Strategic Technologies, Inc. offers a full range of Courses and Implementation assistance in Reliability - centered Maintenance.

For additional Information on:

Training...

A series of highly developed courses designed to ensure your people develop the relevant skills quickly.

Technical Support...

Assistance in ensuring that your facilitators and review groups apply RCM 2 correctly, and so that your organization derives the most benefit from the RCM 2 process.

Project Management...

Assisting your management to assess the resources required to implement RCM 2 and monitor progress during Implementation.

Obtaining additional copies of this book.

Please contact:

Strategic Technologies, Inc.
Suite 400 • 2400 Ardmore Blvd.
Pittsburgh, PA 15221

(412) 731-4200 • 800-552-7262 • Fax (412) 731-4466

Reliability-centered Maintenance

Second Edition

Reliability-centered Maintenance

Second Edition

John Moubray

Industrial Press Inc.

Library of Congress Cataloging-in-Publication Data

Moubray, John
    Reliability-centered maintenance / John Moubray.--2nd ed.
    440p. 15.6 X 23.5 cm
    Includes bibliographical references and index.
    ISBN 0-8311-3078-4
      1. Plant Maintenance. 2. Reliability (Engineering)
    3. Maintainability (Engineering) I. Title.
    TS192.M68 1997   97-8176
    658.2'02--dc21   CIP

INDUSTRIAL PRESS INC.
200 Madison Avenue
New York, NY 10016-4078

Published in Great Britain by Butterworth Heinemann, Oxford, and
in the United States of America since 1992 by Industrial Press Inc.

Printed and bound in the United States of America
by Quinn Woodbine, Woodbine, New Jersey

Typeset by the author
John Moubray

10 9 8 7 6

For Edith

Contents

Preface

Humanity continues to depend to an ever-increasing extent on the wealth generated by highly mechanized and automated businesses. We also depend more and more on services such as the uninterrupted supply of electricity or trains which run on time. More than ever, these depend in turn on the continued integrity of physical assets.

Yet when these assets fail, not only is this wealth eroded and not only are these services interrupted, but our very survival is threatened. Equipment failure has played a part in some of the worst accidents and environmental incidents in industrial history – incidents which have become bywords, such as *Amoco Cadiz*, Chernobyl, Bhopal and Piper Alpha. As a result, the processes by which these failures occur and what must be done to manage them are rapidly becoming very high priorities indeed, especially as it becomes steadily more apparent just how many of these failures are caused by the very activities which are supposed to prevent them.

The first industry to confront these issues was the international civil aviation industry. On the basis of research which challenges many of our most firmly and widely-held beliefs about maintenance, this industry evolved a completely new strategic framework for ensuring that any asset continues to perform as its users want it to perform. This framework is known within the aviation industry as MSG3, and outside it as Reliability-centered Maintenance, or RCM.

Reliability-centered Maintenance was developed over a period of thirty years. One of the principal milestones in its development was a report commissioned by the United States Department of Defense from United Airlines and prepared by Stanley Nowlan and the late Howard Heap in 1978. The report provided a comprehensive description of the development and application of RCM by the civil aviation industry. It forms the basis of both editions of this book and of much of the work done in this field outside the airline industry in the last fifteen years.

Since the early 1980's, the author and his associates have helped companies to apply RCM in hundreds of industrial locations around the world – work which led to the development of RCM2 for industries other than aviation in 1990.

The first edition of this book (published in the UK in 1991 and the USA in 1992) provided a comprehensive introduction to RCM2.

Since then, the RCM philosophy has continued to evolve, to the extent that it became necessary to publish a second edition incorporating the new developments. Several new chapters were added, while others were revised and extended. Foremost among the changes were:

• a more comprehensive review of the role of functional analysis and the definition of failed states in Chapters 2 and 3

• a much broader and deeper look at failure modes and effects analysis in the context of RCM, with special emphasis on the question of levels of analysis and the degree of detail required in Chapter 4

• new material on how to establish acceptable levels of risk in Chapter 5 and Appendix 3

• the addition of more rigorous approaches to the determination of failure-finding task intervals in Chapter 8

• more about the implementation of RCM recommendations in Chapter 11, with extra emphasis on the RCM auditing process

• more information on how RCM should – and should not – be applied in Chapter 13, including a more comprehensive look at the role of the RCM facilitator

• new material on the measurement of the overall performance of the maintenance function in Chapter 14

• a brief review of asset hierarchies in Appendix 1, together with a summary of the (often overstated) role played by functional hierarchies and functional block diagrams in the application of RCM

• a review of different types of human error in Appendix 2, together with a look at the part they play in the failure of physical assets

• the addition of no fewer than 50 new techniques to the appendix on condition monitoring (now Appendix 4).

In the first revision of the second edition (identified by the number 2.1 on the cover), the word 'tolerable' was substituted for 'acceptable' in discussions about risk in Chapters 5 and 8 and in Appendix 3, in order to align this book more closely with standard terminology used in the world of risk. It also included further material on the practicality of failure-finding task intervals in Chapter 8, and slightly revised material on RCM implementation strategies in Chapter 13.

In the second revision of the second edition (identified by the number 2.2 on the cover), the word 'injured' was substituted for the word 'hurt' when discussing safety consequences because the former was felt to be more appropriate in the context of RCM. More precise definitions of the terms *technically feasible* and *worth doing* were added at the beginning of Chapters 6 and 7, while the discussion about scheduled restoration and scheduled discard tasks in Part 4 of Chapter 6 and Part 9 of Chapter 7 was revised to take cognizance of the growing number of cases where the latter may be more cost-effective than the former in situations where both are technically feasible. Finally, a glossary of the most commonly used technical terms in RCM was added just before the bibliography.

The third revision of the second edition makes reference to SAE Standard JA1011: *"Evaluation Criteria for Reliability-centered Maintenance Processes"*. This standard was published in August 1999, and it is mentioned in Chapters 1 and 15.

The book is intended for maintenance, production and operations managers who wish to learn what RCM is, what it achieves and how it is applied. It will also provide students on business or management studies courses with a comprehensive textbook on the formulation of strategies for the management of physical (as opposed to financial) assets. Finally, the book will be invaluable for any students of any branch of engineering who seek an understanding of the state-of-the-art in modern maintenance strategy formulation. It is designed to be read at three levels:

- Chapter 1 is written for those who only wish to review the key elements of Reliability-centred Maintenance.

- Chapters 2 to 10 describe the main elements of the technology of RCM, and will be of most value to those who seek a reasonable technical grasp of the subject.

- The remaining chapters are for those who want to know more about the technical and historical background to RCM (Chapters 12 and 15), about the key steps involved in the implementation of RCM recommendations (Chapter 11), about how the RCM process should be applied (Chapter 13) and about what RCM achieves (Chapter 14).

JOHN MOUBRAY

Asheville
North Carolina
May 2000

Acknowledgments

It has only been possible to write both editions of this book with the help of a great many people around the world. In particular, I would like to record my continuing gratitude to every one of the hundreds of people with whom I have been privileged to work over last fifteen years, each of whom has contributed something to the material in these pages.

In addition, I would like to pay special tribute to a number of people who played a major role in helping to develop and refine the RCM philosophy to the point discussed in this edition of this book.

Firstly, special thanks are due to the late Stan Nowlan for laying the foundations for both editions of this book so thoroughly, both through his own writings and in person, and to all his colleagues in the commercial aviation industry for their pioneering work in this field.

Special thanks are also due to Dr Mark Horton, for his help in developing many of the concepts embodied in Chapters 5 and 8, and to Peter Stock for researching and helping to co-author Appendix 4.

I am also indebted to all members of the Aladon network for their help in applying the concepts and for their continuous feedback about what works and what doesn't work, much of which is also reflected in these pages. Foremost among these are my colleagues Joel Black, Chris James, Hugh Colman and Ian Hipkin, and my associates Alan Katchmar, Frat Amarra, Phil Clarke, Alun Roberts, Michael Hawdon, Ray Peden, Simon Deakin, Tony Landi, Paul Mills and Theuns Koekemoer.

Among the many clients who have proved and are continuing to prove that RCM is a viable force in industry, I am especially indebted to the following:

Gino Palarchio and Ron Thomas of Dofasco Steel
Mike Hopcraft, Terry Belton and Barry Camina of Ford of Europe
Joe Campbell of the British Steel Corporation
Vincent Ryan and Frank O'Connor of the Irish Electricity Supply Board
Francis Cheng of Hong Kong Electric
Nancy Regan of the US Naval Air Command
Denis Udy, Roger Crouch, Kevin Weedon and Malcolm Regler of the Royal Navy

Stan May of Northern Electric
Eric Zindel of Siemens
Don Turner and Trevor Ferrer, formerly of China Light & Power
Dick Pettigrew of Rohm & Haas
Pat McRory of BP Exploration
Ron Doucet of the Iron Ore Company of Canada
Al Weber of Eastman Kodak.
The roles played by Don Humphrey, Richard Hall, Brian Davies, Tom Edwards, David Willson, Brian Oxenham, Sandy Dunn and the late Joe Versteeg in helping to develop or to propagate the concepts discussed in this book are also acknowledged with gratitude.

Perhaps the most important recent development in the field of RCM was the publication in August 1999 of SAE Standard JA1011: *"Evaluation Criteria for Reliability-centered Maintenance Processes"*. This standard is playing a pivotal role in clarifying what is - and what is not - RCM. These acknowledgments would be incomplete without gratefully recognizing on behalf of myself and everyone else with an interest in the responsible custodianship of physical assets, the pivotal role of Dana Netherton, the chairman of the SAE RCM Committee, in bringing it to fruition.

Finally, a special word of thanks to my family for creating an environment in which it was possible to write both editions of this book, and to Aladon Ltd for permission to reproduce the RCM Information and Decision Worksheets and the RCM 2 Decision Diagram.

1 Introduction to Reliability-centered Maintenance

1.1 The Changing World of Maintenance

Over the past twenty years, maintenance has changed, perhaps more so than any other management discipline. The changes are due to a huge increase in the number and variety of physical assets (plant, equipment and buildings) that must be maintained throughout the world, much more complex designs, new maintenance techniques and changing views on maintenance organization and responsibilities.

Maintenance is also responding to changing expectations. These include a rapidly growing awareness of the extent to which equipment failure affects safety and the environment, a growing awareness of the connection between maintenance and product quality, and increasing pressure to achieve high plant availability and to contain costs.

The changes are testing attitudes and skills in all branches of industry to the limit. Maintenance people are having to adopt completely new ways of thinking and acting, as engineers and as managers. At the same time the limitations of maintenance systems are becoming increasingly apparent, no matter how much they are computerised.

In the face of this avalanche of change, managers everywhere are looking for a new approach to maintenance. They want to avoid the false starts and dead ends that always accompany major upheavals. *Instead they seek a strategic framework that synthesizes the new developments into a coherent pattern, so that they can evaluate them sensibly and apply those likely to be of most value to them and their companies.*

This book describes a philosophy that provides just such a framework. It is called Reliability-centered Maintenance, or RCM.

If it is applied correctly, RCM transforms the relationships between the undertakings that use it, their existing physical assets and the people who operate and maintain those assets. It also enables new assets to be put into effective service with great speed, confidence and precision.

This chapter provides a brief introduction to RCM, starting with a look at how maintenance has evolved over the past sixty years.

Since the 1930's, the evolution of maintenance can be traced through three generations. RCM is rapidly becoming a cornerstone of the Third Generation, but this generation can only be viewed in perspective in the light of the First and Second Generations.

The First Generation

The First Generation covers the period up to World War II. In those days industry was not very highly mechanized, so downtime did not matter much. This meant that the prevention of equipment failure was not a very high priority in the minds of most managers. At the same time, most equipment was simple and much of it was over-designed. This made it reliable and easy to repair. As a result, there was no need for systematic maintenance of any sort beyond simple cleaning, servicing and lubrication routines. The need for skills was also lower than it is today.

The Second Generation

Things changed dramatically during World War II. Wartime pressures increased the demand for goods of all kinds while the supply of industrial manpower dropped sharply. This led to increased mechanization. By the 1950's machines of all types were more numerous and more complex. Industry was beginning to depend on them.

As this dependence grew, downtime came into sharper focus. This led to the idea that equipment failures could and should be prevented, which led in turn to the concept of *preventive maintenance*. In the 1960's, this consisted mainly of equipment overhauls done at fixed intervals.

The cost of maintenance also started to rise sharply relative to other operating costs. This led to the growth of *maintenance planning and control systems*. These have helped greatly to bring maintenance under control, and are now an established part of the practice of maintenance.

Finally, the amount of capital tied up in fixed assets together with a sharp increase in the cost of that capital led people to start seeking ways in which they could maximize the life of the assets.

The Third Generation

Since the mid-seventies, the process of change in industry has gathered even greater momentum. The changes can be classified under the headings of *new expectations*, *new research* and *new techniques*.

New expectations

Figure 1.1 shows how expectations of maintenance have evolved.

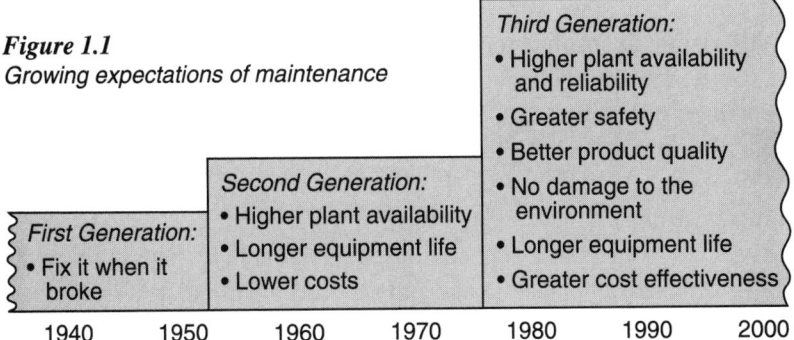

Figure 1.1
Growing expectations of maintenance

Third Generation:
• Higher plant availability and reliability
• Greater safety
• Better product quality
• No damage to the environment
• Longer equipment life
• Greater cost effectiveness

Second Generation:
• Higher plant availability
• Longer equipment life
• Lower costs

First Generation:
• Fix it when it broke

1940   1950   1960   1970   1980   1990   2000

Downtime affects the productive capability of physical assets by reducing output, increasing operating costs and affecting customer service. By the 1960's and 1970's, this was already a major concern in the manufacturing, mining and transport sectors. The effects of downtime are being aggravated by the worldwide move towards just-in-time systems, where reduced stocks of materials throughout the supply chain mean that quite small equipment failures are now increasingly likely to interfere with the operation of an entire facility. In recent times, the growth of mechanization and automation has meant that *reliability* and *availability* are now also key issues in sectors as diverse as health care, data processing, telecommunications and building management.

Greater automation also means that more and more failures affect our ability to sustain satisfactory *quality standards*. This applies as much to standards of service as it does to product quality. For instance, equipment failures can affect climate control in buildings and the punctuality of transport networks as much as they can interfere with the consistent achievement of specified tolerances in manufacturing.

More and more failures have serious *safety* or *environmental* consequences, at a time when standards in these areas are rising rapidly. In some parts of the world, the point is approaching where organizations either conform to society's safety and environmental expectations, or they cease to operate. This adds an order of magnitude to our dependence on the integrity of our physical assets – one that goes beyond cost and becomes a simple matter of organizational survival.

At the same time as our dependence on physical assets is growing, so too is their *cost – to operate* and *to own*. To secure the maximum return on the investment that they represent, they must be kept working efficiently for as long as we want them to. Finally, the *cost of maintenance* itself is still rising, in absolute terms and as a proportion of total expenditure. In some industries, it is now the second highest or even the highest element of operating costs. As a result, in only thirty years it has moved from almost nowhere to the top of the league as a cost control priority.

New research
Quite apart from greater expectations, new research is changing many of our most basic beliefs about age and failure. In particular, it is apparent that there is less and less connection between the operating age of most assets and how likely they are to fail.

Figure 1.2 shows how the earliest view of failure was simply that as things got older, they were more likely to fail. A growing awareness of 'infant mortality' led to widespread Second Generation belief in the 'bathtub' curve.

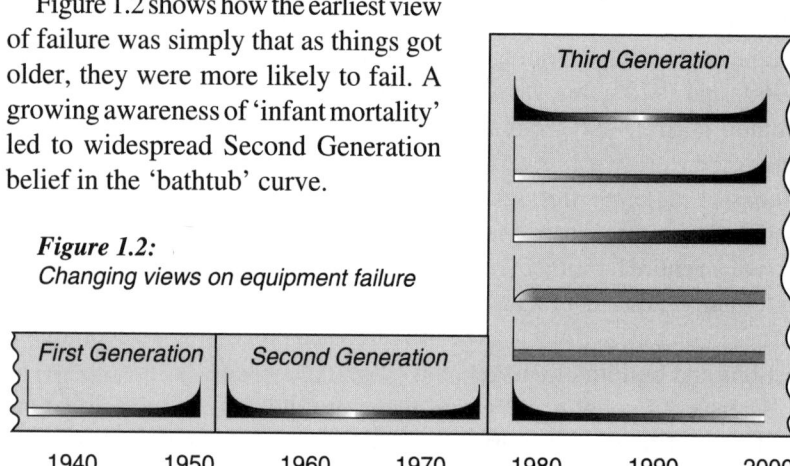

Figure 1.2:
Changing views on equipment failure

However, Third Generation research has revealed that not one or two but *six* failure patterns actually occur in practice. As discussed in more detail later in this chapter, one of the most important conclusions to emerge from this research is a growing realization that although they may be done exactly as planned, a great many traditionally-derived maintenance tasks achieve nothing, while some are actively counterproductive and even dangerous. This is especially true of many tasks done in the name of preventive maintenance. On the other hand, many more maintenance tasks that are essential to the safe operation of modern, complex industrial systems do not appear in the associated maintenance programs.

In other words, industry in general is devoting a great deal of attention to doing maintenance work correctly (doing the job right), but much more needs to be done to ensure that the jobs that are being planned are the jobs that should be planned (doing the right job).

New techniques
There has been explosive growth in new maintenance concepts and techniques. Hundreds have been developed over the past twenty years, and more are emerging every week.

Figure 1.3 shows how the classical emphasis on overhauls and administrative systems has grown to include many new developments in a number of different fields.

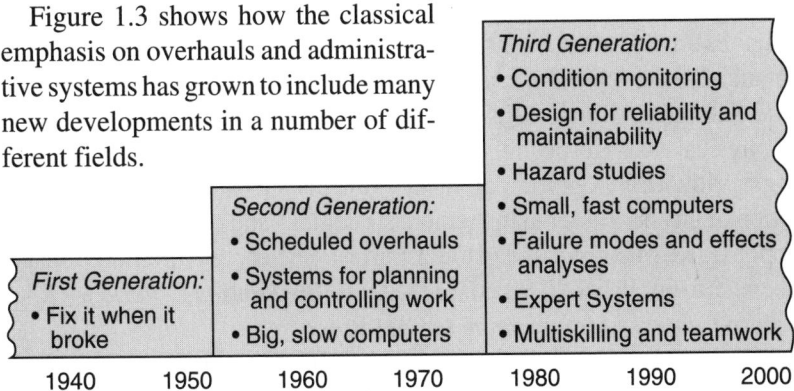

Figure 1.3: Changing maintenance techniques

The new developments include:

• *decision support tools,* such as hazard studies, failure modes and effects analyses and expert systems
• *new maintenance techniques,* such as condition monitoring
• *designing equipment* with a much greater emphasis on reliability and maintainability
• *a major shift in organizational thinking* towards participation, team-working and flexibility.

As mentioned earlier, a major challenge facing maintenance people nowadays is not only to learn what these techniques are, but to decide which are worthwhile and which are not in their own organizations. If we make the right choices, it is possible to improve asset performance and *at the same time* contain and even reduce the cost of maintenance. If we make the wrong choices, new problems are created while existing problems only get worse.

The challenges facing maintenance
The first industry to confront these challenges systematically was the commercial aviation industry. A crucial element of its response was the realization that as much effort needs to be devoted to ensuring that maintainers are doing the right job as to ensuring that they are doing the job right. This realization led in turn to the development of the comprehensive decision-making process known within aviation as MSG3, and outside it as Reliability-centered Maintenance, or RCM.

In nearly every field of organized human endeavour, RCM is now becoming as fundamental to the responsible custodianship of physical assets as double-entry bookkeeping is to the responsible custodianship of financial assets. No other comparable technique exists for identifying the true, safe minimum of tasks that must be done to preserve the functions of physical assets, especially in critical or hazardous situations. A growing worldwide recognition of the key role played by RCM in the formulation of physical asset management strategies - and of the importance of applying RCM correctly - led the American Society of Automotive Engineers[1999] to publish SAE Standard JA1011: *"Evaluation Criteria for Reliability-Centered Maintenance (RCM) Processes"*.

The process described in Chapters 2 to 10 of this book complies with this standard. The rest of the book discusses how RCM should be applied and how RCM-based failure management policies should be implemented, in addition to providing more background on key technical issues.

The remainder of this chapter introduces RCM in more detail.

1.2 Maintenance and RCM

From the engineering viewpoint, there are two elements to the management of any physical asset. It must be maintained and from time to time it may also need to be modified.

The major dictionaries define *maintain* as *cause to continue* (Oxford) or *keep in an existing state* (Webster). This suggests that maintenance means preserving something. On the other hand, they agree that to *modify* something means to *change* it in some way. This distinction between maintain and modify has profound implications which are discussed at length in later chapters. However, we focus on maintenance at this point.

When we set out to maintain something, what is it that we wish to *cause to continue?* What is the *existing state* that we wish to preserve?

The answer to these questions can be found in the fact that every physical asset is put into service because someone wants it to do something. In other words, they expect it to fulfil a specific function or functions. So it follows that when we maintain an asset, the state we wish to preserve must be one in which it continues to do whatever its users want it to do.

Maintenance: Ensuring that physical assets continue to do what their users want them to do

What the users want will depend on exactly where and how the asset is being used (the operating context). This leads to the following formal definition of Reliability-centered Maintenance:

Reliability-centered Maintenance: a process used to determine what must be done to ensure that any physical asset continues to do what its users want it to do in its present operating context.

1.3   RCM: The seven basic questions

The RCM process entails asking seven questions about the asset or system under review, as follows:

• *what are the functions and associated performance standards of the asset in its present operating context?*
• *in what ways does it fail to fulfil its functions?*
• *what causes each functional failure?*
• *what happens when each failure occurs?*
• *in what way does each failure matter?*
• *what can be done to predict or prevent each failure?*
• *what should be done if a suitable proactive task cannot be found?*

These questions are introduced briefly in the following paragraphs, and then considered in detail in Chapters 2 to 10.

Functions and Performance Standards

Before it is possible to apply a process used to determine what must be done to ensure that any physical asset continues to do whatever its users want it to do in its present operating context, we need to do two things:

• determine what its users want it to do

• ensure that it is capable of doing what its users want to start with.

This is why the first step in the RCM process is to define the functions of each asset in its operating context, together with the associated desired standards of performance. What users expect assets to be able to do can be split into two categories:

• *primary functions*, which summarize why the asset was acquired in the first place. This category of functions covers issues such as speed, output, carrying or storage capacity, product quality and customer service.

• *secondary functions*, which recognize that every asset is expected to do more than simply fulfil its primary functions. Users also have expectations in areas such as safety, control, containment, comfort, structural integrity, economy, protection, efficiency of operation, compliance with environmental regulations and even the appearance of the asset,

The users of the assets are usually in the best position by far to know exactly what contribution each asset makes to the physical and financial well-being of the organization as a whole, so it is essential that they are involved in the RCM process from the outset.

Done properly, this step alone usually takes up about a third of the time involved in an entire RCM analysis. It also usually causes the team doing the analysis to learn a remarkable amount – often a frightening amount – about how the equipment actually works.

Functions are explored in more detail in Chapter 2.

Functional Failures

The objectives of maintenance are defined by the functions and associated performance expectations of the asset under consideration. But how does maintenance achieve these objectives?

The only occurrence which is likely to stop any asset performing to the standard required by its users is some kind of failure. This suggests that maintenance achieves its objectives by adopting a suitable approach to the management of failure. However, before we can apply a suitable blend of failure management tools, we need to identify what failures can occur.

The RCM process does this at two levels:

- firstly, by identifying what circumstances amount to a failed state
- then by asking what events can cause the asset to get into a failed state.

In the world of RCM, failed states are known as *functional failures* because they occur when an asset is *unable to fulfil a function to a standard of performance which is acceptable to the user.* In addition to the total inability to function, this definition encompasses partial failures, where the asset still functions but at an unacceptable level of performance (including situations where the asset cannot sustain acceptable levels of quality or accuracy). Clearly these can only be identified after the functions and performance standards of the asset have been defined.

Functional failures are discussed at greater length in Chapter 3.

Failure Modes

As mentioned in the previous paragraph, once each functional failure has been identified, the next step is to try to identify all the *events which are reasonably likely to cause each failed state.* These events are known as *failure modes.* 'Reasonably likely' failure modes include those which have occurred on the same or similar equipment operating in the same context, failures which are currently being prevented by existing maintenance regimes, and failures which have not happened yet but which are considered to be real possibilities in the context in question.

Most traditional lists of failure modes incorporate failures caused by deterioration or normal wear and tear. However, the list should include failures caused by human errors (on the part of operators and maintainers) and design flaws so that all reasonably likely causes of equipment failure can be identified and dealt with appropriately. It is also important to identify the cause of each failure in enough detail to ensure that time and effort are not wasted trying to treat symptoms instead of causes. On the other hand, it is equally important to ensure that time is not wasted on the analysis itself by going into too much detail.

Failure Effects

The fourth step in the RCM process entails listing *failure effects*, which describe what happens when each failure mode occurs. These descriptions should include all the information needed to support the evaluation of the consequences of the failure, such as:

- what evidence (if any) that the failure has occurred
- in what ways (if any) it poses a threat to safety or the environment
- in what ways (if any) it affects production or operations
- what physical damage (if any) is caused by the failure
- what must be done to repair the failure.

Failure modes and effects are discussed at greater length in Chapter 4.

The process of identifying functions, functional failures, failure modes and failure effects yields surprising and often very exciting opportunities for improving performance and safety, and also for eliminating waste

Failure Consequences

A detailed analysis of an average industrial undertaking is likely to yield between three and ten thousand possible failure modes. Each of these failures affects the organization in some way, but in each case, the effects are different. They may affect operations. They may also affect product quality, customer service, safety or the environment. They will all take time and cost money to repair.

It is these consequences which most strongly influence the extent to which we try to prevent each failure. In other words, if a failure has serious consequences, we are likely to go to great lengths to try to avoid it. On the other hand, if it has little or no effect, then we may decide to do no routine maintenance beyond basic cleaning and lubrication.

A great strength of RCM is that it recognizes that the consequences of failures are far more important than their technical characteristics. In fact, it recognizes that the only reason for doing any kind of proactive maintenance is not to avoid failures *per se*, but to avoid or at least to reduce the *consequences* of failure. The RCM process classifies these consequences into four groups, as follows:

- *Hidden failure consequences:* Hidden failures have no direct impact, but they expose the organization to multiple failures with serious, often catastrophic, consequences. (Most of these failures are associated with protective devices which are not fail-safe.)

- *Safety and environmental consequences:* A failure has safety consequences if it could injure or kill someone. It has environmental consequences if it could lead to a breach of any corporate, regional, national or international environmental standard.

- *Operational consequences:* A failure has operational consequences if it affects production (output, product quality, customer service or operating costs in addition to the direct cost of repair)

- *Non-operational consequences:* Evident failures which fall into this category affect neither safety nor production, so they involve only the direct cost of repair.

We will see later how the RCM process uses these categories as the basis of a strategic framework for maintenance decision-making. By forcing a structured review of the consequences of each failure mode in terms of the above categories, it integrates the operational, environmental and safety objectives of the maintenance function. This helps to bring safety and the environment into the mainstream of maintenance management.

The consequence evaluation process also shifts emphasis away from the idea that *all* failures are bad and must be prevented. In so doing, it focuses attention on the maintenance activities which have most effect on the performance of the organization, and diverts energy away from those which have little or no effect. It also encourages us to think more broadly about different ways of managing failure, rather than to concentrate only on failure prevention. Failure management techniques are divided into two categories:

- *proactive tasks:* these are tasks undertaken before a failure occurs, in order to prevent the item from getting into a failed state. They embrace what is traditionally known as 'predictive' and 'preventive' maintenance, although we will see later that RCM uses the terms *scheduled restoration, scheduled discard* and *on-condition maintenance*

- *default actions:* these deal with the failed state, and are chosen when it is not possible to identify an effective proactive task. Default actions include *failure-finding, redesign* and *run-to-failure.*

The consequence evaluation process is discussed again briefly later in this chapter, and in much more detail in Chapter 5. The next section of this chapter looks at proactive tasks in more detail

Proactive Tasks

Many people still believe that the best way to optimize plant availability is to do some kind of proactive maintenance on a routine basis. Second Generation wisdom suggested that this should consist of overhauls or component replacements at fixed intervals. Figure 1.4 illustrates the fixed interval view of failure.

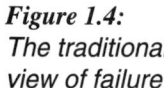

Figure 1.4:
The traditional
view of failure

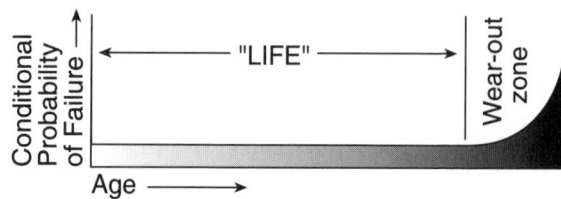

Figure 1.4 is based on the assumption that most items operate reliably for a period 'X', and then wear out. Classical thinking suggests that extensive records about failure will enable us to determine this life and so make plans to take preventive action shortly before the item is due to fail in future.

This model is true for certain types of simple equipment, and for some complex items with dominant failure modes. In particular, wear-out characteristics are often found where equipment comes into direct contact with the product. Age-related failures are also often associated with fatigue, corrosion, abrasion and evaporation.

However, equipment in general is far more complex than it was twenty years ago. This has led to startling changes in the patterns of failure, as shown in Figure 1.5. The graphs show conditional probability of failure against operating age for a variety of electrical and mechanical items.

Pattern A is the well-known bathtub curve. It starts with a high incidence of failure (known as *infant mortality*) followed by constant or slowly increasing conditional probability of failure, then by a wear-out zone.

Pattern B shows constant or slowly increasing conditional probability of failure, ending in a wear-out zone – the same as Figure 1.4.

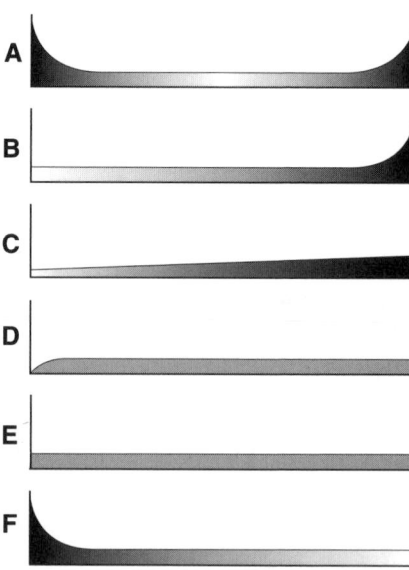

Figure 1.5:
Six patterns of failure

Pattern C shows slowly increasing conditional probability of failure, but there is no identifiable wear-out age. Pattern D shows low conditional probability of failure when the item is new or just out of the shop, then a rapid increase to a constant level, while pattern E shows a constant conditional probability of failure at all ages (random failure). Pattern F starts with high infant mortality, which drops eventually to a constant or very slowly increasing conditional probability of failure.

Studies done on civil aircraft showed that 4% of the items conformed to pattern A, 2% to B, 5% to C, 7% to D, 14% to E and no fewer than 68% *to pattern F*. (The number of times these patterns occur in aircraft is not necessarily the same as in industry. But there is no doubt that as assets become more complex, we see more and more of patterns E and F.)

These findings contradict the belief that there is always a connection between reliability and operating age. This belief led to the idea that the more often an item is overhauled, the less likely it is to fail. Nowadays, this is seldom true. Unless there is a dominant age-related failure mode, age limits do little or nothing to improve the reliability of complex items. In fact scheduled overhauls can actually *increase* overall failure rates by introducing infant mortality into otherwise stable systems.

An awareness of these facts has led some organizations to abandon the idea of proactive maintenance altogether. In fact, this can be the right thing to do for failures with minor consequences. But when the failure consequences are significant, *something* must be done to prevent or predict the failures, or at least to reduce the consequences.

This brings us back to the question of proactive tasks. As mentioned earlier, RCM divides proactive tasks into three categories, as follows:
• scheduled restoration tasks
• scheduled discard tasks
• scheduled on-condition tasks.

Scheduled restoration and scheduled discard tasks
Scheduled restoration entails remanufacturing a component or overhauling an assembly at or before a specified age limit, regardless of its condition at the time. Similarly, scheduled discard entails discarding an item at or before a specified life limit, regardless of its condition at the time.

Collectively, these two types of tasks are now generally known as *preventive* maintenance. They used to be by far the most widely used form of proactive maintenance. However for the reasons discussed above, they are much less widely used than they were twenty years ago.

On-condition tasks

The continuing need to prevent certain types of failure, and the growing inability of classical techniques to do so, are behind the growth of new types of failure management. The majority of these techniques rely on the fact that most failures give some warning of the fact that they are about to occur. These warnings are known as *potential failures*, and are defined as *identifiable physical conditions which indicate that a functional failure is about to occur or is in the process of occurring.*

The new techniques are used to detect potential failures so that action can be taken to avoid the consequences which could occur if they degenerate into functional failures. They are called *on-condition* tasks because items are left in service *on the condition* that they continue to meet desired performance standards. (On-condition maintenance includes *predictive maintenance, condition-based maintenance* and *condition monitoring.*)

Used appropriately, on-condition tasks are a very good way of managing failures, but they can also be an expensive waste of time. RCM enables decisions in this area to be made with particular confidence.

Default Actions

RCM recognizes three major categories of default actions, as follows:

- *failure-finding:* Failure-finding tasks entail checking hidden functions periodically to determine whether they have failed (whereas condition-based tasks entail checking if something is failing).

- *redesign:* redesign entails making any one-off change to the built-in capability of a system. This includes modifications to the hardware and also covers once-off changes to procedures.

- *no scheduled maintenance:* as the name implies, this default entails making no effort to anticipate or prevent failure modes to which it is applied, and so those failures are simply allowed to occur and then repaired. This default is also called *run-to-failure,*

The RCM Task Selection Process

A great strength of RCM is the way it provides simple, precise and easily understood criteria for deciding which (if any) of the proactive tasks is *technically feasible* in any context, and if so for deciding how often they should be done and who should do them. These criteria are discussed in more detail in Chapters 6 and 7.

Whether or not a proactive task is technically feasible is governed by the *technical characteristics* of the task and of the failure which it is meant to prevent. Whether it is *worth doing* is governed by how well it deals with the *consequences* of the failure. If a proactive task cannot be found which is both technically feasible and worth doing, then suitable default action must be taken. The essence of the task selection process is as follows:

- for *hidden failures*, a proactive task is worth doing if it reduces the risk of the multiple failure associated with that function to a tolerably low level. If such a task cannot be found then a scheduled *failure-finding task* must be performed. If a suitable failure-finding task cannot be found, the secondary default decision is that the item may have to be redesigned (depending on the consequences of the multiple failure).

- for failures with *safety* or *environmental* consequences, a proactive task is only worth doing if it reduces the risk of that failure on its own to a very low level indeed, if it does not eliminate it altogether. If a task cannot be found that reduces the risk of the failure to a tolerably low level, *the item must be redesigned or the process must be changed.*

- if the failure has *operational* consequences, a proactive task is only worth doing if the total cost of doing it *over a period of time* is less than the cost of the operational consequences and the cost of repair over the same period. In other words, the task must be *justified on economic grounds*. If it is not justified, the initial default decision is *no scheduled maintenance*. (If this occurs and the operational consequences are still unacceptable then the secondary default decision is again redesign).

- if a failure has *non-operational* consequences a proactive task is only worth doing if the cost of the task over a period of time is less than the cost of repair over the same period. So these tasks must also be *justified on economic grounds*. If it is not justified, the initial default decision is again *no scheduled maintenance*, and if the repair costs are too high, the secondary default decision is once again redesign.

This approach means that proactive tasks are only specified for failures which really need them, which in turn leads to substantial reductions in routine workloads. Less routine work also means that the remaining tasks are more likely to be done properly. This together with the elimination of counterproductive tasks leads to more effective maintenance.

Compare this with the traditional approach to the development of maintenance policies. Traditionally, the maintenance requirements of each asset are assessed in terms of its real or assumed technical characteristics, without considering the consequences of failure. The resulting schedules are used for all similar assets, again without considering that different consequences apply in different operating contexts. This results in large numbers of schedules which are wasted, not because they are 'wrong' in the technical sense, but because they achieve nothing.

Note also that the RCM process considers the maintenance requirements of each asset before asking whether it is necessary to reconsider the design. This is simply because the maintenance engineer who is on duty *today* has to maintain the equipment as it exists *today*, not what should be there or what might be there at some stage in the future.

1.4 Applying the RCM process

Before setting out to analyze the maintenance requirements of the assets in any organization, we need to know what these assets are and to decide which of them are to be subjected to the RCM review process. This means that a plant register must be prepared if one does not exist already. In fact, the vast majority of industrial organizations nowadays already possess plant registers which are adequate for this purpose, so this book only touches on the most desirable attributes of such registers in Appendix 1.

Planning
If it is correctly applied, RCM leads to remarkable improvements in maintenance effectiveness, and often does so surprisingly quickly. However, the successful application of RCM depends on meticulous planning and preparation. The key elements of the planning process are as follows:

• decide which assets are most likely to benefit from the RCM process, and if so, exactly how they will benefit

• assess the resources required to apply the process to the selected assets

• in cases where the likely benefits justify the investment, decide in detail who is to perform and who is to audit each analysis, when and where, and arrange for them to receive appropriate training

• ensure that the operating context of the asset is clearly understood.

Review groups

We have seen how the RCM process embodies seven basic questions. In practice, maintenance people simply cannot answer all these questions on their own. This is because many (if not most) of the answers can only be supplied by production or operations people. This applies especially to questions concerning functions, desired performance, failure effects and failure consequences.

For this reason, a review of the maintenance requirements of any asset should be done by small teams which include *at least* one person from the maintenance function and one from the operations function. The seniority of the group members is less important than the fact that they should have a thorough knowledge of the asset under review. Each group member should also have been trained in RCM. The make-up of a typical RCM review group is shown in Figure 1.6:

The use of these groups not only enables management to gain access to the knowledge and expertise of each member of the group on a systematic basis, but the members themselves gain a greatly enhanced understanding of the asset in its operating context.

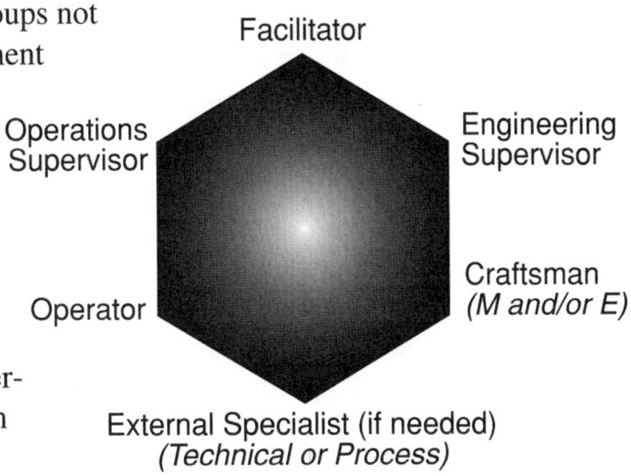

Facilitator

Operations Supervisor

Engineering Supervisor

Operator

Craftsman *(M and/or E)*

External Specialist (if needed) *(Technical or Process)*

Figure 1.6: A typical RCM review group

Facilitators

RCM review groups work under the guidance of highly trained specialists in RCM, known as facilitators. The facilitators are the most important people in the RCM review process. Their role is to ensure that:

• the RCM analysis is carried out at the right level, that system boundaries are clearly defined, that no important items are overlooked and that the results of the analysis are properly recorded

• RCM is correctly understood and applied by the group members

• the group reaches consensus in a brisk and orderly fashion, while retaining the enthusiasm and commitment of individual members

• the analysis progresses reasonably quickly and finishes on time.
Facilitators also work with RCM project managers or sponsors to ensure that each analysis is properly planned and receives appropriate managerial and logistic support.
Facilitators and RCM review groups are discussed in more detail in Chapter 13.

The outcomes of an RCM analysis
If it is applied in the manner suggested above, an RCM analysis results in three tangible outcomes, as follows:

• maintenance schedules to be done by the maintenance department

• revised operating procedures for the operators of the asset

• a list of areas where one-off changes must be made to the design of the asset or the way in which it is operated to deal with situations where the asset cannot deliver the desired performance in its current configuration.

Two less tangible outcomes are that participants in the process learn a great deal about how the asset works, and also tend to function better as teams.

Auditing and implementation
Immediately after the review has been completed for each asset, senior managers with overall responsibility for the equipment must satisfy themselves that decisions made by the group are sensible and defensible.

After each review is approved, the recommendations are implemented by incorporating maintenance schedules into maintenance planning and control systems, by incorporating operating procedure changes into the standard operating procedures for the asset, and by handing recommendations for design changes to the appropriate design authority. Key aspects of auditing and implementation are discussed in Chapter 11.

1.5 What RCM Achieves

Desirable as they are, the outcomes listed above should only be seen as a means to an end. Specifically, they should enable the maintenance function to fulfil all the expectations listed in Figure 1.1 at the beginning of this chapter. How they do so is summarized in the following paragraphs, and discussed again in more detail in Chapter 14.

- *Greater safety* and *environmental integrity:* RCM considers the safety and environmental implications of every failure mode before considering its effect on operations. This means that steps are taken to minimize all identifiable equipment-related safety and environmental hazards, if not eliminate them altogether. By integrating safety into the mainstream of maintenance decision-making, RCM also improves attitudes to safety.

- *Improved operating performance (output, product quality and customer service):* RCM recognizes that *all* types of maintenance have some value, and provides rules for deciding which is most suitable in every situation. By doing so, it helps ensure that only the most effective forms of maintenance are chosen for each asset, and that suitable action is taken in cases where maintenance cannot help. This much more tightly focused maintenance effort leads to quantum jumps in the performance of *existing assets* where these are sought.

  RCM was developed to help airlines draw up maintenance programs for new types of aircraft *before* they enter service. As a result, it is an ideal way to develop such programs for *new assets*, especially complex equipment for which no historical information is available. This saves much of the trial and error which is so often part of the development of new maintenance programs – trial which is time-consuming and frustrating, and error which can be very costly.

- *Greater maintenance cost-effectiveness:* RCM continually focuses attention on the maintenance activities which have most effect on the performance of the plant. This helps to ensure that everything spent on maintenance is spent where it will do the most good.

  In addition, if RCM is correctly applied to existing maintenance systems, it reduces the amount of *routine* work (in other words, maintenance tasks to be undertaken on a *cyclic* basis) issued in each period, usually by 40% to 70%. On the other hand, if RCM is used to develop a new maintenance program, the resulting scheduled workload is much lower than if the program is developed by traditional methods.

- *Longer useful life of expensive items,* due to a carefully focused emphasis on the use of on-condition maintenance techniques.

- A *comprehensive database:* An RCM review ends with a comprehensive and fully documented record of the maintenance requirements of all the significant assets used by the organization. This makes it possible *to adapt to changing circumstances* (such as changing shift patterns or new technology) without having to reconsider all maintenance policies from scratch. It also enables equipment users to demonstrate that their maintenance programs are built on rational foundations (the *audit trail* required by more and more regulators). Finally, the information stored on RCM worksheets *reduces the effects of staff turnover* with its attendant loss of experience and expertise.

An RCM review of the maintenance requirements of each asset also provides a much clearer view of the *skills required to maintain each asset*, and for deciding *what spares should be held in stock*. A valuable by-product is also *improved drawings and manuals*.

- *Greater motivation of individuals,* especially people who are involved in the review process. This leads to greatly improved general understanding of the equipment in its operating context, together with wider 'ownership' of maintenance problems and their solutions. It also means that solutions are more likely to endure.

- *Better teamwork:* RCM provides a common, easily understood technical language for everyone who has anything to do with maintenance. This gives maintenance and operations people a better understanding of what maintenance can (and cannot) achieve and what must be done to achieve it.

All of these issues are part of the mainstream of maintenance management, and many are already the target of improvement programs. A major feature of RCM is that it provides an effective step-by-step framework for tackling *all* of them at once, and for involving everyone who has anything to do with the equipment in the process.

RCM yields results very quickly. In fact, if they are correctly focused and correctly applied, RCM reviews can pay for themselves in a matter of months and sometimes even a matter of weeks, as discussed in Chapter 14. The reviews transform both the perceived maintenance requirements of the physical assets used by the organization and the way in which the maintenance function as a whole is perceived. The result is more cost-effective, more harmonious and much more successful maintenance.

2 Functions

Most people become engineers because they feel at least some affinity for things, be they mechanical, electrical or structural. This affinity leads them to derive pleasure from assets in good condition, but to feel offended by assets in poor condition.

These reflexes have always been at the heart of the concept of preventive maintenance. They have given rise to concepts such as 'asset care', which as the name implies, seeks to care for assets *per se*. They have also led some maintenance strategists to believe that maintenance is all about preserving the inherent reliability or built-in capability of any asset.

In fact, this is not so.

As we gain a deeper understanding of the role of assets in business, we begin to appreciate the significance of the fact that any physical asset is put into service because someone wants it to do something. So it follows that when we maintain an asset, *the state which we wish to preserve must be one in which it continues to do whatever its users want it to do*. Later in this chapter, we will see that this state – what the users want – is fundamentally different from the built-in capability of the asset.

This emphasis on what the asset *does* rather than what it *is* provides a whole new way of defining the objectives of maintenance for any asset – one which focuses on what the user wants. This is the most important single feature of the RCM process, and is why many people regard RCM as 'TQM applied to physical assets'.

Clearly, in order to define the objectives of maintenance in terms of user requirements, we must gain a crystal clear understanding of the functions of each asset together with the associated performance standards. This is why the RCM process starts by asking:

- **what are the functions and associated performance standards of the asset in its present operating context?**

This chapter considers this question in more detail. It describes how functions should be defined, explores the two main types of performance standards, reviews different categories of functions and shows how functions should be listed.

2.1 Describing functions

It is a well established principle of value engineering that a function statement should consist of a verb and an object. It is also helpful to start such statements with the word 'to' ('to pump water', 'to transport people', etc).

However, as explained at length in the next part of this chapter, users not only expect an asset to fulfil a function. They also expect it to do so to an acceptable level of performance. So a function definition – and by implication the definition of the objectives of maintenance for the asset – is not complete unless it specifies as precisely as possible the level of performance desired by the user (as opposed to the built-in capability).

For instance, the primary function of the pump in Figure 2.1 would be listed as:
• To pump water from Tank X to Tank Y at not less than 800 liters per minute.

This example shows that a complete function statement consists of a verb, an object and the standard of performance desired by the user.

***A function statement should consist of a verb,
an object and a desired standard of performance***

2.2 Performance standards

The objective of maintenance is to ensure that assets continue to do what their users want them to do. The extent to which any user wants any asset to do anything can be defined by a minimum standard of performance. If we could build an asset which could deliver that minimum performance without deteriorating in any way, then that would be the end of the matter. The machine would run continuously with no need for maintenance.

However, in the real world, things are not that simple.

The laws of physics tell us that any organized system which is exposed to the real world will deteriorate. The end result of this deterioration is total disorganization (also known as 'chaos' or 'entropy'), unless steps are taken to arrest whatever process is causing the system to deteriorate.

For instance, the pump in Figure 2.1 is pumping water into a tank from which the water is drawn at a rate of 800 liters/minute. One process that causes the pump to deteriorate *(failure mode)* is impeller wear. This happens regardless of whether it is pumping acid or lubricating oil, and regardless of whether the impeller is made of titanium or mild steel. The only question is how fast it will wear to the point that it can no longer deliver 800 liters/minute.

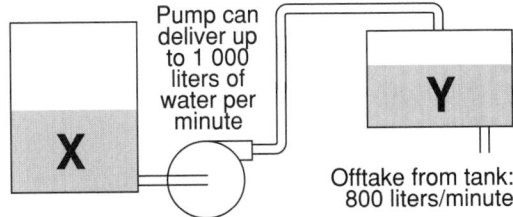

Figure 2.1:
Initial capability vs desired performance

Pump can deliver up to 1 000 liters of water per minute

Offtake from tank: 800 liters/minute

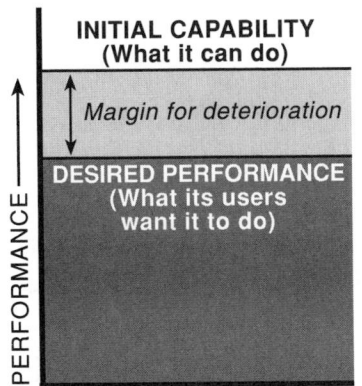

INITIAL CAPABILITY
(What it can do)

Margin for deterioration

DESIRED PERFORMANCE
(What its users want it to do)

PERFORMANCE

So if deterioration is inevitable, it must be allowed for. This means that when any asset is put into service, it must be able to deliver *more* than the minimum standard of performance desired by the user. What the asset is able to deliver is known as its *initial capability* (or inherent reliability). Figure 2.2 illustrates the right relationship between this capability and desired performance.

Figure 2.2: *Allowing for deterioration*

For instance, in order to ensure that the pump shown in Figure 2.1 does what its users want *and to allow for deterioration*, the system designers must specify a pump which has an initial built-in capability of something greater than 800 liters/minute. In the example shown, this initial capability is 1 000 liters per minute.

This means that performance can be defined in *two* ways, as follows:

• desired performance *(what the user wants the asset to do)*

• built-in capability *(what it can do)*

Later chapters look at how maintenance helps ensure that assets continue to fulfil their intended functions, either by ensuring that their capability remains above the minimum standard desired by the user or by restoring something approaching the initial capability if it drops below this point. When considering the question of restoration, bear in mind that:

• the initial capability of any asset is established by its design and by how it is made

• maintenance can only restore the asset to this initial level of capability – it cannot go beyond it.

In practice, most assets *are* adequately designed and built, so it is usually possible to develop maintenance programs which ensure that such assets continue to do what their users want.

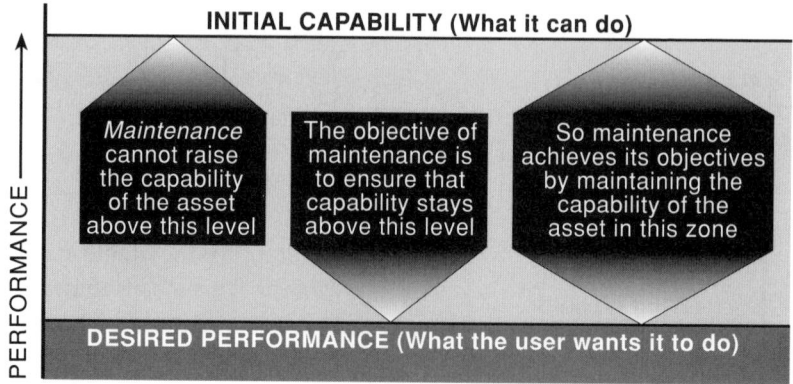

Figure 2.3: A maintainable asset

In short, such assets are maintainable, as shown in Figure 2.3.

On the other hand, if the desired performance exceeds the initial capability, no amount of maintenance can deliver the desired performance. In other words, such assets are not maintainable, as shown in Figure 2.4.

For instance, if the pump shown in Figure 2.1 had an initial capability of 750 liters/minute, it would not be able to keep the tank full. Since the maintenance program does not exist which makes pumps bigger, maintenance cannot deliver the desired performance in this context. Similarly, if we make a habit of trying to draw 15 kW *(desired performance)* from a 10 kW electric motor *(initial capability)*, the motor will keep tripping out and will eventually burn out prematurely. No amount of maintenance will make this motor big enough. It may be perfectly adequately designed and built in its own right - it just cannot deliver the desired performance in the context in which it is being used.

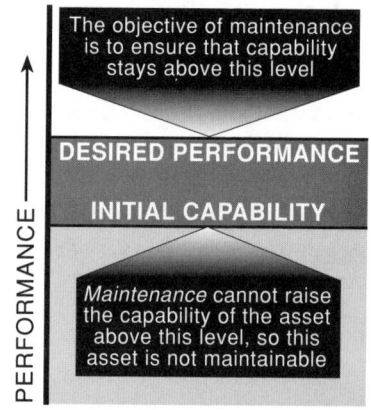

Figure 2.4:
A non-maintainable situation

Two conclusions which can be drawn from the above examples are that:

- for any asset to be maintainable, the desired performance of the asset must fall within the envelope of its initial capability

- in order to determine whether this is so, we not only need to know the initial capability of the asset, but we also need to know exactly what minimum performance the user is prepared to accept in the context in which the asset is being used.

This underlines the importance of identifying precisely *what the users want* when starting to develop a maintenance program. The following paragraphs explore key aspects of performance standards in more detail.

Multiple performance standards
Many function statements incorporate more than one and sometimes several performance standards.

For example, one function of a chemical reactor in a batch-type chemical plant might be listed as:
• To heat up to 500 kg of product X from ambient temperature to boiling point (125°C) in one hour.
In this case, the weight of product, the temperature range and the time all present different performance expectations. Similarly, the primary function of a motor car might be defined as:
• To transport up to 5 people along made roads at speeds of up to 90 mph
Here the performance expectations relate to speed and number of passengers.

Quantitative performance standards
Performance standards should be quantified where possible, because quantitative standards are inherently much more precise than qualitative standards. Special care should be taken to avoid qualitative statements like 'to produce as many widgets as required by production', or 'to go as fast as possible'. Function statements of this type are meaningless, if only because they make it impossible to define exactly when the item is failed.

In reality, it can be extraordinarily difficult to define precisely what is required, but just because it is difficult does not mean that it cannot or should not be done. One major user of RCM summed up this point by saying 'If the users of an asset cannot specify precisely what performance they want from an asset, they cannot hold the maintainers accountable for sustaining that performance.'

Qualitative standards
In spite of the need to be precise, it is sometimes impossible to specify quantitative performance standards so we have to live with qualitative statements.

For instance, the primary function of a painting is usually 'to look acceptable' (if not 'attractive'). What is meant by 'acceptable' varies hugely from person to person and is impossible to quantify. As a result, user and maintainer need to take care to ensure that they share a common understanding of what is meant by words like 'acceptable' before setting up a system intended to preserve that acceptability.

Absolute performance standards

A function statement which contains no performance standard at all usually implies an absolute.

For instance, the concept of containment is associated with nearly all enclosed systems. Function statements covering containment are often written as follows:
• To contain liquid X
The absence of a performance standard suggests that the system must contain *all* the liquid, and that any leakage at all amounts to a failed state. In cases where an enclosed system can tolerate some leakage, the amount which can be tolerated should be incorporated as a performance standard in the function statement.

Variable performance standards

Performance expectations (or applied stress) sometimes vary infinitely between two extremes.

Consider for example a truck used to deliver loads of assorted goods to urban retailers. Assume that the actual loads vary between (say) 0 (empty) and 5 tons, with an average of 2.5 tons, and the distribution of loads is as shown in Figure 2.5. To allow for deterioration, the initial capability of the truck must be more than the 'worst case' load, which in this example is 5 tons. The maintenance program in turn must ensure that the capability does not drop below this level, in which case it would automatically satisfy the full range of performance expectations.

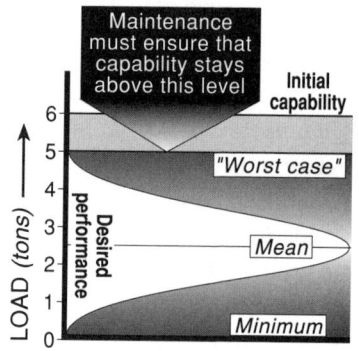

Figure 2.5:
Variable performance standards

Upper and lower limits

In contrast to variable performance expectations, some systems exhibit variable capability. These are systems which simply cannot be set up to function to exactly the same standard every time they operate.

For example, a grinding machine used to finish grind a crankshaft will not produce *exactly* the same finished diameter on every journal. The diameters will vary, if only by a few microns. Similarly, a filling machine in a food factory will not fill two successive containers with exactly the same weight of food. The weights will vary, if only by a few milligrams.

Figure 2.6 indicates that capability variations of this nature usually vary about a mean. In order to accommodate this variability, the associated desired standards of performance incorporate an upper and lower limit.

For instance, the primary function of a sweet-packing machine might be:
• To pack 250±1 gm of sweets into bags at a minimum rate of 75 bags per minute.
The primary function of the grinding machine might be:
• To finish grind main bearing journals in a cycle time of 3.00 ±0.03 minutes to a diameter of 75 ±0.1 mm with a surface finish of Ra0.2.

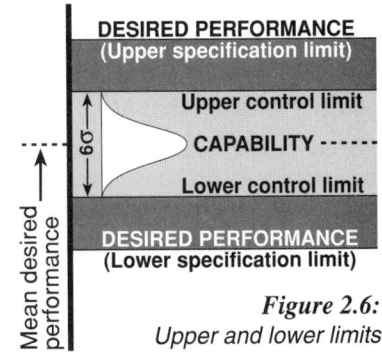

Figure 2.6:
Upper and lower limits

(In practice, this kind of variability is usually unwelcome for a number of reasons. Ideally, processes should be so stable that there is no variation at all and hence no need for two limits. In pursuit of this ideal, many industries are spending a great deal of time and energy on designing processes that vary as little as possible. However, this aspect of design and development is beyond the scope of this book. Right now we are concerned purely with variability from the viewpoint of maintenance.)

How much variability can be tolerated in the specification of any product is usually governed by external factors.

For instance, the lower limit which can be tolerated on the crankshaft journal diameter is governed by factors such as noise, vibration and harshness, and the upper limit by the clearances needed to provide adequate lubrication. The lower limit of the weight of the bag of sweets (relative to the advertised weight) is usually governed by trading standards legislation, while the upper limit is governed by the amount of product which the company can afford to give away.

In cases like these, the desired performance limits are known as the upper and lower *specification* limits. The limits of capability (usually defined as being three standard deviations either side of the mean) are known as the upper and lower *control* limits. Quality management theory suggests that in a well managed process, the difference between the control limits should ideally be half the difference between the specification limits. This multiple should allow a more than adequate margin for deterioration from a maintenance viewpoint.

Upper and lower limits not only apply to product quality. They also apply to other functional specifications such as the accuracy of gauges and the settings of control systems and protective devices. This issue is discussed further in Chapter 3.

2.3 The Operating Context

In Chapter 1, RCM was defined as 'a process used to determine the maintenance requirements of any physical asset in its operating context'. This context pervades the entire maintenance strategy formulation process, starting with the definition of functions.

For example, consider a situation where a maintenance program is being developed for a truck used to transport material from Startsville to Endburg. Before the functions and associated performance standards of this vehicle can be defined, the people developing the program need to ensure that they thoroughly understand the operating context.

For instance, how far is Startsville from Endburg? Over what sort of roads and what sort of terrain? What are the 'typical worst case' weather and traffic conditions on this route? What load is the truck carrying (fragile? corrosive? abrasive? explosive?) What speed limits and other regulatory constraints apply to the route? What fuel facilities exist along the way?

The answers to these questions might lead us to define the primary function of this vehicle as follows: 'To transport up to 40 tons of steel slabs at speeds of up to 60 mph (average 45 mph) from Startsville to Endburg on one tank of fuel'.

The operating context also profoundly influences the requirements for secondary functions. In the case of the truck, the climate may demand air conditioning, regulations may demand special lighting, the remoteness of Endburg may demand that special spares be carried on board, and so on.

Not only does the context drastically affect functions and performance expectations, but it also affects the nature of the failure modes which could occur, their effects and consequences, how often they happen and what must be done to manage them.

For instance, consider again the pump shown in Figure 2.1. If it were moved to a location where it pumps mildly abrasive slurry into a Tank B from which the slurry is being drawn at a rate of 900 liters per minute, the primary function would be:
• To pump slurry into Tank B at not less than 900 liters per minute.
This is a higher performance standard than in the previous location, so the standard to which it has to be maintained rises accordingly. Because it is now pumping slurry instead of water, the nature, frequency and severity of the failure modes also change. As a result, although the pump itself is unchanged, it is likely to end up with a completely different maintenance program in the new context.

All this means that anyone setting out to apply RCM to any asset or process must ensure that they have a crystal clear understanding of the operating context before they start. Some of the most important factors which need to be considered are discussed in the following paragraphs.

Batch and flow processes
In manufacturing plants, the most important feature of the operating context is the type of process. This ranges from flow process operations where nearly all the equipment is interconnected, to jobbing operations where most of the machines are independent.

In flow processes, the failure of a single asset can either stop the entire plant or significantly reduce output, unless surge capacity or standby plant is available. On the other hand, in batch or jobbing plants, most failures only curtail the output of a single machine or line. The consequences of such failures are determined mainly by the duration of the stoppage and the amount of work-in-process queuing in front of subsequent operations.

These differences mean that the maintenance strategy applied to an asset which is part of a flow process could be radically different from the strategy applied to an identical asset in a batch environment.

Redundancy
The presence of redundancy – or alternative means of production – is a feature of the operating context which must be considered in detail when defining the functions of any asset.

The importance of redundancy is illustrated by the three identical pumps shown in Figure 2.7. Pump B has a stand-by, while pump A does not.

Figure 2.7:
Different
operating
contexts

Stand Alone | Duty | Stand-by

This means that the primary function of pump A is to transfer liquid from one point to another *on its own*, and that of pump B to do it *in the presence of a standby*. This difference means that the maintenance requirements of these pumps will be different (just how different we see later), even though the pumps are identical.

Quality standards
Quality standards and standards of customer service are two more aspects of the operating context which can lead to differences between the descriptions of the functions of otherwise identical machines.

For example, identical milling stations on two transfer machines might have the same basic function – to mill a workpiece. However, depth of cut, cycle time, flatness tolerance and surface finish specifications might all be different. This could lead to quite different conclusions about their maintenance requirements.

Environmental standards
An increasingly important aspect of the operating context of any asset is the impact which it has (or could have) on the environment. Growing worldwide interest in environmental issues means that when we maintain any asset, we actually have to satisfy two sets of 'users'. The first is the people who operate the asset itself. The second is society as a whole, which wants both the asset and the process of which it forms part not to cause undue harm to the environment.

What society wants is expressed in the form of increasingly stringent environmental standards and regulations. These are international, national, regional, municipal or even corporate standards. They cover an extraordinarily wide range of issues, from the biodegradability of detergents to the content of exhaust gases. In the case of processes, they tend to concentrate on unwanted liquid, solid and gaseous by-products.

Most industries are responding to society's environmental expectations by ensuring that equipment is designed to comply with the associated standards. However, it is not enough simply to ensure that a plant or process is environmentally sound at the moment it is commissioned. Steps also have to be taken to ensure that it remains in compliance throughout its life.

Taking the right steps is becoming a matter of urgency, because all over the world, more and more incidents which seriously affect the environment are occurring because some physical asset did not behave as it should – in other words, because something failed. The associated penalties are becoming very harsh indeed, so long-term environmental integrity is now a particularly important issue for maintenance people.

Safety hazards
An increasing number of organizations have either developed themselves or subscribe to formal standards concerning acceptable levels of risk. In some cases, these apply at corporate level, in others to individual sites and in yet others to individual processes or assets. Clearly, wherever such standards exist, they are an important part of the operating context.

Shift arrangements
Shift arrangements profoundly affect the operating context. Some plants operate for eight hours per day five days a week (and even less in bad times). Others operate continuously for seven days a week, and yet others somewhere in between.

In a single shift plant, production lost due to failures can usually be made up by working overtime. This overtime leads to increased production costs, so maintenance strategies are evaluated in the light of these costs.

On the other hand, if an asset is working 24 hours per day, seven days per week, it is seldom possible to make up for lost time, so downtime causes lost sales. This costs a great deal more than extra overtime, so it is worth trying much harder to prevent failures under these circumstances. However, it is also more difficult to make equipment available for maintenance in a fully-loaded plant, so maintenance strategies need to be formulated with special care.

As products move through their life cycles or as economic conditions change, organizations can move from one end of this spectrum to the other surprisingly quickly. For this reason, it is wise to review maintenance policies every time this aspect of the operating context changes.

Work-in-process
Work-in-process refers to any material which has not yet been through all the steps of the manufacturing process. It may be stored in tanks, in bins, in hoppers, on pallets, on conveyors or in special stores. The consequences of the failure of any machine are greatly influenced by the amount of this work-in-process between it and the next machines in the process.

Consider an example where the volume of work in the queue is sufficient to keep the next operation working for six hours and it only takes four hours to repair the failure mode under consideration. In this case, the failure would be unlikely to affect overall output. Conversely, if it took eight hours to repair, it could affect overall output because the next operation would come to a halt. The severity of these consequences in turn depends on
• the amount of work-in-process between that operation and the next and so on down the line, and
• the extent to which any of the operations affected is a bottleneck operation (in other words an operation which governs the output of the whole line).

Although plant stoppages cost money, it also costs money to hold stocks of work-in-process. Nowadays stock-holding costs of any kind are so high that reducing them to an absolute minimum is a top priority. This is a major objective of 'just-in-time' systems and their derivatives.

These systems reduce work-in-process stocks, so the cushion that the stocks provided against failure is rapidly disappearing. This is a vicious circle, because the pressure on maintenance departments to reduce failures in order to make it possible to do without the cushion is also increasing.

So from the maintenance viewpoint, a balance has to be struck between the economic implications of operational failures, and:

- the cost of holding work-in-process stocks in order to mitigate the effects of those failures, or

- the cost of doing proactive maintenance tasks with a view to anticipating or preventing the failures.

To strike this balance successfully, this aspect of the operating context must be particularly clearly understood in manufacturing operations.

Repair time

Repair times are influenced by the *speed of response* to the failure, which is a function of failure reporting systems and staffing levels, and the *speed of repair* itself, which is a function of the availability of spares and appropriate tools and of the capability of the person doing the repairs.

These factors heavily influence the effects and the consequences of failures, and they vary widely from one organization to another. As a result, this aspect of the operating context also needs to be clearly understood.

Spares

It is possible to use a derivative of the RCM process to optimize spares stocks and the associated failure management policies. This derivative is based on the fact that the *only* reason for keeping a stock of spare parts is to avoid or reduce the consequences of failure.

The relationship between spares and failure consequences hinges on the time it takes to procure spares from suppliers. If it could be done instantly there would be no need to stock any spares at all. But in the real world procuring spares takes time. This is known as the lead time, and it ranges from a matter of minutes to several months or years. If the spare is not a stock item, the lead time often dictates how long it takes to repair the failure, and hence the severity of its consequences. On the other hand, holding spares in stock also costs money, so a balance needs to be struck, on a case-by-case basis, between the cost of holding a spare in stock and the total cost of not holding it. In some cases, the weight and/or dimensions of the spares also need to be taken into account because of load and space restrictions, especially in facilities like oil platforms and ships.

This spares optimization process is beyond the scope of this book. However, when applying RCM to an existing facility, one has to start somewhere. In most cases, the best way to deal with spares is as follows:

- use RCM to develop a maintenance strategy based on existing spares holding policies,

- review the failure modes associated with key spares on an exception basis, by establishing what impact (if any) a change in the present stockholding policy would have on the initial maintenance strategy, and then picking the most cost-effective maintenance strategy/spares holding policy.

If this approach is adopted, then the existing spares holding policy can be seen as part of the (initial) operating context.

Market demand
The operating context sometimes features cyclic variations in demand for the products or services provided by the organization.

For example, soft drink companies experience greater demand for their products in summer than in winter, while urban transport companies experience peak demand during rush hours.

In cases like these, the operational consequences of failure are much more serious at the times of peak demand, so in this type of industry, this aspect of the operating context needs to be especially clearly understood when defining functions and assessing failure consequences.

Raw material supply
Sometimes the operating context is influenced by cyclic fluctuations in the supply of raw materials. Food manufacturers often experience periods of intense activity during harvest times and periods of little or no activity at other times. This applies especially to fruit processors and sugar mills. During peak periods, operational failures not only affect output, but can lead to the loss of large quantities of raw materials if these cannot be processed before they deteriorate.

Documenting the operating context
For all the above reasons, it is essential to ensure that everyone involved in the development of a maintenance program for any asset fully understands the operating context of that asset. The best way to do so is to document the operating context, if necessary up to and including the overall mission statement of the entire organization, as part of the RCM process.

Figure 2.8 overleaf shows a hypothetical operating context statement for the grinding machine mentioned earlier. The crankshaft is used in a type of engine used in motor car model X.

Make car model X
(Corresponding asset: Model X Car Division)

Model X division employs 4 000 people to produce 220 000 cars this year. Sales forecasts indicate that this could rise to 320 000 per year within 3 years. We are now number 18 in national customer satisfaction rankings, and intend to reach 15th place next year and 10th place the following year. The target for lost time injuries throughout the division is one per 500 000 paid hours. The probability of a fatality occurring anywhere in the division should be less than one in 50 years. The division plans to conform to all known environmental standards.

Make engines
(Corresponding asset: Motown Engine Plant)

The Motown Engine Plant produces all the engines for model X cars. 140 000 Type 1 and 80 000 Type 2 engines are produced per year. In order to achieve the customer satisfaction targets for the entire vehicle, warranty claims for engines must drop from the present level of 20 per 1000 to 5 per 1000. The plant suffered three reportable environmental excursions last year – our target is not more than one in the next three years. The plant shuts down for two weeks per year to allow production workers to take their main annual vacations.

Make Type 2 engines
(Corresponding asset:Type 2 Engine Line)

The Type 2 engine line presently works 110 hours per week (2 x 10 hr shifts 5 days per week and one 10 hour shift on Saturdays). The assembly line could produce 140 000 engines per year in these hours if it ran continuously with no defects, but overall output of engines is limited by the speed of the crankshaft manufacturing line. The company would like as much maintenance as possible to be done during normal hours without interfering with production.

Machine crankshafts
(Corresponding asset: Crankshaft machining line 2)

The crankshaft line consists of 25 operations, and is nominally able to produce 20 crankshafts per hour (2 200 per week, 110 000 per 50 week year). It currently sometimes fails to produce the requirement of 1 600 per week in normal time. When this happens, the line has to work overtime at an additional cost of £800 per hour. (Since most of the forecast growth will be for Type 2 engines, stoppages on this line could eventually lead to lost sales of model X cars unless the performance is improved.) There should be no crankshafts stored between the end of the crankshaft line and the engine assembly line, but operations in fact keep a pallet of about 60 crankshafts to provide some 'insurance' against stoppages. This enables the crankshaft line to stop for up to 3 hours without stopping assembly. Crankshaft machining defects have not caused any warranty claims, but the scrap rate on this line is 4%. The initial target is 1.5%.

Finish grind crankshaft main and big end journals
(Corresponding asset: Ajax Mark 5 grinding machine)

The finish grinding machine grinds 5 main and 4 big end journals. It is the bottleneck operation on the crankshaft line, and the cycle time is 3.0 minutes. The finished diameter of the main journals is 75mm ± 0.1mm, and of the big ends 53mm ± 0.1 mm. Both journals have a surface finish of Ra0.2. The grinding wheels are dressed every cycle, a process which takes 0.3 minutes out of each 3 minute cycle. The wheels need to be replaced after 3 500 crankshafts, and replacement takes 1.8 hours. There are usually about ten crankshafts on the conveyor between this machine and the next operation, so a stoppage of 25 minutes can be tolerated without interfering with the next operation. Total buffer stocks on the conveyors between this machine and the end of the line mean that this machine can stop for about 45 minutes before the line as a whole stops. Finish grinding contributes 0.4% to the present overall scrap rate.

Figure 2.8: An operating context statement

The hierarchy starts with the division of the corporation which produces this model, but it could have gone up one level further to include the entire corporation. Note also that a context statement at any level should apply to *all* the assets below it in the hierarchy, not just the asset under review.

The context statements at the higher levels in this hierarchy are simply broad function statements. Performance standards at the highest levels quantify expectations from the viewpoint of the overall business. At lower levels, performance standards become steadily more specific until one reaches the asset under review. The primary and secondary functions of the asset at this level are defined as described in the rest of this chapter.

2.4 Different Types of Functions

Every physical asset has more than one – often several – functions. If the objective of maintenance is to ensure that the asset can continue to fulfil these functions, then they must *all* be identified together with their current desired standards of performance. At first glance, this may seem to be a fairly straightforward exercise. However in practice it nearly always turns out to be the single most challenging and time-consuming aspect of the maintenance strategy formulation process,

This is especially true of older facilities. Products change, plant configurations change, people change, technology changes and performance expectations change – but still we find assets in service that have been there since the plant was built. Defining precisely what they are supposed to be doing *now* requires very close cooperation between maintainers and users. It is also usually a profound learning experience for everyone involved.

Functions are divided into two main categories (***primary*** and ***secondary*** functions) and then further divided into various sub-categories. These are reviewed on the following pages, starting with primary functions.

Primary functions

Organizations acquire physical assets for one, possibly two, seldom more than three main reasons. These 'reasons' are defined by suitably worded function statements. Because they are the 'main' reasons why the asset is acquired, they are known as ***primary functions***. They are the reasons why the asset exists at all, so care should be taken to define them as precisely as possible.

Primary functions are usually fairly easy to recognize. In fact, the names of most industrial assets are based on their primary functions.

For instance the primary function of a packing machine is to pack things, of a crusher to crush something and so on.

As mentioned earlier, the real challenge lies in defining the current performance expectations associated with these functions. For most types of equipment, the performance standards associated with primary functions concern speeds, volumes and storage capacities. Product quality also usually need to be considered at this stage.

Chapter 1 mentioned that our ability to achieve and sustain satisfactory quality standards depends increasingly on the capability and condition of the assets which produce the goods. These standards are usually associated with primary functions. As a result, take care to incorporate product quality criteria into primary function statements where relevant. These include *dimensions* for machining, forming or assembly operations, *purity standards* for food, chemicals and pharmaceuticals, *hardness* in the case of heat treatment, filling *levels* or *weights* for packaging, and so on.

Functional block diagrams

If an asset is very complex or if the interaction between different systems is poorly understood, it is sometimes helpful to clarify the operating context by drawing up functional block diagrams. These are simply diagrams showing *all* the primary functions of an enterprise at any given level. They are discussed in more detail in Appendix 1.

Multiple independent primary functions

An asset can have more than one primary function. For instance, the very name of a military fighter/bomber suggests that it has two primary functions. In such cases, both should be listed in the functional specification.

A similar situation is often found in manufacturing, where the same asset may be used to perform different functions at different times. For instance, a single reactor vessel in a chemical plant might be used at different times to reflux (boil continuously) three different products under three different sets of conditions, as follows:

Product	1	2	3
Pressure	2 bar	10 bar	6 bar
Temperature	180°C	120°C	140°C
Batch size	500 liters	600 liters	750 liters

(It could be said that this vessel is not performing three different functions, but that it is performing the same function to different standards of performance. In fact, the distinction does not matter because we arrive at the same conclusion either way.)

In cases like this, one could list a separate function statement for each product. This would logically lead to three separate maintenance programs for the same asset. Three programs may be feasible – perhaps even desirable – if each product runs continuously for very long periods.

However, if the interval between long-term maintenance tasks is longer than the change-over intervals, then it is impractical to change the tasks every time the machine is changed over to a different product.

One way around this problem is to combine the 'worst case' standards associated with each product into one function statement.

In the above example, a combined function statement could be 'to reflux up to 750 liters of product at temperatures up to 180°C and pressures up to 10 bar.'

This will lead to a maintenance program which might embody some over-maintenance some of the time, but which will ensure that the asset can handle the worst stresses to which it will be exposed.

Serial or dependent primary functions
One often encounters assets which must perform two or more primary functions in series. These are known as serial functions.

For instance, the primary functions of a machine in a food factory may be 'to fill 300 cans with food per minute' and then 'to seal 300 cans per minute'.

The distinction between multiple primary functions and serial primary functions is that in the former case, each function can be performed independently of the other, while in the latter, one function must be performed before the other. In other words, for the canning machine to work properly it must fill the cans before it seals them.

Secondary Functions

Most assets are expected to fulfil one or more functions in addition to their primary functions. These are known as *secondary functions*.

For example, the primary function of a motor car might be described as follows:
• to transport up to 5 people at speeds of up to 90 mph along made roads
If this was the only function of the vehicle, then the only objective of the maintenance program for this car would be to preserve its ability to carry up to 5 people at speeds of up 90 mph along made roads. However, this is only part of the story, because most car owners expect far more from their vehicles, ranging from the ability to carry luggage to the ability to indicate how much fuel is in the fuel tank.

To help ensure that none of these functions are overlooked, they are divided into seven categories as follows:

- environmental integrity
- safety/structural integrity
- control/containment/comfort
- appearance
- protection
- economy/efficiency
- superfluous functions.

The first letters of each line in this list form the word ESCAPES. Although secondary functions are usually less obvious than primary functions, the loss of a secondary function can still have serious consequences – sometimes more serious than the loss of a primary function. As a result, secondary functions often need as much if not more maintenance than primary functions, so they too must be clearly identified. The following pages explore the main categories of these functions in more detail.

Environmental integrity
Part 2 of this chapter explained how society's environmental expectations have become a critical feature of the operating context of many assets. RCM begins the process of compliance with the associated standards by incorporating them in appropriately worded function statements.

For instance, one function of a car exhaust or a factory smoke stack might be 'to contain no more than X micrograms of a specified chemical per cubic meter'. The car exhaust system might also be the subject of environmental restrictions dealing with noise, and the associated functional specification might be 'to emit no more than X dB measured at a distance of Y meters behind the exhaust outlet'

Safety
Most users want to be reasonably sure that their assets will not hurt or kill them. In practice, most safety hazards emerge later in the RCM process as failure modes. However, in some cases it is necessary to write function statements which deal with specific threats to safety.

For instance, two safety-related functions of a toaster are 'to prevent users from touching electrically live components' and 'not to burn the users'.

Many processes and components are unable to fulfil the safety expectations of users on their own. This has given rise to additional functions in the form of protective devices. These devices pose some of the most difficult and complex challenges facing the maintainers of modern industrial plant. As a result, they are dealt with separately below.

A further subset of safety-related functions are those which deal with product contamination and hygiene. These are most often found in the food and pharmaceutical industries. The associated performance standards are usually tightly specified, and lead to rigorous and comprehensive maintenance routines (cleaning and testing/validation).

Structural integrity

Many assets have a structural secondary function. This usually involves supporting some other asset, sub-system or component.

For example, the primary function of the wall of a building might be to protect people and equipment from the weather, but it might also be expected to support the roof (and bear the weight of shelves and pictures).

Large, complex structures with multiple load bearing paths and high levels of redundancy need to be analyzed using a specialized version of RCM. Typical examples of such structures are airframes, the hulls of ships and the structural elements of offshore oil platforms.

Structures of this type are rare in industry in general, so the relevant analytical techniques are not covered in this book. However, straightforward, single-celled structural elements can be analyzed in the same way as any other function described in this chapter.

Control

In many cases, users not only want assets to fulfil functions to a given standard of performance, but they also want to be able to regulate the performance. This expectation is summarized in separate function statements.

For instance, the primary function of a car as suggested earlier was 'to transport up to 5 people at speeds of up to 90 mph along made roads'. One control function associated with this function could be 'to enable driver to regulate speed at will between -10 mph (reverse) and +90 mph'.

Indication or feedback forms an important subset of the control category of functions. This includes functions which provide operators with real-time information about the process (gauges, indicators, telltales, VDU's and control panels), or which record such information for later analysis (digital or analog recording devices, cockpit voice recorders in aircraft, etc). Performance standards associated with these functions not only relate to the ease with which it should be possible to read and assimilate or to playback the information, but also cover its accuracy.

For instance, the function of the speedometer of a car might be described as 'to indicate the road speed to the driver to within +5 -0% of the actual speed'.

Containment

In the case of assets used to *store* things, a primary function is to contain whatever is being stored. However, containment should also be acknowledged as a secondary function of all devices used to *transfer* material of any sort – especially fluids. This includes pipes, pumps, conveyors, chutes, hoppers and pneumatic and hydraulic systems.

Containment is also an important secondary function of items like gearboxes and transformers. (In this context, note again the remarks on Page 26 about performance standards and containment.)

Comfort

Most people expect their assets not to cause them anxiety, grief or pain. These expectations are listed under the heading of 'comfort' because the major English dictionaries define comfort as being freedom from anxiety, pain, grief and so on. (These expectations can also be classified under the heading of 'ergonomics'.)

Too much discomfort affects morale, so it is undesirable from a human viewpoint. It is also bad business because people who are anxious or in pain are more likely to make incorrect decisions. Anxiety is caused by poorly explained, unreliable or unintelligible control systems, be they for domestic appliances or for oil refineries. Pain is caused by assets – especially clothing and furniture – which are incompatible with the people using them.

The best time to deal with these problems is of course at the design stage. However, deterioration and/or changing expectations can cause this category of functions to fail like any other. The best way to set about ensuring that this doesn't happen is to define appropriate functional specifications.

For example, one function of a control panel might be 'To indicate clearly to a color-blind operator up to five feet away whether pump A is operating or shut down'. A control-room chair might be expected 'To allow operators to sit comfortably for up to one hour at a time without inducing drowsiness'.

Appearance

The appearance of many items embodies a specific secondary function. For instance, the primary function of the paintwork on most industrial equipment is to protect it from corrosion, but a bright color might be used to enhance its visibility for safety reasons. Similarly, the main func-tion of a sign outside a factory is to show the name of the company which occupies the premises, but a secondary function is to project an image.

Protective devices

As physical assets become more complex, the number of ways they can fail is growing almost exponentially. This has led to corresponding growth in the variety and severity of failure consequences. In an attempt to eliminate (or at least to reduce) these consequences, increasing use is being made of automatic protective devices. These work in one of five ways:

* to draw the attention of the operators to abnormal conditions *(warning lights and audible alarms which respond to failure effects. The effects are monitored by a variety of sensors including level switches, load cells, overload or overspeed devices, vibration or proximity sensors, temperature or pressure switches, etc)*

* to shut down the equipment in the event of a failure *(these devices also respond to failure effects, using the same types of sensors and often the same circuits as alarms, but with different settings)*

* to eliminate or relieve abnormal conditions which follow a failure and which might otherwise cause much more serious damage *(fire-fighting equipment, safety valves, rupture discs or bursting discs, emergency medical equipment)*

* to take over from a function which has failed *(standby plant of any sort, redundant structural components)*

* to prevent dangerous situations from arising in the first place *(guards)*.

The purpose of these devices is to protect people from failures or to protect machines or to protect products – in some cases all three.

Protective devices ensure that the failure of the function being protected is much less serious than it would be if there were no protection. The presence of protection also means that the maintenance requirements of a protected function are often less stringent than they would be otherwise.

Consider a milling machine whose milling cutter is driven by a toothed belt. If the belt were to break in the absence of any protection, the feed mechanism would drive the stationary cutter into the workpiece (or vice versa) and cause serious secondary damage. This can be avoided in two ways:
• by implementing a comprehensive proactive maintenance routine designed to prevent the failure of the belt
• by providing protection such as a broken belt detector to shut down the machine as soon as the belt breaks. In this case, the only consequence of a broken belt is a brief stoppage while it is replaced, so the most cost-effective maintenance policy might simply be to let the belt fail. But *this policy is only valid if the broken belt detector is working*, and steps must be taken to ensure that this is so.

The maintenance of protective devices – especially devices which are not fail-safe – is discussed in much more detail in Chapters 5 and 8. However, this example demonstrates two fundamental points:

- that protective devices often need more routine maintenance attention than the devices they are protecting
- that we cannot develop a sensible maintenance program for a protected function without also considering the maintenance requirements of the protective device.

It is only possible to consider the maintenance requirements of protective devices if we understand their functions. So when listing the functions of any asset, we must list the functions of *all* protective devices.

A final point about protective devices concerns the way their functions should be described. These devices act by exception (in other words when something else goes wrong), so it is important to describe them correctly. In particular, protective function statements should include the words 'if' or 'in the event of', followed by a very brief summary of the circumstances or the event which would activate the protection.

For instance, if we were to describe the function of a tripwire as being 'to stop the machine', anyone reading this description could be forgiven for thinking that the tripwire is the normal stop/start device. To remove any ambiguity, the function of a tripwire should be described as follows:
• *to be capable of* stopping the machine *in the event of* an emergency at any point along its length
The function of a safety valve may be described as follows:
• *to be capable of* relieving the pressure in the boiler *if* it exceeds 250 psi.

Economy/efficiency

Anyone who uses assets of any sort only has finite financial resources. This leads them to put a limit on what they are prepared to spend on operating and maintaining it. How much they are prepared to spend is governed by a combination of three factors:
• the actual extent of their financial resources
• how much they want whatever the asset will do for them
• the availability and cost of competitive ways of achieving the same end.
At the operating context level, functional expectations concerning costs are usually spelled out in the form of expenditure budgets.

At the asset level, economic issues can be addressed directly by function statements which define what users expect in areas such as fuel economy and loss of process materials.

For instance, a car might be expected 'to travel at least 35 miles per gallon of fuel at a constant 65 mph, and at least 50 miles per gallon of fuel at 35 mph'. A fossil fuel power station might be expected 'to export at least 45% of the latent energy in the fuel as electrical power.' A plant using an expensive solvent might want 'to lose no more than 0.5% of solvent X per month'.

Superfluous Functions

Items or components are sometimes encountered which are completely superfluous. This usually happens when equipment has been modified frequently over a period of years, or when new equipment has been over-specified. (These comments do not apply to redundant components built in for safety reasons, but to items which serve no purpose at all in the context under consideration.)

For example, a pressure reducing valve was built into the supply line between a gas manifold and a gas turbine. The original function of the valve was to reduce the gas pressure from 120 psi to 80 psi. The system was later modified to reduce the manifold pressure to 80 psi, after which the valve served no useful purpose.

It is sometimes argued that items like these do no harm and it costs money to remove them, so the simplest solution may be to leave them alone until the whole plant is decommissioned. Unfortunately, this is seldom true in practice. Although these items have no positive function, they can still fail and so reduce overall system reliability. To avoid this, they need maintenance, which means that they still consume resources.

It is not unusual to find that between 5% and 20% of the components of complex systems are superfluous in the sense described above. If they are eliminated, it stands to reason that the same percentage of maintenance problems and costs will also be eliminated. However, before this can be done with confidence, the functions of these components first need to be identified and clearly understood.

A Note on Reliability

There is often a temptation to write 'reliability' function statements such as 'to operate 7 days a week, 24 hours per day'. In fact, reliability is not a function in its own right. It is a performance expectation which pervades all the other functions. It is properly dealt with by dealing appropriately with each of the failure modes which could cause each loss of function. This issue is discussed further in Chapter 13.

Using the ESCAPES Categories
There will often be doubt about which of the ESCAPES categories some functions belong to. For instance, should the function of a seat reclining mechanism be classified under the heading of 'control' or 'comfort'?

In practice the precise classification does not matter. What does matter is that we identify and define all the functions which are likely to be expected by the user. The list of categories merely serves as an aide memoire to help ensure that none of these expectations are overlooked.

2.5 How Functions should be Listed

A properly written functional specification – especially one which is fully quantified – precisely defines the objectives of the enterprise. This ensures that everyone involved knows exactly what is wanted, which in turn ensures that maintenance activities remain focused on the real needs of the users (or 'customers'). It also makes it easier to absorb changes triggered by changing expectations without derailing the whole enterprise.

Functions are listed on RCM Information Worksheets in the left hand column. Primary functions are listed first and the functions are numbered numerically, as shown in Figure 2.9. (These functions apply to the exhaust system of a 5 megawatt gas turbine.)

A complete Information Worksheet is shown at the end of Chapter 4.

Figure 2.9:
Describing functions

RCM II INFORMATION WORKSHEET © 1996 ALADON LTD	SYSTEM / SUB-SYSTEM

	FUNCTION
1	To channel all the hot turbine exhaust gas without restriction to a fixed point 10 meters above the roof of the turbine hall
2	To reduce exhaust noise levels to ISO Noise Rating 30 at 150 meters
3	To ensure that the surface temperature of the ducting inside the turbine hall does not exceed 60°C
4	To transmit a warning signal to the turbine control system if the exhaust gas temperature exceeds 475°C and a shutdown signal if it exceeds 500°C at a point 4 meters from the turbine
5	To allow free movement of the ducting in response to temperature changes

3 Functional Failures

Chapter 1 explained that the RCM process entails asking seven questions about selected assets, as follows:

- *what are the functions and associated performance standards of the asset in its present operating context?*
- *in what ways does it fail to fulfil its functions?*
- *what causes each functional failure?*
- *what happens when each failure occurs?*
- *in what way does each failure matter?*
- *what can be done to predict or prevent each failure?*
- *what should be done if a suitable proactive task cannot be found?*

Chapter 2 dealt at length with the first question. After a brief introduction to the general concept of failure, this chapter considers the second question, which deals with *functional failures*.

3.1 Failure

In the previous chapter, we saw that people or organizations acquire assets because they want the assets to do something. Not only that, but they also expect their assets to fulfil the intended functions to an acceptable stan-dard of performance.

Chapter 2 went on to explain that for any asset both to do what its users want and to allow for deterioration, the initial capability of the asset must exceed the desired standard of performance. Thereafter, as long as the capability of the asset continues to exceed the desired standard of performance, the user will be satisfied.

On the other hand, if for any reason the asset is unable to do what the user wants, the user will consider it to have failed.

This leads to a basic definition of failure:

'Failure' is defined as the inability of any asset to do what its users want it to do.

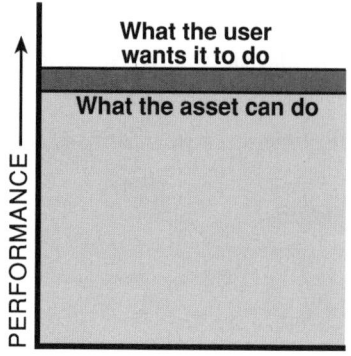

This is illustrated in Figure 3.1.

For instance, if the pump shown in Figure 2.1 on Page 22 is unable to pump 800 liters per minute, it will not be able to keep the tank full and so its users will regard it as 'failed'.

Figure 3.1:
The general failed state

3.2 Functional Failures

The above definition treats the concept of failure as if it applies to an asset as a whole. In practice, this definition is vague because it does not distinguish clearly between the failed state (functional failure) and the events which cause the failed state (failure modes). It is also simplistic, because it does not take into account the fact that each asset has more than one function, and each function often has more than one desired standard of performance. The implications are explored in the following paragraphs.

Functions and Failures

We have seen that an asset is failed if it doesn't do what its users want it to do. We have also seen that what anything must do is defined as a function and that every asset has more than one and often several different functions. Since it is possible for each one of these functions to fail, it follows that any asset can suffer from a variety of different failed states.

For instance, the pump in Figure 2.1 has at least two functions. One is to pump water at not less than 800 liters/minute, and the other is to contain the water. It is perfectly feasible for such a pump to be capable of pumping the required amount (*not failed* in terms of its primary function) while leaking excessively (*failed* in terms of the secondary function).

Conversely, it is equally possible for the pump to deteriorate to the point where it cannot pump the required amount (*failed* in terms of its primary function), while it still contains the liquid (*not failed* in terms of the secondary function).

This shows why it is more accurate to define failure in terms of the loss of specific functions rather than the failure of an asset as a whole. It also shows why the RCM process uses the term 'functional failure' to describe failed states, rather than 'failure' on its own. However, to complete the definition of failure, we also need to look more closely at the question of performance standards.

Performance Standards and Failure

As discussed in the first part of this chapter, the boundary between satisfactory performance and failure is specified by a performance standard. Given that performance standards apply to individual functions, 'failure' can be defined precisely by defining a functional failure as follows:

A functional failure is defined as the inability of any asset to fulfil a function to a standard of performance which is acceptable to the user

The following paragraphs discuss different aspects of functional failure under the following headings:
• partial and total failure
• upper and lower limits
• gauges and indicators
• the operating context.

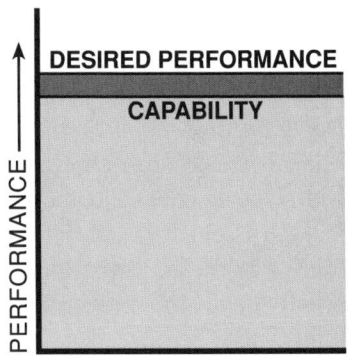

Partial and total failures
The above definition of functional failure covers complete loss of function. It also covers situations where the asset still functions, but performs outside acceptable limits.

Figure 3.2:
Functional failure

For example, the primary function of the pump discussed earlier is 'to pump water from tank X to Tank Y at not less than 800 liters/minute'. This function could suffer from two functional failures, as follows:
• fails to pump any water at all
• pumps water at less than 800 liters per minute.

Partial failure is nearly always caused by different failure modes from total failure, and the consequences are different. This is why *all* the functional failures which could affect each function should be recorded.

Record all the functional failures associated with each function.

Note that partial failure should not be confused with the situation where the asset deteriorates slightly but its capability remains above the level of performance required by the user.

For example, the initial capability of the pump in Figure 2.1 is 1 000 liters per minute. Impeller wear is inevitable, so this capability will decline. As long as it does not decline to the point where the pump is unable to pump 800 liters per minute, it will still be able to fill the tank and so keep the users satisfied in the context described.

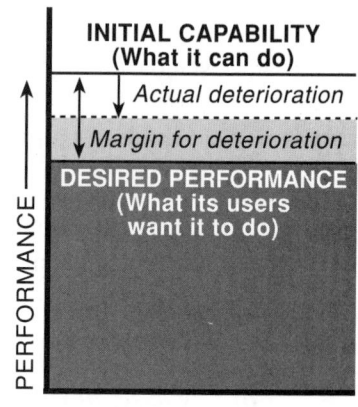

Figure 3.3:
Asset still OK despite
some deterioration

However if the capability of the asset deteriorates so much that it falls below the desired performance, its users will consider it to have failed.

Upper and lower limits
The previous chapter explained that the performance standards associated with some functions incorporate upper and lower limits. Such limits mean that the asset has failed if it produces products which are over the upper limit *or* below the lower limit. In these cases, the breach of the upper limit usually needs to be identified *separately* from the breach of the lower limit. This is because the failure modes and/or the consequences associated with going over the upper limit are usually different from those associated with going below the lower limit.

For example, the primary function of a sweet-packing machine was listed in Chapter 2 as being 'To pack 250±1 gm of sweets into bags at a minimum rate of 75 bags per minute'. This machine has failed:
• if it stops altogether
• if it packs more than 251 gm of sweets into any bags
• if it packs less than 249 gm into any bags
• if it packs at a rate of less than 75 bags per minute.

The function of a crankshaft grinding machine was listed as 'To finish grind main bearing journals in a cycle time of 3.00 ±0.03 minutes to a diameter of 75 ±0.1 mm with a surface finish of Ra 0.2'.
• Completely unable to grind workpiece
• Grinds workpiece in a cycle time longer than 3.03 minutes
• Grinds workpiece in a cycle time less than 2.97 minutes
• Diameter exceeds 75.1 mm
• Diameter is below 74.9 mm
• Surface finish too rough.

Of course, if only one limit applies to a particular parameter, then only one failed state is possible. For instance, the absence of a lower limit on the roughness specification in the above example suggests that it is not possible to make the item too smooth. In some circumstances, this may not actually be true, so care needs to be taken to verify this point when analyzing functions of this type.

In practice, the failed states associated with upper and lower limits can manifest themselves in two ways. Firstly, the spread of capability could breach the specification limits in one direction only. This is illustrated in Figure 3.4, which shows that this type of failed state can be likened to a number of shots hitting a target in a tight group but way off center.

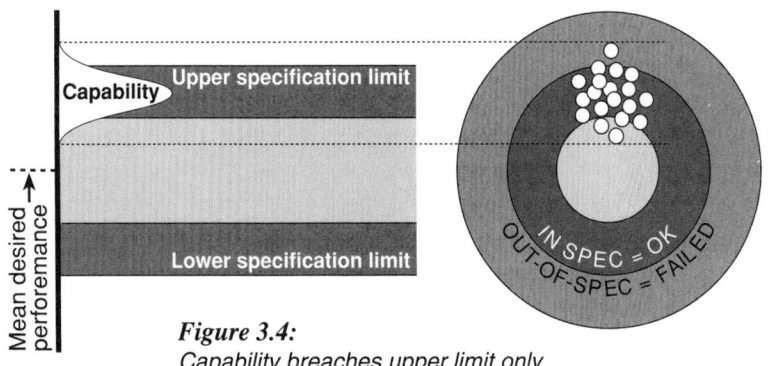

Figure 3.4:
Capability breaches upper limit only

The second failed state occurs when the spread of capability is so broad that it breaches both the upper and the lower specification limits. Figure 3.5 shows that this can be likened to shots scattered all over the target.

Note that in both of the above cases, not all of the products produced by the processes in question will be failed. If the breach is minor, only a small percentage of out-of-spec products will be produced. However, the further off centre the grouping in the first case, or the broader the spread in the second case, the higher will be the percentage of failures.

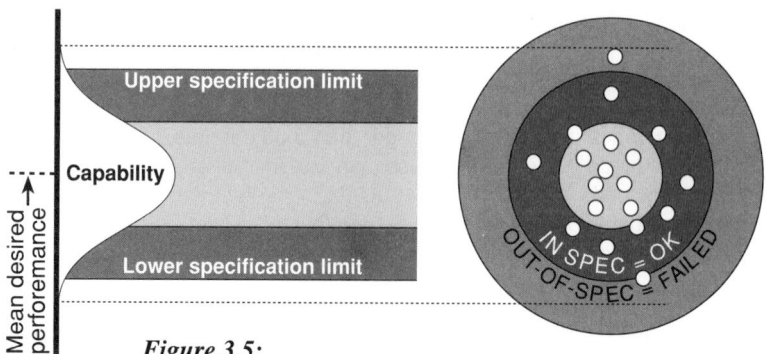

Figure 3.5:
Capability breaches upper and lower limits

Figure 2.6 illustrated a process which is in control and in specification. Figures 3.4 and 3.5 show that processes which are out of control and out of spec are in a failed state. The failure modes which can cause these failed states are discussed in the next chapter. (Chapter 7 deals with the implications of a process which is out-of-control but within specification.)

Gauges and indicators

The above discussion has tended to focus on product quality. Chapter 2 mentioned that upper and lower limits also apply to the performance standards associated with gauges, indicators, protection and control systems. Depending on failure modes and consequences, it may also be necessary to treat the breach of these limits separately when listing functional failures.

For instance, the function of a temperature gauge could be listed as 'to display the temperature of process X to within *(say)* 2% of the actual process temperature'. This gauge can suffer from three functional failures, as follows:
• fails altogether to display process temperature
• displays a temperature more than 2% higher than the actual temperature
• displays a temperature more than 2% lower than the actual temperature.

Functional failures and the operating context

The exact definition of failure for any asset depends very much on its operating context. This means that in the same way that we should not generalize about the functions of identical assets, so we should take care not to generalize about functional failures.

For example, we saw how the pump shown in Figure 2.1 fails if it is completely unable to pump water, and if it is able to pump less than 800 liters/minute. If the same pump is used to fill a tank from which water is drawn at 900 liters/minute, the second failed state occurs if the throughput drops below 900 liters/minute.

Who should set the standard?

An issue which needs careful consideration when defining functional failures is the 'user'. To this day, most maintenance programs in use around the world are compiled by maintenance people working on their own. These people usually decide for themselves what is meant by 'failed'.

In practice, their view of failure often turns out to be quite different from that of the users, with sometimes disastrous consequences for the effectiveness of their programs.

For example, one function of a hydraulic system is to contain oil. How well it should fulfil this function can be subject to widely differing points of view. There are production managers who believe that a hydraulic leak only amounts to a functional failure if it is so bad that the equipment stops working altogether. On the other hand, a maintenance manager might suggest that a functional failure has occurred if the leak causes excessive consumption of hydraulic oil over a long period of time. Then again, a safety officer might say that a functional failure has occurred if the leak creates a pool of oil on the floor in which people could slip and fall or which might create a fire hazard. This is illustrated in Figure 3.6.

Figure 3.6:
Different views
about failure

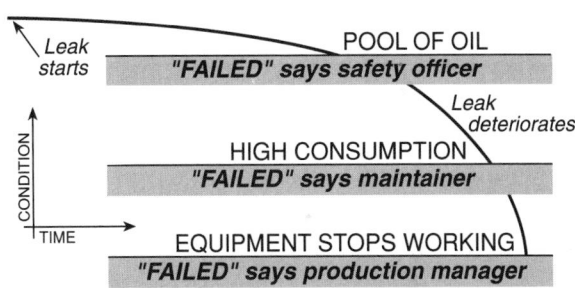

The maintenance manager (who controls the hydraulic oil budget) may ask the operators for access to the hydraulic system to repair leaks 'because oil consumption is excessive'. However access may be denied because the operators think the machine 'is still working OK'. When this happens, the maintenance people (1) record that the machine 'was not released for preventive maintenance', and (2) form the opinion that their production colleagues 'don't believe in PM'. For similar reasons, the maintenance manager might not release a maintenance person to repair a small leak when requested to do so by the safety officer.

In fact, all three parties almost certainly do believe in prevention. The real problem is that they have not taken the trouble to agree exactly what is meant by 'failed' so they do not share a common understanding of what they are seeking to prevent.

This example illustrates three key points:

- the performance standard used to define functional failure – in other words, the point where we say 'so far and no further' – defines the level of proactive maintenance needed to avoid that failure (in other words, to sustain the required level of performance)

- much time and energy can be saved if these performance standards are clearly established *before the failures occur*

- the performance standards used to define failure *must* be set by operations and maintenance people working together with anyone else who has something legitimate to say about how the asset should behave.

How Functional Failures should be Listed

Functional failures are listed in the second column of the RCM Information Worksheet. They are coded alphabetically, as shown in Figure 3.7.

RCM II INFORMATION WORKSHEET © 1996 ALADON LTD	SYSTEM	*5 MW Turbine*	
	SUB-SYSTEM	*Exhaust System*	
FUNCTION			**FUNCTIONAL FAILURE**
1	To channel all the hot turbine exhaust gas without restriction to a fixed point 10 meters above the roof of the turbine hall	A	Unable to channel gas at all
		B	Gas flow restricted
		C	Fails to contain the gas
		D	Fails to convey gas to a point 10 m above the roof
2	To reduce exhaust noise levels to ISO Noise Rating 30 at 150 meters	A	Noise level exceeds ISO Noise Rating 30 at 150 meters
3	To ensure that duct surface temperature inside turbine hall does not rise above 60°C	A	Duct surface temperature exceeds 60°C
4	To transmit a warning signal to the control system if exhaust gas temperature exceeds 475°C and a shutdown signal if it exceeds 500°C at a point 4 metres from the turbine	A	Incapable of sending a warning signal if exhaust temperature exceeds 475°C
		B	Incapable of sending a shutdown signal if exhaust temperature exceeds 500°C
5	To allow free movement of ducting in response to temperature changes	A	Does not allow free movement of ducting

Figure 3.7: Describing functional failures

4 Failure Modes and Effects Analysis (FMEA)

We have seen that by defining the functions and desired standards of performance of any asset, we are defining the objectives of maintenance with respect to that asset. We have also seen that defining functional failures enables us to spell out exactly what we mean by 'failed'. These two issues were addressed by the first two questions of the RCM process.

The next two questions seek to identify the *failure modes* which are reasonably likely to cause each functional failure, and to ascertain the *failure effects* associated with each failure mode. This is done by performing a *failure modes and effects analysis* (FMEA) for each functional failure.

This chapter describes the main elements of an FMEA, starting with a definition of the term 'failure mode'.

4.1 What is a Failure Mode?

A failure mode could be defined as any event which is likely to cause an asset (or system or process) to fail. However, Chapter 3 explained that it is both vague and simplistic to apply the term 'failure' to an asset as a whole. It is much more precise to distinguish between a 'functional failure' (a failed *state)* and a 'failure mode' (an *event* which could cause a failed state). This distinction leads to the following more precise definition of a failure mode:

> *A failure mode is any event*
> *which causes a functional failure*

The best way to show the connection and the distinction between failed states and the events which could cause them, is to list functional failures first, then to record the failure modes which could cause each functional failure, as shown in Figure 4.1 overleaf.

RCM II INFORMATION WORKSHEET © 1996 ALADON LTD	SYSTEM	*Cooling Water Pumping System*	
	SUB-SYSTEM		

FUNCTION		FUNCTIONAL FAILURE (Loss of Function)		FAILURE MODE (Cause of Failure)		
1	To transfer water from tank X to tank Y at not less than 800 liters/minute	A	Unable to transfer any water at all	1	Bearing seizes	
				2	**Impeller comes adrift**	
				3	**Impeller jammed by foreign object**	
				4	Coupling hub shears due to fatigue	
				5	Motor burns out	
				6	Inlet valve jams closed	
					... etc	
		B	Transfers less than 800 liters per minute	1	**Impeller worn**	
				2	Partially blocked suction line	
					... etc	

Figure 4.1: Failure modes of a pump

Figure 4.1 also indicates that at the very least, a description of a failure mode should consist of a noun and a verb. The description should contain enough detail for it to be possible to select an appropriate failure management strategy, but not so much detail that excessive amounts of time are wasted on the analysis process itself.

In particular, the verbs used to describe failure modes should be chosen with care, because they strongly influence the subsequent failure management policy selection process. For instance, verbs such as 'fails' or 'breaks' or 'malfunctions' should be used sparingly, because they give little or no indication as to what might be an appropriate way of managing the failure. The use of more specific verbs makes it possible to select from the full range of failure management options.

For example, a term like 'coupling fails' provides no clue as to what might be done to anticipate or prevent the failure. However, if we say 'coupling bolts come loose' or 'coupling hub fails due to fatigue', then it becomes much easier to identify a possible proactive task.

In the case of valves or switches, one should also indicate whether the loss of function is caused by the item failing in the open or closed position – 'valve jams closed' says more than 'valve fails'. In the interests of complete clarity, it may sometimes be necessary to take this one step further.

For instance, 'valve jams closed due to rust on lead screw' is clearer than 'valve jams closed'. Similarly, one might need to distinguish between 'bearing seizes due to normal wear and tear' and 'bearing seizes due to lack of lubrication'.

These issues are discussed at length later in this chapter, but first we look at why we need to analyze failure modes at all.

4.2 Why Analyze Failure Modes?

A single machine can fail for dozens of reasons. A group of machines or system such as a production line can fail for hundreds of reasons. For an entire plant, the number can rise into the thousands or even tens of thousands. Most managers shudder at the thought of the time and effort likely to be involved in identifying all these failure modes. Many decide that this type of analysis is just too much work, and abandon the whole idea entirely. In doing so, these managers overlook the fact that on a day-to-day basis, *maintenance is really managed at the failure mode level*. For instance:

- work orders or job requests are raised to cover specific failure modes
- day-to-day maintenance planning is all about making plans to deal with specific failure modes
- in most industrial undertakings, maintenance and operations people hold meetings every day. The meetings usually consist almost entirely of discussions about what has failed, what caused it (and who is to blame), what is being done to repair it and – sometimes – what can be done to stop it happening again. In short, the entire meeting is spent discussing failure modes
- to a large extent, technical history recording systems record individual failure modes (or at least, what was done to rectify them).

Sadly, in too many cases, these failure modes are discussed, recorded or otherwise dealt with *after* they have occurred. Dealing with failures after they have happened is of course the essence of *reactive* maintenance.

Proactive management on the other hand, means dealing with events *before* they occur – or at least, deciding how they should be dealt with if they were to occur. In order to do this, we need to know beforehand what events are likely to occur. The 'events' in this context are failure modes. So if we wish to apply truly proactive maintenance to any physical asset, we must try to identify all the failure modes which are reasonably likely to affect that asset. Ideally, they should be identified before they occur at all, or if this is not possible, before they occur again.

Once each failure mode has been identified, it then becomes possible to consider what happens when it occurs, to assess its consequences and to decide what (if anything) should be done to anticipate, prevent, detect or correct it – or perhaps even to design it out.

So the maintenance task selection process – and much of the subsequent management of these tasks – is carried out at the failure mode level. This is briefly illustrated in the following example, and then discussed at length in the remaining chapters of this book:

Consider again the RCM Information Worksheet shown in Figure 4.1. This applies to the primary function of the pump first shown in Figure 2.1. Figure 4.2 shows that the pump is a direct-coupled single-stage back-pull-out end-suction volute pump sealed by a mechanical seal. In this example, we look more closely at the three failure modes which are thought to be likely to affect the impeller only. These are discussed in some detail below and summarized in Figure 4.2:

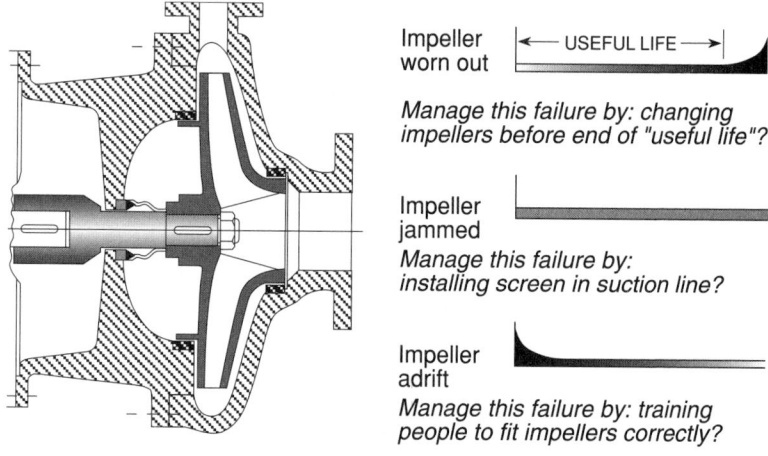

Impeller worn out |◄── USEFUL LIFE ──►|

Manage this failure by: changing impellers before end of "useful life"?

Impeller jammed

Manage this failure by: installing screen in suction line?

Impeller adrift

Manage this failure by: training people to fit impellers correctly?

Figure 4.2: *Failures of the impeller of a centrifugal pump*

- *Impeller wear* is likely to be an age-related phenomenon. As shown in Figure 4.1, this means that it is likely to conform to the second of the six failure patterns introduced in Figure 1.5 on Page 12 (Failure Pattern B). So if we know roughly what the useful life of the impeller is, and if the consequences of the failure are serious enough, then we may decide to *prevent this failure* by changing the impeller just before the end of the useful life.

- *Impeller jammed by foreign object:* The likelihood of a foreign object appearing in the suction line will almost certainly have nothing to do with how long the impeller has been in service. As a result, it stands to reason that this failure mode will occur on a random basis (Pattern E in Figure 1.5). There would also be no warning of the fact that the failure is about to occur. So if the consequences were serious enough, and the failure happened often enough, we would be likely to consider *modifying the system*, perhaps by installing some sort of filter or screen in the suction line.

- *Impeller adrift:* If the impeller fastening mechanism is adequately designed and it still keeps coming adrift, this would almost certainly be because it wasn't put on properly in the first place. (If we knew that this was so, then perhaps the failure mode should actually be described as 'Impeller fitted incorrectly'.) This in turn means that the failure mode is most likely to occur soon after start up, as shown in Figure 4.2 (Pattern F in Figure 1.5), and we would probably deal with it by improving the relevant *training or procedures.*

This example reinforces the point that the level at which we manage the maintenance of any asset is not at the level of the asset as a whole (in this case, the pump), and not even at the level of any component (in this case, the impeller), but at the level of each failure mode. So before we can develop a systematic, proactive maintenance management strategy for any asset, *we must identify what these failure modes are* (or could be).

The example also suggests that one of the failure modes could be eliminated by a design change and another by improving training or procedures. So *not every failure mode is dealt with by scheduled maintenance.* Chapters 5 to 9 describe an orderly approach to deciding what *is* likely to be the most suitable way of dealing with each failure.

Note also that the failure management solutions proposed in Figure 4.2 represent only one of several possibilities in each case.

For instance we could monitor impeller wear by monitoring the pump performance and only change the impeller when it needs it. We also need to bear in mind that adding a screen to the suction line adds three more failure possibilities, which need to be analyzed in turn (it could block up, it could be holed and therefore cease to screen, and it could disintegrate and damage the impeller.)

Chapters 6 to 9 examine these alternatives in more detail.

These points all indicate that the identification of failure modes is one of the most important steps in the development of any program intended to ensure that any asset continues to fulfil its intended functions. In practice, depending on the complexity of the item, its operating context and the level at which it is being analyzed, between one and thirty failure modes are usually listed per functional failure.

The next two sections of this chapter consider two of the key issues in this area under the following headings:

- categories of failure modes
- level of detail.

Thereafter, the last three parts of the chapter consider failure effects, sources of information for an FMEA, and how failure modes and effects should be listed.

4.3 Categories of Failure Modes

Some people regard maintenance as being all about – and only about – dealing with deterioration. Some even go so far as to specify that FMEA's carried out on their assets should deal only with failure modes caused by deterioration, and should ignore other categories of failure modes (such as human errors and design flaws). This is unfortunate, because it often transpires that deterioration causes a surprisingly small proportion of failures. In these cases, restricting the analysis to deterioration only can lead to a woefully incomplete maintenance strategy.

On the other hand, if one accepts that maintenance means ensuring that physical assets continue to do whatever their users want them to do, then a comprehensive maintenance program must address *all* the events that are reasonably likely to threaten that functionality. Failure modes can be classified into one of three groups, as follows:
• when capability falls below desired performance
• when desired performance rises above initial capability
• when the asset is not capable of doing what is wanted from the outset.
Each of these categories is discussed in the following paragraphs.

Falling Capability

The first category of failure modes covers situations where capability is above desired performance to begin with, but then drops below desired performance after the asset is put into service, as illustrated in Figure 4.3.

The five principal causes of reduced capability are listed below:
• deterioration
• lubrication failures
• dirt
• disassembly
• 'capability reducing' human errors.

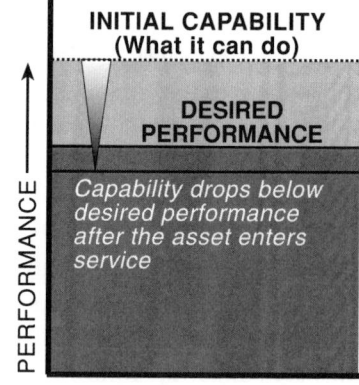

Figure 4.3:
Failure Mode Category 1

Deterioration
Any physical asset that fulfils a function which brings it into contact with the real world is subject to a variety of stresses. These stresses cause the

asset to deteriorate by lowering its capability, or more accurately, its *resistance to stress*. Eventually resistance drops so much that the asset can no longer deliver the desired performance – in other words, it fails. Deterioration covers all forms of 'wear and tear' (fatigue, corrosion, abrasion, erosion, evaporation, degradation of insulation, etc). These failure modes should of course be included in a list of failure modes wherever they are thought to be reasonably likely. The level of detail with which they need to be recorded is discussed in the next part of this chapter.

Lubrication failure
Lubrication is associated with two types of failure modes. The first concerns lack of lubricant, and the second the failure of the lubricant itself.

With regard to lack of lubrication, things have changed considerably in the last two decades. Twenty years ago, the majority of lubrication points were replenished manually. The cost of lubricating each of these points was tiny compared to the cost of not doing so. It was also tiny compared to the cost of analyzing the lubrication requirements of each point in detail. This meant that it was just not worth carrying out an in-depth analytical exercise to set up a lubrication program. Instead, these programs were usually set up on the basis of a quick survey by a lubrication specialist.

Nowadays however, 'sealed-for-life' components and centralized lubrication systems have become the norm in most industries. This has led to a massive reduction in the number of points where a human has to apply oil or grease to a machine, and a massive increase in the consequences of failure (especially the failure of centralized lubrication systems). From the analytical viewpoint, this means that it is now cost-effective to:

• use RCM to analyze centralized lubrication systems in their own right

• consider the loss of lubricant in the few remaining manually lubricated points as individual failure modes.

The second category of failures associated with lubrication concerns deterioration of the lubricant itself. It is caused by phenomena such as shearing of the oil molecules, oxidation of the base oil and additive depletion. In some cases, deterioration of the oil may be aggravated by the build-up of sludge or the presence of water or other contaminants. A lubricant may also fail to do its job simply because the wrong lubricant has been used. If any or all of these failures modes are considered to be likely in the context under consideration, then they should be recorded and subjected to further analysis. (This also applies to transformer oil and hydraulic oil.)

Dirt

Dirt or dust is a very common cause of failure. It interferes directly with machines by causing them to block, stick or jam. It is also a principal cause of the failure of functions which deal with the appearance of assets (things which should look clean look dirty). Dirt can also cause product quality problems, either by getting into the clamping mechanisms of machine tools and causing misalignment, or by getting directly into products such as food, pharmaceuticals or the oilways of engines. As a result, failures caused by dirt should be listed in the FMEA whenever they are likely to interfere with a significant function of the asset.

Disassembly

If components fall off machines, assemblies fall apart or whole machines come adrift, the consequences are usually very serious, so the relevant failure modes should be listed. These are usually the failure of welds, soldered joints or rivets due to fatigue or corrosion, or the failure of threaded components such as bolts, electrical connections or pipefittings which can also fail due to fatigue or corrosion, or which simply come undone.

Also take care to record the functions and associated failure modes of locking mechanisms such as split pins and lock nuts when considering the integrity of assemblies.

Human errors which reduce capability

The final subset of the 'falling capability' category of failure modes are those caused by human error. As the name implies, these refer to errors which reduce the capability of the process to the extent that it is unable to function as required by the user.

Examples include manually operated valves left shut causing a process to be unable to start, parts incorrectly fitted by maintenance craftsmen or sensors set in such a way that a machine trips out when nothing is wrong.

If failure modes of this type are known to occur, they should be recorded in the FMEA so that appropriate failure management decisions can be made later in the process. However, when listing failure modes caused by people, take care simply to record *what* went wrong and not *who* caused it. If too much emphasis is placed on 'who' at this stage, the analysis could become unnecessarily adversarial, and people begin to lose sight of the fact that it is an exercise in avoiding or solving problems, not attaching blame. For instance, it is enough to say 'control valve set too high', not 'control valve incorrectly set by instrument technician'.

Increase in Desired Performance *(or Increase in Applied Stress)*

The second category of failure modes occurs when desired performance is within the envelope of the capability of the asset when it is first put into service, but then the desired performance increases until it falls outside the capability envelope. This causes the asset to fail in one of two ways:

- the desired performance rises until the asset can no longer deliver it, or
- the increase in stress causes deterioration to accelerate to the extent that the asset becomes so unreliable that it is effectively useless.

An example of the first case occurs if the users of the pump shown in Figure 2.1 were to increase the offtake from the tank to 1 050 liters per minute. Under these circumstances, the pump is unable to keep the tank full. (Note that in this case, the users are not forcing the pump to work any faster – they have simply opened a valve a bit wider somewhere else in the system.)

The second case occurs for instance if the owner of a motor car whose engine is 'red-lined' at 6 000 rpm persists in revving the motor to 7 000 rpm. This causes the engine to deteriorate more quickly than if the user keeps the revs within the prescribed limits, so it fails more often.

This phenomenon is illustrated in Figure 4.4. It occurs for four reasons, the first three of which embody some kind of human error:

- sustained, deliberate overloading
- sustained, unintentional overloading
- sudden, unintentional overloading
- incorrect process material.

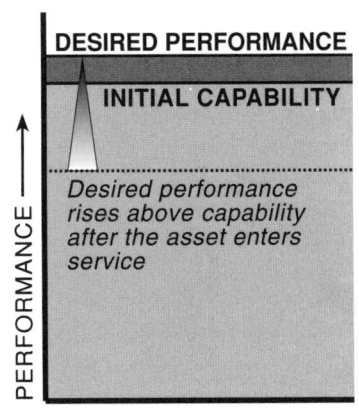

Figure 4.4:
Failure Mode Category 2

Sustained, deliberate overloading
In many industries, users quickly give in to the temptation simply to speed up equipment in response to increased demand for existing products. In other cases, people use assets acquired for one product to process a product with different characteristics (such as larger, heavier unit sizes or higher quality standards). People do this in the belief that they will get more out of their facilities without any increase in capital investment. This may even be true in the short term. However, this solution carries long-term penalties in terms of reduced reliability and/or availability, especially when the increased stresses begin to approach or exceed the ability of the asset to withstand them.

(This phenomenon causes some of the most ferocious disputes between maintenance and operations people. When it occurs, operations people tend to claim that 'there must be something wrong with our maintenance', while maintenance accuses operations of 'flogging the machine to death'. These disputes occur because operations people usually focus on what they want out of each asset, while maintenance people tend to think in terms of what it can do. Neither of them are 'wrong' – they are simply considering the problem from two different points of view.)

In these cases, implementing 'better' maintenance procedures will do little or nothing to solve the problem. In fact, maintaining a machine which cannot deliver the desired performance has been likened to rearranging the deck chairs on the *Titanic*. In such cases we need to look beyond maintenance for solutions. The two options are to modify the asset to improve its inherent capability, or to lower our expectations and operate the machine within its existing capabilities.

Sustained, unintentional overloading
Many industries respond to increased demand by undertaking formal 'debottlenecking' programs. These programs entail increasing the capability of a production facility – such as a production line – to accommodate a new level of desired performance. However, much to the chagrin of their sponsors, these programs often seem to end up causing more problems than they solve. This usually happens because a few small subsystems or components get left out of the overall upgrade program, with sometimes devastating results. How this occurs is illustrated in Figure 4.5.

Demand for the products produced by the facility illustrated in the example has increased to the extent that its users wish to increase output from 400 to 500 tons per week. The dotted lines represent the capability of each operation. They show that most of the operations are already capable of meeting the new requirement. However, operations 3, 8 and 10 are capable of less than 500 tons, so they are the 'bottlenecks'. To achieve the new target, the users 'debottleneck' these operations by installing new machines or components which are capable of producing well over 500 tons per week. They also upgrade the power supplies to match.

However, in this example the need to upgrade the instrument air supply was overlooked, so the plant begins to suffer intermittent instrument problems when demand for instrument air is at a maximum. (Note also that although the unchanged operations were already capable of more than 500 tons, their margin for deterioration is reduced by the upgrade program, so they also begin to fail more often.)

Clearly, if a plant is suffering from failure modes of this type, they should be recorded in the FMEA so that they can be dealt with appropriately.

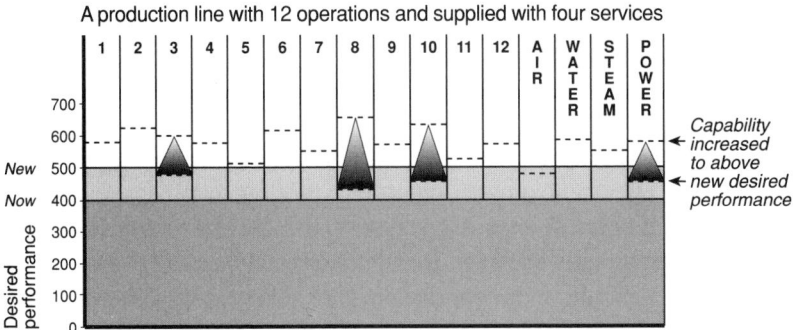

Figure 4.5: The Destabilizing Impact of 'Debottlenecking'

(Some industrial organizations have found that despite the best efforts of their engineers, debottlenecking usually causes so much instability that it is forbidden in all but the most tightly controlled and heavily restricted circumstances. In these cases, growth is handled by allowing for it in the design of the original plant and/or by building new plants.)

Sudden, unintentional overloading
Many failures are caused by sudden and (usually) unintentional increases in applied stress, usually caused in turn by one of the following:

- incorrect operation (for instance, if a machine is put into reverse while moving forward)
- incorrect assembly (for instance, overtorquing a bolt)
- external damage (for instance, if a fork lift truck smashes into a pump or lightning strikes a poorly protected electrical installation).

These are not actually increases in *desired* performance, because no-one wants the operator to put the machine into reverse at the wrong moment or the fork lift to smash the pump. However, they belong in this category because applied stress rises above the ability of the asset to withstand it.

If any of these failure modes are thought to be reasonably likely in the context under consideration, they should be incorporated in the FMEA.

Incorrect process or packaging materials
Manufacturing processes often suffer functional failures caused by process materials which are out of specification (in terms of such variables as consistency, hardness or pH). Similarly, packaging plants often suffer from inadequate or incompatible packaging materials.

In both cases, the machines fail or run badly because they cannot handle the out-of-spec material. This can be seen as an increase in applied stress. In practice, these 'failure modes' are seldom the result of a failure of the asset under review, but are nearly always the effect of a failure elsewhere in the system. This means that remedial action has to be applied to a different asset. However, acknowledging these failures in the analysis of the affected asset helps to ensure that they will receive attention when the system which is really causing the problem is analyzed. As a result, these failure modes should be incorporated in the FMEA where they are known to affect the asset under review, with a comment in the failure effects column which directs attention to the real source of the problem.

Initial incapability

Chapter 2 explained that for any asset to be maintainable, its desired performance must fall within the envelope of its initial capability. It went on to mention that the majority of assets are in fact built this way. However, situations do arise where desired performance is outside the envelope of the initial capability right from the outset, as shown in Figure 4.6.

This incapability problem seldom affects entire assets. It usually affects just one or two functions of one or two components, but these weak links upset the operation of the whole chain. The first step towards rectifying design problems of this nature is to list them as failure modes in an FMEA.

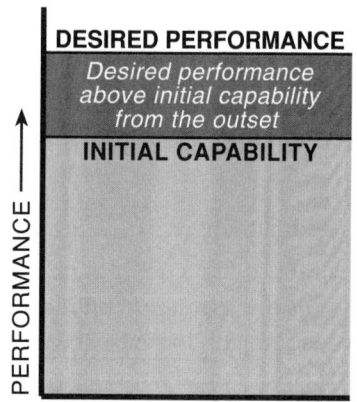

Figure 4.6: Failure Mode Category 3

4.4 How Much Detail?

Earlier in this chapter, it was mentioned that failure modes should be described in enough detail for it to be possible to select an appropriate failure management strategy, but not in so much detail that excessive amounts of time are wasted on the analysis process itself.

> *Failure modes should be defined in enough detail for it to be possible to select a suitable failure management policy*

In practice, it can be surprisingly difficult to find an appropriate level of detail. However, it is important to do so, because the level of detail profoundly affects the validity of the FMEA and the amount of time needed to do it. Too little detail and/or too few failure modes lead to superficial and sometimes dangerous analyses. Too many failure modes and/or too much detail causes the entire RCM process to take much longer than it needs to. In extreme cases, excessive detail can cause the process to take two or even three times longer than necessary (a phenomenon known as *analysis paralysis*). This means that it is essential to try to strike the right balance. Some of the key factors which need to be taken into account are discussed in the following paragraphs.

Causation

The causes of any functional failure can be defined to almost any level of detail, and different levels are appropriate in different situations. At one extreme, it is sometimes enough to summarize the causes of a functional failure in one statement, such as 'machine fails'. At the other, we may need to consider what goes wrong at the molecular level and/or explore the remoter corners of the psyche of the operators and maintainers in a bid to define so-called root causes of failure.

The extent to which failure modes can be described at different levels of detail is illustrated in Figure 4.7 on the next three pages.

Figure 4.7 is based on the pump set shown in Figure 4.2, some of whose failure modes were listed in Figure 4.1. Figure 4.7 lists ways in which the pump set might suffer from the functional failure 'unable to transfer any water at all'. These failure modes are considered at seven different levels of detail.

The top level (Level 1) is failure of the pump set as a whole. Level 2 recognizes the failure of the five major components of the pump set - the pump, the drive shaft, the motor, the switchgear and inlet/outlet. Thereafter failures are considered in progressively more detail. When considering this example, please note that
- levels have been defined and failure modes allocated to each level for the purpose of this example only. They are not any kind of universal classification.
- Figure 4.7 does not show all failure possibilities at each level so don't use this example as a definitive model
- it is possible to analyze some of the failure modes at even lower levels than level 7, but it would very seldom be necessary to do so in practice
- the failure modes listed only apply to the functional failure 'unable to transfer water at all'. Figure 4.7 does not show failure modes which would cause other functional failures, such as loss of containment or loss of protection.

LEVEL 1	LEVEL 2	LEVEL 3	LEVEL 4	LEVEL 5	LEVEL 6	LEVEL 7
Pump set fails	Pump fails	Impeller fails	Impeller comes adrift	Mounting nut undone	Nut not tightened correctly	Assembly error
				Mounting nut worn away	Nut eroded/corroded away	
					Nut made of wrong material	Wrong material specified
						Wrong material supplied
				Impeller nut cracked	Impeller nut overtightened	Assembly error
					Nut made of wrong material	Wrong material specified
						Wrong material supplied
				Impeller key sheared	Wrong key steel specified	Design error
						Procurement error
					Wrong key steel supplied	Storekeeping error
						Requisitioning error
			Object smashes impeller	Part in system after maintenance	Assembly error	*See Appendix 2*
				Foreign object enters system	Suction strainer not installed	Assembly error
					Strainer holed by corrosion	
		Casing ruptured	Casing bolts come loose	Casing bolts undertightened	Assembly error	*See Appendix 2*
				Bolt loosened by vibration		
				Casing bolts corroded away		
				Bolts fail due to fatigue		
			Casing joint fails	Joint incorrectly fitted	Assembly error	*See Appendix 2*
				Joint fails due to fretting		
			Casing smashed	Casing smashed by vehicle	Operating error	*See Appendix 2*
					Pump in vulnerable position	Design error
				Smashed by object from sky	Casing hit by meteorite	
					Casing hit by part of aircraft	
		Pump seal fails	Normal wear and tear	Seal abraded		
			Pump runs dry	*See "water supply fails" below*		
			Seal misaligned	Assembly error	*See Appendix 2*	
			Seal faces dirty	Assembly error	*See Appendix 2*	
			Wrong seal fitted	Wrong seal supplied	Procurement error	*See Appendix 2*
					Storekeeping error	*See Appendix 2*
				Wrong seal specified	Design error	*See Appendix 2*
			Damaged seal installed	Pump seal dropped in stores	Storekeeping error	*See Appendix 2*
				Pump seal damaged in transit	Procurement error	*See Appendix 2*

Figure 4.7:
Failure modes at different levels of detail

LEVEL 1	LEVEL 2	LEVEL 3	LEVEL 4	LEVEL 5	LEVEL 6	LEVEL 7
Pump set fails	Motor fails	Bearings seize	Normal wear and tear	Subsurface fatigue on outer race		
				Balls worn away		
			Axial thrust too great	Motor undersized		
			Lubrication failure	Bearing seals fail	Seals damaged on installation	Assembly error
					Seals poorly fitted to bearing	Manufacturing error
				Grease fails	Base oil oxidised	
					Grease liquified	
					Additives depleted	
					Manufacturing error	
				Wrong lubricant installed		
			Bearing wrongly installed	Damaged prior to installation	Bearing dropped in store	Storekeeping error
					Bearing damaged in transit	Procurement error
				Damaged during installation	Bearing hit with hammer	Assembly error
				Bearing misaligned	Assembly error	See Appendix 2
				Defective bearing installed	Defective bearing supplied	Manufacturing error
					Bearing corroded in store	Storekeeping error
				Wrong bearing installed	Wrong bearing specified	Design error
					Wrong bearing supplied	Procurement error
		Motor reverses	Motor wired incorrectly	Assembly error	See Appendix 2	
		Stator winding burns out	Motor insulation fails	Insulation deteriorates	Normal wear and tear	
				Motor operated at high load	Operating error	
				Insulation damp	Motor casing gasket fails	Normal deterioration
						Gasket fitted incorrectly
					Motor casing damaged	Motor dropped in store
						Motor hit by foreign object
					Motor stored in damp area	Storekeeping error
					Casing gasket not fitted	Assembly error
					Water sprayed on motor	Operating error
					Motor casing bolts loose	Assembly error
			Motor overheated	Fan grille blocked by dirt		
			Motor fan fails	Fan fitted the wrong way round	Assembly error	See Appendix 2
				Fan not fitted	Assembly error	See Appendix 2
		Not switched on	Operating error	See Appendix 2		

Figure 4.7 (continued):
Failure modes at different levels of detail

LEVEL 1	LEVEL 2	LEVEL 3	LEVEL 4	LEVEL 5	LEVEL 6	LEVEL 7
Pump set fails	Driveline fails	Shaft shears	Shears due to fatigue	Stress raisers in steps in shaft	Sharp radii specified	Design error
					Wrong radii cut	Manufacturing error
				Defective steel supplied	Steel manufacturing error	
		Drive key shears	Wrong key steel specified	Design error	See Appendix 2	
			Wrong key steel supplied	Procurement error	See Appendix 2	
			Key cut too short	Assembly error	See Appendix 2	
	Valve closed	V/v jammed shut	Valve handle missing	Handle cannibalized	Assembly error	See Appendix 2
			Valve spindle seized	Spindle seized due to corrosion	Grease worn off valve spindle	
					Wrong grease used	Assembly error
		Valve left shut	Operating error	See Appendix 3		
	Power fails	Switchgear fails	Contactor fails open	Contactor points worn		
				Contactor coil burns out		
				Contactor spring fails due to fatigue		
				Contactor points dirty	Points dirty on installation	Assembly error
					Switchbox cover lets in dirt	Cover badly fitted
						Cover not closed properly
		Spurious trip	Overload c/b set too low	Assembly error	See Appendix 2	
				Overload setting drifts		
			Spurious failure of fuse	Wrong fuse fitted	Procurement error	See Appendix 2
				Defective fuse fitted	Defective fuse supplied	Manufacturing error
					Fuse damaged on installation	Assembly error
			Switched off accidentally	Operating error	See Appendix 2	
		Power cable fails	Cable insulation fails	Operating error		
				Insulation deteriorates		
				Insulation manufacturing defect		
			Cable damaged	Cable damaged by impact	Operating error	
				Cable abraded	Cable too long	
					Cable poorly restrained	Design error
			Connection fails	Connection loose	Loosens in service	
					Installed too loose	Assembly error
				Connection corroded	Terminal box fails	Box damaged by impact
						Box cover loose
						Box cover seal not fitted
		Incoming power fails				

Figure 4.7 (continued):
Failure modes at different levels of detail

The first point to emerge from this example is the connection between the level of detail and the number of failure modes listed. The example shows the further one 'drills down' in an FMEA, the larger the number of failure modes that can be listed.

For instance, there are five failure modes listed at level 2 for the pump set in Figure 4.7 but 64 at level 6.

Two more key issues which arise from Figure 4.7 concern 'root causes' and human error. They are discussed below.

Root causes
The term 'root cause' is often used in connection with the analysis of failures. It implies that if one drills down far enough, it is possible to arrive at a final and absolute level of causation. In fact, this is seldom the case.

For instance, in Figure 4.7 the failure mode 'impeller nut overtightened' is listed at level 6, which in turn is caused by an 'assembly error' at level 7. If we were to go down one level further, the assembly error might have occurred because the 'fitter was distracted' (level 8). He might have been distracted because his 'child was ill' (level 9). This failure might have occurred because the 'child ate bad food in restaurant' (level 10).

Clearly, this process of drilling down could go on almost forever – way beyond the point at which the organization doing the FMEA has any control over the failure modes. This is why this chapter stresses repeatedly that the level at which any failure mode *should* be identified is the level at which it is possible to identify an appropriate failure management policy. (This is equally true whether one is carrying out an FMEA before failures occur or a 'root cause analysis' after a failure has occurred.)

    The fact that the level which is appropriate varies for different failure modes means that we do not have to list all failure modes at the same level on the Information Worksheet. Some failure modes might be identified at level 2, others at level 7, and the rest somewhere in between.

For instance, in one particular context, it may be appropriate to list only those failure modes shaded in grey in Figure 4.7. In another context, it may be appropriate for an entire FMEA for an identical pump set to consist of the single failure mode 'pump set fails'. Another context may call for yet another selection.

Obviously, in order to be able to stop at an appropriate level, the people doing such analyses need to be aware of the full range of failure manage-ment policy options. These are discussed at length in Chapters 6 to 9.

    Other factors which influence the level of detail are considered in the rest of this part of this chapter and again in part 7.

Human error

Part 3 of this chapter mentioned a number of general ways in which human error could cause machines to fail. It went on to suggest that if the associated failure modes are thought to be reasonably likely, they should be incorporated in the FMEA. This has been done in Figure 4.7, where all the failure modes ending with the word 'error' are some form of human error. Appendix 2 provides a brief summary of key issues involved in the classification and management of such errors.

Probability

Different failure modes occur at different frequencies. Some may occur regularly, at average intervals measured in months, weeks or even days. Others may be extremely improbable, with mean times between occurrences measured in millions of years. When preparing an FMEA, decisions must be made continuously as to what failure modes are so unlikely that they can safely be ignored. This means that we do not try to list every single failure possibility regardless of its likelihood.

> ***When listing failure modes, do not try to list every single failure possibility regardless of its likelihood***

Only failure modes which might reasonably be expected to occur in the context in question should be recorded. A list of 'reasonably likely' failure modes should include the following:

• *failures which have occurred before* on the same or similar assets. These are the most obvious candidates for inclusion in an FMEA unless the asset has been modified so that the failure cannot occur again. As discussed later, sources of information about these failures include people who know the asset well (your own employees, vendors or other users of the same equipment), technical history records and data banks. In this context note the comments in Part 6 of this chapter about the shortcomings of most technical history records, and in Chapter 12 about the danger of too much reliance on historical data.

• *failure modes which are already the subject of proactive maintenance routines*, and so which would occur if no proactive maintenance was being done. One way to ensure that none of these failure modes have been overlooked is to study existing maintenance schedules and ask "what failure mode would occur if we did not do this task?"

However, a review of existing schedules should only be carried out as a final check after the rest of the RCM analysis has been completed in order to reduce the possibility of perpetuating the status quo. (Some users of RCM are tempted to assume that *all* reasonably likely failure modes are covered by their existing PM systems, and hence that these are the only failure modes which need to be considered in the FMEA. This assumption leads these users to develop the entire FMEA by working backwards from their existing maintenance schedules, and then working forwards again through the last three steps of the RCM process. This approach is usually adopted in the belief that it will speed up or 'streamline' the process. In fact this approach is not recommended because among other shortcomings, it leads to dangerously incomplete RCM analyses.)

- *any other failure modes which have not yet occurred but which are considered to be real possibilities.* Identifying and deciding how to deal with failures which have not happened yet is an essential feature of pro-active management in general and of risk management in particular. It is also one of the most challenging aspects of the RCM process, because it calls for a high degree of judgement. On the one hand, we need to list all reasonably likely failure modes, while on the other we don't want to waste time on failures which have never occurred before and which are extremely unlikely (incredible) in the context in question.

For example, 'sealed-for-life' bearings are installed on the motor driving the pump shown in Figure 4.7. This means that the likelihood of lubrication failure is low – so low that it would not be included in most FMEA's. On the other hand, failure due to lack of lubricant probably would be included in FMEA's prepared for manually lubricated components, centralized lube systems and gearboxes.

However, the decision not to list a failure mode should be tempered by careful consideration of the failure consequences.

Consequences

If the consequences are likely to be very severe indeed, then less likely failure possibilities *should* be listed and subjected to further analysis.

For instance, if the pump set in Figure 4.7 was installed in a food factory or a vehicle assembly plant, the failure mode 'casing smashed by an object falling from the sky' would be dismissed immediately as being laughably unlikely. However, if the pump were pumping something really nasty in a nuclear installation, this failure mode is more likely to be taken seriously even though it is still highly improbable. (Appropriate failure management policies might be either to ban aircraft from flying over the facility, or to design a roof which can withstand a crashing aircraft.)

Another example from Figure 4.7 is 'motor not switched on'. This failure mode is likely to be dismissed on the grounds of improbability in most situations. Even if it does occur, the consequences may be so trivial that it is excluded from the FMEA. (On the other hand, if it could occur and it does matter – especially in cases where things must be switched on in a particular sequence and something could be damaged if they are not – then this failure mode *should* be considered.)

Cause vs Effect

Care should be taken not to confuse causes and effects when listing failure modes. This is a subtle mistake most often made by people who are new to the RCM process.

For example, one plant had some 200 gearboxes, all of the same design and all performing more or less the same function on the same type of equipment. Initially, the following failure modes were recorded for one of these gearboxes:
• Gearbox bearings seize
• Gear teeth stripped.
These failure modes were listed to begin with because the people carrying out the review recalled that each failure had happened in the past to their knowledge (some of the gearboxes were twenty years old). The failures did not affect safety but they did affect production. So the implication was that it might be worth doing preventive tasks like 'check gear teeth for wear' or 'check gearbox for backlash', and 'check gearbox bearings for vibration'. However, further discussion revealed that both failures had occurred because the oil level had not been checked when it should have been, so the gearboxes had actually failed due to lack of oil. What is more, no-one could recall that any of the gearboxes had failed if they had been properly lubricated. As a result, the failure mode was eventually recorded as:
• Gearbox fails due to lack of oil.
This underlined the importance of the obvious proactive task, which was to check the oil level periodically. (This is not to suggest that all gearboxes should be analyzed in this way. Some are much more complex or much more heavily loaded, and so are subject to a wider variety of failure modes. In other cases, the failure consequences may be much more severe, which would call for a more defensive view of failure possibilities.)

Failure Modes and the Operating Context

We have seen how the functions and functional failures of any item are influenced by its operating context. This is also true of failure modes in terms of causation, probability and consequences.

For example, consider the three pumps shown in Figure 2.7. The failure modes which are likely to affect the standby pump (such as brinelling of the bearings, stagnation of water in the pump casing and even the 'borrowing' of key components to use elsewhere in an emergency) are different from those which might affect the duty pump, as set out in Figure 4.7.

Similarly, a vehicle operating in the Arctic would be subject to different failure modes from the same make of vehicle operating in the Sahara desert. Similarly, a gas turbine powering a jet aircraft would have different failure modes from the same type of turbine acting as a prime mover on an oil platform.

These differences mean that great care should be taken to ensure that the operating context is identical before applying an FMEA developed in one set of circumstances to an asset which is used in another. (Note also the comments regarding the use of generic FMEA's in part 6 of this chapter.)

The operating context affects levels of analysis as well as the causes and consequences of failure. As discussed earlier, it might be appropriate to identify failure modes for two identical assets at one level in one operating context and at another level in another.

4.5 Failure Effects

The fourth step in the RCM review process entails listing what happens when each failure mode occurs. These are known as *failure effects.*

Failure effects describe what happens when a failure mode occurs

(Note that failure effects are *not* the same as failure consequences. A failure effect answers the question "what happens?", whereas a failure consequence answers the question "(how) does it matter?".)

A description of failure effects should include all the information needed to support the evaluation of the consequences of the failure. Specifically, when describing the effects of a failure, the following should be recorded:

• what evidence (if any) that the failure has occurred
• in what ways (if any) it poses a threat to safety or the environment
• in what ways (if any) it affects production or operations
• what physical damage (if any) is caused by the failure
• what must be done to repair the failure.

These issues are reviewed in the following paragraphs. Note that one of the objectives of this exercise is to establish whether proactive maintenance is necessary. If we are to do this correctly, we cannot assume that some sort of proactive maintenance is being done already, so the effects of a failure should be described as if nothing was being done to prevent it.

Evidence of Failure

Failure effects should be described in a way which enables the team doing the RCM analysis to decide whether the failure will become evident to the operating crew under normal circumstances.

For instance, the description should state whether the failure causes warning lights to come on or alarms to sound (or both), and whether the warning is given on a local panel or in a central control room (or both).

Similarly, the description should state whether the failure is accompanied (or preceded) by obvious physical effects such as loud noises, fire, smoke, escaping steam, unusual smells, or pools of liquid on the floor. It should also state whether the machine shuts down as a result of the failure.

For example, if we are considering the seizure of the bearings of the pump shown in Figure 3.5, the failure effects might be described as follows (the italics describe what would make it evident to the operators that a failure has occurred):

• Motor trips out and *trip alarm sounds in the control room.* Tank Y *low level alarm sounds after 20 minutes,* and *tank runs dry after 30 minutes.* Downtime required to replace the bearings 4 hours.

In the case of a stationary gas turbine, a failure mode that occurred in practice was the gradual build up of combustion deposits on the compressor blades. These deposits could be partially removed by the periodic injection of special materials into the air stream, a process known as 'jet blasting'. The failure effects were described accordingly as follows:

• Compressor efficiency declines and governor compensates to sustain power output, causing exhaust temperature to rise. Exhaust temperature is displayed on the local control panel and in the central control room. If no action is taken, exhaust gas temperature rises above 475°C under full power. *A high exhaust gas temperature alarm annunciates on the local control panel and a warning light comes on in the central control room.* Above 500°C, *the control system shuts down the turbine.* (Running at temperatures above 475°C shortens the creep life of the turbine blades.) The blades can be partially cleaned by jet blasting, and jet blasting takes about 30 minutes.

This is an unusually complex failure mode, so the description of the failure effects is somewhat longer than usual. The average description of a failure effect usually amounts to between twenty and sixty words.

When describing failure effects, do not prejudge the evaluation of the failure consequences by using the words 'hidden' or 'evident' They are part of the consequence evaluation process, and using them prematurely could bias this evaluation incorrectly.

Finally, when dealing with protective devices, failure effect descriptions should state briefly what would happen if the protected device were to fail while the protective device was unserviceable.

Safety and Environmental Hazards

Modern industrial plant design has evolved to the point that only a small proportion of failure modes present a direct threat to safety or the environment. However, if there is a possibility that someone could get injured or killed as a direct result of the failure, or an environmental standard or regulation could be breached, the failure effect should describe how this could happen. Examples include:

• increased risk of fire or explosions
• the escape of hazardous chemicals (gases, liquids or solids)
• electrocution
• falling objects
• pressure bursts (especially pressure vessels and hydraulic systems)
• exposure to very hot or molten materials
• the disintegration of large rotating components
• vehicle accidents or derailments
• exposure to sharp edges or moving machinery
• increased noise levels
• the collapse of structures
• the growth of bacteria
• ingress of dirt into food or pharmaceutical products
• flooding.

When listing these effects, do not make qualitative statements like "this failure has safety consequences" or "this failure affects the environment". Simply state what happens, and leave the evaluation of the consequences to the next stage of the RCM process.

Note also that we are not only concerned about possible threats to our own staff (operators and maintainers), but also about threats to the safety of customers and the community as a whole. This may call for some research by the team doing the analysis into the environmental and safety standards which govern the process under review.

Secondary Damage and Production Effects

Failure effect descriptions should also help with decisions about operational and non-operational failure consequences. To do so, they should indicate how production is affected (if at all), and for how long. This is usually given by the amount of downtime associated with each failure.

In this context, downtime means the total amount of time the asset would normally be out of service owing to this failure, from the moment it fails until the moment it is fully operational again. As indicated in Figure 4.8, this is usually much longer than the repair time.

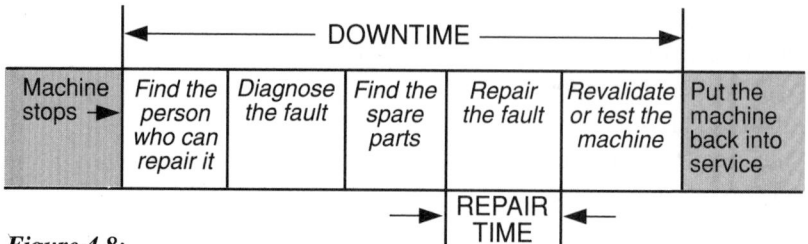

Figure 4.8:
Downtime vs repair time

Downtime as defined above can vary greatly for different occurrences of the same failure, and the most serious consequences are usually caused by the longer outages. Since it is consequences which are of most interest to us, the downtime recorded on the information worksheet should be based on the 'typical worst case'.

For instance, if the downtime caused by a failure which occurs late on a weekend night shift is usually much longer than it is when the failure occurs on a normal day shift, and if such night shifts are a regular occurrence, we list the former.

It is of course possible to reduce the operational consequences of a failure by taking steps to shorten the downtime, most often by reducing the time it takes to get hold of a spare part. However, as discussed in Chapter 2, we are still in the process of defining the problem at this stage so the analysis should be based – at least initially – on current spares holding policies.

Note that if the failure affects operations, it is more important to record downtime than the *mean time to repair* the failure (MTTR), for two reasons:

• in many people's minds, the word 'repair time' has the meaning shown in Figure 4.8. If this is used instead of downtime, it could upset the subsequent assessment of the operational consequences of failure

• we should base the assessment of consequences on the 'typical worst case' and not the 'mean', as discussed above.

If the failure does not cause a process stoppage, then the average amount of time it takes to repair the failure should be recorded, because this can be used help establish manpower requirements.

In addition to downtime, any other ways in which the failure could have a significant effect on the operational capability of the asset should be listed. Possibilities include:

• whether and how product quality or customer service is affected, and if so whether any financial penalties are involved
• whether any other equipment or activity also has to stop (or slow down)
• whether the failure leads to an increase in overall operating costs in addition to the direct cost of repair (such as higher energy costs)
• what secondary damage (if any) is caused by the failure.

Corrective Action

Failure effects should also state what must be done to repair the failure. This can be included in the statement about downtime, as shown in italics in the following examples:
• Downtime *to replace bearings* about four hours
• Downtime *to clear the blockage and reset the trip switch* about 30 minutes
• Downtime to *strip the turbine and replace the disc* about 2 weeks.

4.6 Sources of Information about Modes and Effects

When considering where to get information needed to draw up a reasonably comprehensive FMEA, remember the need to be proactive. This means that as much emphasis should be placed on what could happen as on what has happened. The most common sources of information are discussed in the following paragraphs, together with a brief review of their main advantages and disadvantages.

The manufacturer or vendor of the equipment
When carrying out an FMEA, the source of information which usually springs to mind first is the manufacturer. This is especially so in the case of new equipment. In some industries, this has reached the point where manufacturers or vendors are routinely asked to provide a comprehensive FMEA as part of the equipment supply contract. Apart from anything else, this request implies that manufacturers know everything that needs to be known about how the equipment can fail and what happens when this occurs. This is seldom the case in reality.

In practice, few manufacturers are involved in the day-to-day operation of the equipment. After the end of the warranty period, almost none get regular feedback from the equipment users about what fails and why. The best that many of them can do is try to draw conclusions about how their equipment is performing from a combination of anecdotal evidence and an analysis of spares sales (except when a really spectacular failure occurs, in which case lawyers tend to take over from engineers. At this point, rational technical discussion about root causes often ceases.)

Manufacturers also have little access to information about the operating context of the equipment, desired standards of performance, failure consequences and the skills of the user's operators and maintainers. More often the manufacturers know nothing about these issues. As a result, FMEA's compiled by these manufacturers are usually generic and often highly speculative, which greatly limits their value.

The small minority of equipment manufacturers who are able to produce a satisfactory FMEA on their own usually fall into one of two categories:

• they are involved in maintaining the equipment throughout its useful life, either directly or through closely associated vendors. For instance most privately-owned motor vehicles are maintained by the dealers who sold the vehicles. This enables the dealers to provide the manufacturers with copious failure data.

• they are paid to carry out formal reliability studies on prototypes as part of the initial procurement process. This is a common feature of military procurement, but much less common in industry.

In most cases, the author has found that the best way to access whatever knowledge manufacturers possess about the behavior of the equipment in the field is to ask them to supply experienced field technicians to work alongside the people who will eventually operate and maintain the asset, to develop FMEA's which are satisfactory to both parties. If this suggestion is adopted, the field technicians should of course have unrestricted access to specialist support to help them answer difficult questions.

When adopting this approach, issues such as warranties, copyrights, languages which the participants should be able to speak fluently, technical support, confidentiality, and so on should be handled at the contracting stage, so that everyone knows clearly what to expect of each other.

Note the suggestion to use field technicians rather than designers. Designers are often surprisingly reluctant to admit that their designs can fail, which reduces their ability to help develop a sensible FMEA.

Generic lists of failure modes

'Generic' lists of failure modes are lists of failure modes – or sometimes entire FMEA's – prepared by third parties. They may cover entire systems, but more often cover individual assets or even single components. These generic lists are touted as another method of speeding up or 'streamlining' this part of the maintenance program development process. In fact, they should be approached with great caution, for the following reasons:

- *the level of analysis may be inappropriate:* A generic list may identify failure modes at a level equivalent to (say) level 5 in Figure 4.7, when all that may be needed is level 1. This means that far from streamlining the process, the generic list would condemn the user to analyzing far more failure modes than necessary. Conversely, the generic list may focus on level 3 or 4 in a situation where some of the failure modes really ought to be analyzed at level 5 or 6.

- *the operating context may be different:* The operating context of your asset may have features which make it susceptible to failure modes that do not appear in the generic list. Conversely, some of the modes in the generic list might be extremely improbable (if not impossible) in your context.

- *performance standards may differ:* your asset may operate to standards of performance which mean that your whole definition of failure may be completely different from that used to develop the generic FMEA.

These three points mean that if a generic list of failure modes is used at all, it should only ever be used to supplement a context-specific FMEA, and never used on its own as a definitive list.

Other users of the same equipment

Other users are an obvious and very valuable source of information about what can go wrong with commonly used assets, provided of course that competitive pressures permit the exchange of data. This is often done through industry associations (as in the offshore oil industry), through regulatory bodies (as in civil aviation) or between different branches of the same organization. However, note the above comments about the dan-gers of generic data when considering these sources of information.

Technical history records

Technical history records can also be a valuable source of information. However, they should be treated with caution for the following reasons:

- they are often incomplete
- more often than not, they describe what was done to repair the failure ('replaced main bearing') rather than what caused it
- they do not describe failures which have not yet occurred
- they often describe failure modes which are really the effect of some other failure.

These drawbacks mean that technical history records should only be used as a supplementary source of information when preparing an FMEA, and never as the sole source.

The people who operate and maintain the equipment
In nearly all cases, by far the best sources of information for preparing an FMEA are the people who operate and maintain the equipment on a day-to-day basis. They tend to know the most about how the equipment works, what goes wrong with it, how much each failure matters and what must be done to fix it – and if they don't know, they are the ones who have the most reason to find out.

The best way to capture and to build on their knowledge is to arrange for them to participate formally in the preparation of the FMEA as part of the overall RCM process. The most efficient way to do this is under the guidance of a suitably trained facilitator at a series of meetings. (The most valuable source of additional information at these meetings is a comprehensive set of process and instrumentation drawings, coupled with ready access to process and/or technical specialists on an ad hoc basis.) This approach to RCM was introduced in Chapter 1 and is discussed at much greater length in Chapter 13.

4.7 Levels of Analysis and the Information Worksheet

Part 4 of this chapter showed how failure modes can be described at almost any level of detail. The level of detail which is ultimately selected should enable a suitable failure management policy to be identified. In general, higher levels (less detail) should be selected if the component or sub-system is likely to be allowed to run to failure or subject to failure-finding, while lower levels (more detail) need to be selected if the failure mode is likely to be subjected to some sort of proactive maintenance.

The detail used to describe failure modes on Information Worksheets is also influenced by the level at which the FMEA as a whole is carried out. This in turn is governed by the level at which the entire RCM analysis is performed. For this reason, we review the principal factors which influence the overall level of analysis (which is also known as 'level of indenture') before considering how this affects the detail with which failure modes should be described.

Level of analysis

RCM is defined as a process used to determine what must be done to ensure that any physical asset continues to do whatever its users want it to do in its present operating context. In the light of this definition, we have seen that it is necessary to define the context in detail before we can apply the process. However, we also need to define exactly what the 'physical asset' is to which the process will be applied.

For example, if we apply RCM to a truck, is the entire truck the 'asset'? Or do we subdivide the truck and analyze (say) the drive train separately from the braking system, the steering, the chassis and so on? Or should we further subdivide the drive train and analyze (say) the engine separately from the gearbox, propshaft, differentials, axles and wheels? Or should the engine not be divided into engine block, engine management system, cooling system, fuel system and so on before starting the analysis? What about subdividing the fuel system into tank, pump, pipes and filters?

This issue needs careful thought because an analysis carried out at too high a level becomes too superficial, while one done at too low a level can become unmanageable and unintelligible. The following paragraphs explore the implications of carrying out the analysis at different levels.

Starting at a low level
One of the most common mistakes in the RCM process is carrying out the analysis at too low a level in the equipment hierarchy.

For example, when thinking about the failure modes which could affect a motor vehicle, a possibility which comes to mind is a blocked fuel line. The fuel line is part of the fuel system, so it seems sensible to address this failure mode by raising a Worksheet for the fuel system. Figure 4.9 indicates that if the analysis is carried out at this level, the blocked fuel line might be the seventh failure mode to be identified out of a total of perhaps a dozen which could cause the functional failure 'unable to transfer any fuel at all'.

RCM II INFORMATION WORKSHEET © 1996 ALADON LTD	SYSTEM *Engine*				
	SUB-SYSTEM *Fuel system*				
FUNCTION		**FUNCTIONAL FAILURE** (Loss of Function)		**FAILURE MODE** (Cause of Failure)	
1	To transfer fuel from the fuel tank to the engine at a rate of up to 1 liter per minute	A	Unable to transfer any fuel at all	1 3 7 12	No fuel in tank Fuel filter blocked Fuel line blocked by foreign object Fuel line severed ... etc

Figure 4.9: Failure modes of a fuel system

When the decision worksheet has been completed for this sub-system, the RCM review group proceeds to the next system, and so on until the maintenance requirements of the entire vehicle have been reviewed. This seems to be straightforward enough until we consider that the vehicle can actually be sub-divided into literally dozens – if not hundreds – of sub-systems at this level. If a separate analysis is carried out for each sub-system, the following problems begin to arise:

- the further down the hierarchy one progresses, the more difficult it becomes to conceptualize and define performance standards. (One could also ask who actually cares about the precise amount of fuel passing through the fuel system as long as the fuel economy of the vehicle is within reasonable limits and the vehicle has enough power.)

- at a low level, it becomes equally difficult to visualize and hence to analyze failure consequences.

- the lower the level of the analysis, the more difficult it becomes to decide which components belong to which system (for instance, is the accelerator part of the fuel system or the engine control system?)

- some failure modes can cause many sub-systems to cease to function simultaneously (such as a failure in the supply of power to an industrial plant). If each sub-system is analyzed on its own, failure modes of this type are repeated again and again.

- control and protective loops can become very difficult to deal with in a low-level analysis, especially when a sensor in one sub-system drives an actuator in another through a processor in a third.

For instance, a rev limiter which reads a signal off the flywheel in the 'engine block' sub-system might send a signal through a processor in the 'engine control' sub-system to a fuel shut-off valve in the 'fuel' sub-system.

If special attention is not paid to this issue, the same function ends up being analyzed three times in slightly different ways, and the same failure-finding task prescribed more than once for the same loop.

• a new worksheet has to be raised for each new sub-system. This leads to the generation of vast quantities of paperwork for the analysis of the entire vehicle, or the consumption of equally large amounts of computer memory space. The associated manual or electronic filing systems have to be very carefully structured if the information is to remain manageable. In short, the whole exercise starts to become much more extensive and much more intimidating than it needs to be.

FMEA's are often carried out at too low a level in the equipment hierarchy because of a belief that there is a correlation between the level at which we identify failure modes and the level at which the FMEA (or the RCM analysis as a whole) should be performed. In other words, it is often said that if we want to identify failure modes in detail, then we ought to carry out a separate FMEA for each replaceable component or sub-system.

In fact, this is not so. The level at which failure modes can be identified is independent of the level at which the analysis is performed, as shown in the next section of this chapter.

Starting at the top
Instead of starting the analysis towards the bottom of the equipment hierarchy, we could start at the top.

For example, the primary function of a truck was listed on page 28 as follows: 'To transport up to 40 tons of material at speeds of up to 75 mph (average 60 mph) from Startsville to Endburg on one tank of fuel.' The first functional failure associated with this function is 'Unable to move at all'. The four failure modes shown in Figure 4.9 could all cause this functional failure, so instead of being listed on an Information Worksheet for the fuel system, they could have been listed on a Worksheet covering the entire truck, as shown in Figure 4.10.

RCM II INFORMATION WORKSHEET © 1996 ALADON LTD	SYSTEM *40 ton truck*	
	SUB-SYSTEM	
FUNCTION	**FUNCTIONAL FAILURE** *(Loss of Function)*	**FAILURE MODE** *(Cause of Failure)*
1 To transfer up to 40 tons of material from Startsville to Endburg speeds of up to 75 mph (average 60 mph) on one tank of fuel	A Unable to move at all	18 No fuel in tank 42 Fuel filter blocked 73 **Fuel line blocked by foreign object** 114 Fuel line severed ... etc

Figure 4.10: Failure modes of a truck

The main advantages of starting the analysis at this level are as follows:
- functions and performance expectations are much easier to define
- failure consequences are much easier to assess
- it is easier to identify and analyze control loops and circuits as a whole
- there is less repetition of functions and failure modes
- it is not necessary to raise a new information worksheet for each new sub-system, so analyses carried out at this level consume far less paper.

However, the main disadvantage of performing the analysis at this level is that there are hundreds of failure modes which could render the truck effectively unable to move. These range from a flat front tire to a sheared crankshaft. So if we were to try to list all the failure modes at this level, it is highly likely that several would be overlooked altogether.

For instance, we have seen how the blocked fuel system might have been the seventh failure mode out of twelve to be identified in the analysis carried out at the 'fuel system' level. However, at the truck level, Figure 4.10 shows that it might have been 73rd out of several hundred failure modes.

Intermediate levels

The problems associated with high- and low-level analyses suggest that it may be sensible to carry out the analysis at an intermediate level. In fact, we are almost spoiled for choice, because most assets can be sub-divided into many levels and the RCM process applied at any one of these levels.

For example, Figure 4.11 shows how the 40 ton truck could be divided into at least five levels. It traces the hierarchy from the level of the truck as a whole down to the level of the fuel lines. It goes on to show how the primary function of the asset might be defined at each level on an RCM Information Worksheet, and how the blocked fuel line could be appear at each level.

Given the choice of five (sometimes more) possibilities, how do we select the level at which to perform the analysis?

We have seen that the top level usually embodies too many failure modes per function to permit sensible analysis. In spite of this however, we still need to identify the main *functions* of the asset or system at the highest levels in order to provide a framework for the rest of the analysis.

For example an operator acquires a truck to carry goods from A to B, not to pump fuel along a fuel line. Although the latter function contributes to the former, the overall performance of the asset – and hence of its maintenance – tends to be judged at the highest levels. For instance, the chief executive of a truck fleet is much more likely to ask 'how is truck X performing?' than 'how is the fuel system on truck X performing?' (unless the fuel system is known to be causing a problem).

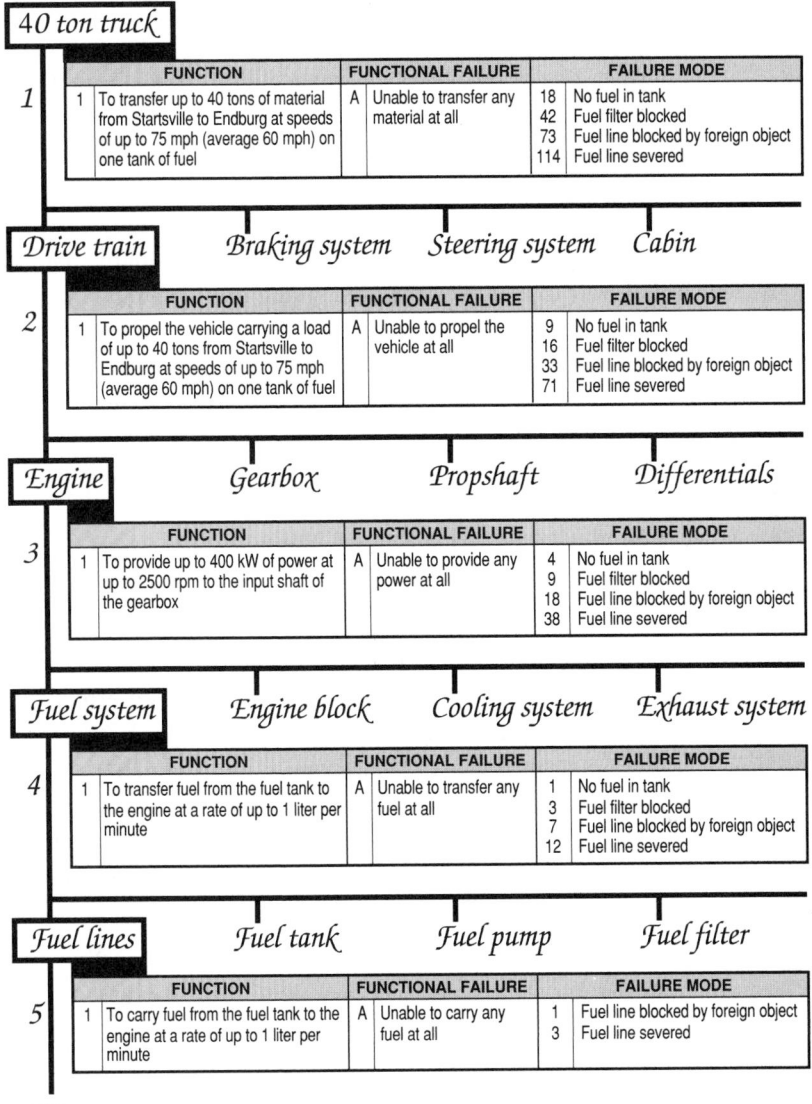

40 ton truck

		FUNCTION		FUNCTIONAL FAILURE		FAILURE MODE
1	1	To transfer up to 40 tons of material from Startsville to Endburg at speeds of up to 75 mph (average 60 mph) on one tank of fuel	A	Unable to transfer any material at all	18 42 73 114	No fuel in tank Fuel filter blocked Fuel line blocked by foreign object Fuel line severed

Drive train   Braking system   Steering system   Cabin

		FUNCTION		FUNCTIONAL FAILURE		FAILURE MODE
2	1	To propel the vehicle carrying a load of up to 40 tons from Startsville to Endburg at speeds of up to 75 mph (average 60 mph) on one tank of fuel	A	Unable to propel the vehicle at all	9 16 33 71	No fuel in tank Fuel filter blocked Fuel line blocked by foreign object Fuel line severed

Engine   Gearbox   Propshaft   Differentials

		FUNCTION		FUNCTIONAL FAILURE		FAILURE MODE
3	1	To provide up to 400 kW of power at up to 2500 rpm to the input shaft of the gearbox	A	Unable to provide any power at all	4 9 18 38	No fuel in tank Fuel filter blocked Fuel line blocked by foreign object Fuel line severed

Fuel system   Engine block   Cooling system   Exhaust system

		FUNCTION		FUNCTIONAL FAILURE		FAILURE MODE
4	1	To transfer fuel from the fuel tank to the engine at a rate of up to 1 liter per minute	A	Unable to transfer any fuel at all	1 3 7 12	No fuel in tank Fuel filter blocked Fuel line blocked by foreign object Fuel line severed

Fuel lines   Fuel tank   Fuel pump   Fuel filter

		FUNCTION		FUNCTIONAL FAILURE		FAILURE MODE
5	1	To carry fuel from the fuel tank to the engine at a rate of up to 1 liter per minute	A	Unable to carry any fuel at all	1 3	Fuel line blocked by foreign object Fuel line severed

etc....

Figure 4.11: *Functions and failures at different levels*

Chapter 2 explained that in practice, a statement of the operating context provides a record of the main functions and associated performance standards of any asset or system at levels above the level at which the RCM analysis is to be carried out.

On the other hand, we have seen the initial inclination is nearly always to start too low in the asset hierarchy. For this reason, a good general rule (especially for people new to RCM) is to carry out the analysis one level – or even two levels – higher than at first seems sensible. This is because it is always easier to break complex sub-systems out of a high-level analysis than it is to go up a level when one has started too low. This is discussed in more detail in the next section of this chapter.

With a bit of practice (especially concerning what is meant by 'a level at which it is possible to identify a suitable failure management policy'), the most suitable level at which to carry out any analysis eventually becomes intuitively obvious. In this context, note that it is not necessary to analyze every system at the same level throughout the asset hierarchy. For instance, the entire braking system could be analyzed at level 2 as shown in Figure 4.11, but it may be necessary to analyze the engine at level 3 or even level 4.

How Failure Modes and Effects Should be Recorded

Once the level of the entire RCM analysis has been established, we then have to decide what degree of detail is necessary to define each failure mode within the framework of that analysis. There is no *technical* reason why all the failure modes cannot be listed (together with their effects) at a level which enables a suitable failure management policy to be selected.

However, even intermediate level analyses sometimes generate too many failure modes per functional failure, especially for primary functions. This usually happens when the asset incorporates complex subassemblies which could themselves suffer from a large number of failure modes. Examples of such subassemblies include small electric motors, small hydraulic systems, small gearboxes, control loops, protective circuits and complex couplings.

Depending as usual on context and consequences, these sub-assemblies can be handled in one of four different ways, as discussed below.

Option 1
List all the reasonably likely failure modes of the subassembly individually as part of the main analysis – in other words, at levels equivalent to level 3, 4, 5 or 6 in Figure 4.7.

For example, consider an asset which could stop completely as a result of the failure of a small gearbox. On the Information Worksheet for this asset, this gearbox failure could be listed as shown below:

FAILURE MODE	FAILURE EFFECT
1 Gearbox bearings seize	Motor trips and alarm sounds in control room. 3 hours downtime to replace gearbox with spare. New bearings fitted in workshop
2 Gear teeth stripped	Motor does not trip but machine stops. 3 hours downtime to replace gearbox with spare. New gears fitted in workshop
3 Gearbox seizes due to lack of oil	Motor trips and alarm sounds in control room. 3 hours downtime to replace gearbox with spare. Seized gearbox would be scrapped
......etc	

In general, the failure modes which could affect a subassembly should be incorporated in a higher level analysis if the subassembly is likely to suffer from no more than about 6 failure modes which are considered to be worth identifying and which will cause any one functional failure of the higher level system.

Option 2
List the failure of the subassembly as a single failure mode on the Information Worksheet to begin with, then raise a new worksheet to analyze the functions, functional failures, failure modes and effects of the subassembly as a separate exercise.

For example, the failure of the gearbox discussed above could have been listed as follows:

FAILURE MODE	FAILURE EFFECT
1 Gearbox fails ......etc	Gearbox analyzed separately

A subassembly is usually worth treating in this way if more than 10 failure modes of the subassembly could cause the loss of any one function of the main assembly.

(If there are between 7 and 9 failure modes per functional failure, use option one or option two, bearing in mind that separate analyses mean more analyses, but fewer failure modes per analysis.)

Option 3
List the failure of the subassembly on the Information Worksheet as a single failure mode – in other words, at a level equivalent to level one or two in Figure 4.7 – record its effects, and leave it at that.

For example, if it was considered appropriate to treat the failure of the gearbox discussed above in this fashion, it would be listed as shown overleaf:

	FAILURE MODE	FAILURE EFFECT
1	Gearbox fails	Motor trips and alarm sounds in control room. Downtime to replace gearbox 3 hours
	......*etc*	

This approach should only be adopted for a component or subassembly which has the following characteristics:

• it is not subject to detailed diagnostic and repair routines when it fails, but is simply replaced and either discarded or subjected to later repair
• it is quite small but quite complex
• it does not have any dominant failure modes
• it is not likely to be susceptible to any form of proactive maintenance.

Option 4

In some cases, a complex subassembly might suffer from one or two dominant failure modes which are readily preventable, and a number of less common failures which may not be worth preventing because the frequency and/or the consequences of the failures do not warrant it.

For example, a small electric motor operating in a dusty environment might be certain to fail due to overheating if the grille covering its cooling fan gets blocked, but failures for other reasons might be few, far between and not very serious if they do occur. In this case, the failure modes for this motor might be listed as follows:
• motor fan blocked by dust
• motor fails (for other reasons).

This option is really a combination of options 1 and 3.

Services

The failure of services (power, water, steam, air, gases, vacuum, etc) are treated as a single failure mode from the point of view of the asset which is supplied by that service, because detailed analysis of these failures is usually beyond the scope of the asset in question. Such failures are noted for information purposes ('Power supply fails'), their effects recorded and they are then analyzed in detail when the service is analyzed as a whole.

A Completed Information Worksheet

Failure effects are listed in the last column of the Information Worksheet alongside the relevant failure mode, as shown in Figure 4.13.

RCM II INFORMATION WORKSHEET © 1996 ALADON LTD	SYSTEM	5 MW Gas Turbine		SYSTEM Nº 216 - 05		Facilitator: N Smith		Date 07 - 07 - 1996	Sheet Nº 1
	SUB-SYSTEM	Exhaust System		SUB-SYSTEM Nº 216 - 05 - 11		Auditor: P Jones		Date 07 - 08 - 1996	of 3

FUNCTION		FUNCTIONAL FAILURE (Loss of function)		FAILURE MODE (Cause of failure)		FAILURE EFFECT (What happens when it fails)
1	To channel all the hot turbine gas without restriction to a fixed point 10m above the roof of the turbine hall	A	Unable to channel gas at all	1	Silencer mountings corroded away	Silencer assembly collapses and falls to bottom of stack. Back pressure causes the turbine to surge violently and shut down on high exhaust gas temperature. Downtime to replace silencer up to four weeks
		B	Gas flow restricted	1	Part of silencer falls off due to fatigue	Depending on nature of blockage, exhaust temperature may rise to where it shuts down the turbine. Debris could damage parts of the turbine. Downtime to repair silencer 4 weeks.
		C	Fails to contain the gas	1	Flexible joint holed by corrosion	The joint is inside turbine hood, so leaking exhaust gases would be extracted by the hood extraction system. Fire and gas detection equipment inside hood is unlikely to detect an exhaust gas leak, and temperatures are unlikely to rise enough to trigger the fire wire. A severe leak may cause gas demister to overheat, and may also melt control wires near the leak with unpredictable effects. Pressure balances inside the hood are such that little or no gas is likely to escape from a small leak, so a small leak is unlikely to be detected by smell or hearing. Downtime to replace joint up to 3 days
				2	Gasket in ducting improperly fitted	Gas escapes into turbine hall and ambient temperature rises. Hall ventilation system would expel gases through louvres to atmosphere, so concentration of gases is unlikely to reach noxious levels. A small leak at this point may be audible. Downtime to repair up to 4 days
				3	Upper bellows holed by corrosion	The upper bellows are outside turbine hall, so a leak here discharges to atmosphere. Ambient noise levels may rise. Downtime to repair up to 1 week.
		D	Fails to convey gas to a point 10 m above roof	1	Exhaust stack mounting bolts shear due to rust	Exhaust stack is likely to be held up by the guy ropes for a while before it falls over, but would lean at an angle. If it did fall over, there is a high probability that it could crush a structure containing people. Downtime to repair a few days to several weeks.
				2	Exhaust stack blown over in gale	The stack structure is designed to withstand winds up to 200 mph, so it is only likely to fall over in a gale if the guy ropes were already weakened, perhaps by corrosion. If it went, it could be blown onto an accommodation module. Downtime to repair up to several weeks.
2	To reduce exhaust noise level to ISO Noise Rating 30 at 50 meters	A	Noise level exceeds ISO Noise Rating 30 at 50 m	1	Silencer material retaining mesh corroded away	Most of the material would be blown out, but some might fall to the bottom of stack and obstruct the turbine outlet, causing high EGT and possible turbine shutdown. Noise levels would rise gradually. Downtime to repair about 2 weeks.
				2	Duct leaks outside turbine hall	.... etc

Figure 4.13: The RCM Information Worksheet

5 Failure Consequences

Previous chapters have explained how the RCM process asks the following seven questions about each asset:

• *what are the functions and associated performance standards of the asset in its present operating context?*
• *in what ways does it fail to fulfil its functions?*
• *what causes each functional failure?*
• *what happens when each failure occurs?*
• *in what way does each failure matter?*
• *what can be done to predict or prevent each failure?*
• *what if a suitable proactive task cannot be found?*

The answers to the first four questions were discussed at length in Chapters 2 to 4. These showed how RCM Information Worksheets are used to record the functions of the asset under review, and to list the associated functional failures, failure modes and failure effects.

The last three questions are asked about each individual failure mode. This chapter considers the fifth question:
• *in what way does each failure matter?*

5.1 Technically Feasible and Worth Doing

Every time a failure occurs, the organization which uses the asset is affected in some way. Some failures affect output, product quality or customer service. Others threaten safety or the environment. Some increase operating costs, for instance by increasing energy consumption, while a few have an impact in four, five or even all six of these areas. Still others may appear to have no effect at all if they occur on their own, but may expose the organization to the risk of much more serious failures.

If any of these failures are not prevented, the time and effort which need to be spent correcting them also affects the organization, because repairing failures consumes resources which might be better used elsewhere.

The nature and severity of these effects govern the *consequences* of the failure. In other words, they govern the extent to which the owners or users of the asset will believe that each failure matters. (Note that failure effects describe *what happens* when a failure occurs, while consequences describe how – and how much – it *matters*. Clearly, if we can reduce the effects of any failure in terms of frequency and/or severity, then it follows that we will also reduce the associated consequences.)

If a failure matters very much, then considerable efforts will be made to avoid, eliminate or minimise the consequences. This is especially true if the failure could injure or kill someone, or if it could have a serious effect on the environment. It is also true of failures that interfere with production or operations, or which cause significant secondary damage.

On the other hand, if the failure only has minor consequences, it is possible that no proactive action will be taken and the failure simply corrected each time it occurs.

This focus on consequences means that RCM starts the task selection process by assessing the effects of each failure mode, and classifying it into one of four broad categories of consequences. The second step is to find out if it is physically possible to perform a proactive task that reduces, or enables action to be taken to reduce, the consequences of failure to an extent that might be acceptable to the owner or user of the asset. If such a task is found, it is said to be *technically feasible*. The criteria governing technical feasibility are discussed in more detail in Chapters 6 and 7.

If a task is technically feasible, then the third step is to ask whether the task actually reduces the failure consequences to an extent that justifies the direct and indirect costs of doing the task. (Direct costs are the costs of labor or materials needed to do the task and to do any associated rectification work. Indirect costs include the cost of any downtime which may be needed to do the task.) If the answer is yes, then the task is said to be *worth doing*.

*A proactive task is worth doing if it reduces the conse-
quences of the associated failure mode to an extent that
justifies the direct and indirect costs of doing the task.*

If it is not possible to find a suitable proactive task, the nature of the failure consequences also indicate what default action should be taken. Default actions are reviewed in Chapters 8 and 9.

The remainder of this chapter considers the criteria used to evaluate the consequences of failure, and hence to decide whether any form of proactive task is *worth doing*. These consequences are divided in two stages into four categories. The first stage separates hidden functions from evident functions.

5.2 Hidden and Evident Functions

We have seen that every asset has more than one and sometimes dozens of functions. When most of these functions fail, it will inevitably become apparent to someone that the failure has occurred.

For instance, some failures cause warning lights to flash or alarms to sound, or both. Others cause machines to shut down or some other part of the process to be interrupted. Others lead to product quality problems or increased use of energy, and yet others are accompanied by obvious physical effects such as loud noises, escaping steam, unusual smells or pools of liquid on the floor.

For example, Figure 2.7 in Chapter 2 showed three pumps which are shown again in Figure 5.1 below. If a bearing on Pump A seizes, pumping capability is lost. This failure on its own will inevitably become apparent to the operators, either as soon as it happens or when some downstream part of the process is interrupted. (The operators might not know immediately that the problem was caused by the bearing, but they would eventually and inevitably become aware of the fact that something unusual had happened.)

Figure 5.1:
Three pumps

Stand Alone   **Duty**   **Stand-by**

Failures of this kind are classed as evident because someone will eventually find out about it when they occur on their own. This leads to the following definition of an evident function:

> ***An evident function is one whose failure will on its own eventually and inevitably become evident to the operating crew under normal circumstances***

However, some failures occur in such a way that nobody knows that the item is in a failed state unless or until some other failure also occurs.

For instance, if Pump C in Figure 5.1 failed, no-one would be aware of the fact because under normal circumstances Pump B would still be working. In other words, the failure of Pump C on its own has no direct impact unless or until Pump B also fails (an abnormal circumstance).

Pump C exhibits one of the most important characteristics of a hidden function, which is that the failure of this pump *on its own* will not become evident to the operating crew under normal circumstances. In other words, it will not become evident unless pump B also fails. This leads to the following definition of a hidden function:

> **A *hidden function* is one whose failure will not become evident to the operating crew under normal circumstances if it occurs on its own.**

The first step in the RCM consequence evaluation process is to separate hidden functions from evident functions because hidden functions need special handling. As explained in Part 6 of this chapter, these functions are associated with protective devices that are not fail safe. Since they can account for *up to half the failure modes that could affect modern, complex equipment,* hidden functions could well become *the* dominant issue in maintenance over the next ten years. However, to place these functions in perspective, we first consider evident failures.

Categories of Evident Failures

Evident failures are classified into three categories in descending order of importance, as follows:

• *safety* and *environmental consequences.* A failure has safety consequences if it could injure or kill someone. It has environmental consequences if it could lead to a breach of any corporate, regional or national environmental standard

• *operational consequences.* A failure has operational consequences if it affects production or operations (output, product quality, customer service or operating costs in addition to the direct cost of repair)

• *non-operational consequences.* Evident failures in this category affect neither safety nor production, so they involve only the direct cost of repair.

By ranking evident failures in this order, RCM ensures that the safety and environmental implications of *every* evident failure mode are considered. This unequivocally puts people ahead of production.

This approach also means that the safety, environmental and economic consequences of each failure are assessed in one exercise, which is much more cost-effective than considering them separately.

The next four sections of this chapter consider each of these categories in detail, starting with the evident categories and then moving on to the rather more complex issues surrounding hidden functions.

5.3   Safety and Environmental Consequences

Safety First

As we have seen, the first step in the consequence evaluation process is to identify hidden functions so that they can be dealt with appropriately. All remaining failure modes – in other words, failures which are not classified as hidden – must by definition be evident. The above paragraphs explained that the RCM process considers the safety and environmental implications of each evident failure mode first. It does so for two reasons:

• a more and more firmly held belief among employers, employees, customers and society in general that injuring or killing people in the course of business is simply not tolerable, and hence that everything possible should be done to minimize the possibility of any sort of safety-related incident or environmental excursion.

• the more pragmatic realization that the probabilities which are tolerated for safety-related incidents tend to be several orders of magnitude lower than those which are tolerated for failures which have operational consequences. As a result, in most of the cases where a proactive task is worth doing from the safety viewpoint, it is also likely to be more than adequate from the operational viewpoint.

At one level, safety refers to the safety of individuals in the workplace. Specifically, RCM asks whether anyone could get injured or killed either as a direct result of the failure mode itself or by other damage which may be caused by the failure.

A failure mode has safety consequences
if it causes a loss of function or other
damage which could injure or kill someone

At another level, 'safety' refers to the safety or well-being of society in general. Nowadays, failures which affect society tend to be classed as 'environmental' issues. In fact, in many parts of the world the point is fast approaching where organizations either conform to society's environmental expectations, or they will no longer be allowed to operate. So quite apart from any personal feelings which anyone may have on the issue, environmental probity is becoming a prerequisite for corporate survival.

Chapter 2 explained how society's expectations take the form of municipal, regional and national environmental standards. Some organizations also have their own sometimes even more stringent corporate standards. A failure mode is said to have environmental consequences if it could lead to the breach of any of these standards.

> *A failure mode has environmental consequences*
> *if it causes a loss of function or other damage*
> *which could lead to the breach of any known*
> *environmental standard or regulation*

Note that when considering whether a failure mode has safety or environmental consequences, we are considering whether one failure mode on its own could have the consequences. This is different from part 6 of this chapter, in which we consider the failure of both elements of a protected system.

The Question of Risk

Much as most people would like to live in an environment where there is no possibility at all of death or injury, it is generally accepted that there is an element of risk in everything we do. In other words, absolute zero is unattainable, even though it is a worthy target to keep striving for. This immediately leads us to ask what *is* attainable.

To answer this question, we first need to consider the question of risk in more detail.

Risk assessment consists of three elements. The first asks what could happen if the event under consideration did occur. The second asks how likely it is for the event to occur at all. The combination of these two elements provides a measure of the degree of risk. The third – and often the most contentious element – asks whether this risk is tolerable.

For example, consider a failure mode which could result in death or injury to ten people *(what could happen)*. The probability that this failure mode could occur is one in a thousand in any one year *(how likely it is to occur)*. On the basis of these figures, the risk associated with this failure is:

10 x (1 in 1000) = 1 casualty per 100 years

Now consider a second failure mode which could cause 1000 casualties, but the probability that this failure could occur is one in 100 000 in any one year. The risk associated with this failure is:

1000 x (1 in 100 000) = 1 casualty per 100 years.

In these examples, the risk is the same although the figures upon which it is based are quite different. Note also that these examples do not indicate whether the risk is tolerable – they merely quantify it. Whether or not the risk is tolerable is a separate and much more difficult question which is dealt with later.

Note that throughout this book, the terms 'probability' (1 in 10 chance of a failure in any one period) and 'failure rate' (once in ten periods on average, corresponding to a mean time between failures of 10 periods) are used as if they are interchangeable when applied to random failures. Strictly speaking, this is not true. However, if the MTBF is greater than about 4 periods, the difference is so small that it can usually be ignored.

The following paragraphs consider each of the three elements of risk in more detail.

What could happen if the failure occurred?

Two issues need to be considered when considering what could happen if a failure were to occur. These are *what actually happens* and *whether anyone is likely to be hurt or killed* as a result.

What actually happens if any failure mode occurs should be recorded on the RCM Information Worksheet as its failure effects, as explained at length in Chapter 4. Part 5 of Chapter 4 also listed a number of typical effects which pose a threat to safety or the environment.

The fact that these effects *could* injure or kill someone does not necessarily mean that they *will* do so every time they occur. Some may even occur quite often without doing so. However, the issue is not whether such consequences are inevitable, but whether they are possible.

For example, if the hook were to fail on a travelling crane used to carry steel coils, the falling load would only injure or kill anyone who happened to be standing under it or very close to it at the time. If no-one was nearby, then no-one would be injured. However, the possibility that someone could be injured means that this failure mode should be treated as a safety hazard and analyzed accordingly.

This example demonstrates the fact that the RCM process assesses safety consequences at the most conservative level. If it is reasonable to assume that any failure mode *could* affect safety or the environment, we assume that it *can*, in which case it must be subjected to further analysis. (We see later that the likelihood that someone will get injured is taken into consideration when evaluating the tolerability of the risk.)

A more complex situation arises when dealing with safety hazards that are already covered by some form of built-in protection. We have seen that one of the main objectives of the RCM process is to establish the most effective way of managing each failure in the context of its consequences. This can only be done if these consequences are evaluated to begin with as if nothing was being done to manage the failure (in other words, to predict or prevent it or to mitigate its consequences).

Protective devices which are designed to deal with the failed or the failing state (alarms, shutdowns and relief systems) are nothing more than built-in failure management systems. As a result, to ensure that the rest of the analysis is carried out from an appropriate zero-base, the consequences of the failure of protected functions should ideally be assessed as if protective devices of this type are not present.

For example, a failure which could cause a fire is always regarded as a safety hazard, because the presence of a fire-extinguishing system does not necessarily *guarantee* that the fire will be controlled and extinguished.

The RCM process can then be used to validate (or revalidate) the suitability of the protective device itself from three points of view:

- its *ability to provide the required protection*. This is done by defining the function of the protective device, as explained in Chapter 2

- whether the protective device responds *fast enough* to avoid the consequences, as discussed in Chapter 7

- what must be done to ensure that the protective device *continues to function* in its turn, as discussed in part 6 of this chapter and Chapter 8.

How likely is the failure to occur?

Part 4 of Chapter 4 mentions that only failure modes which are reasonably likely to occur in the context in question should be listed on the RCM Information Worksheet. As a result, if the Information Worksheet has been prepared on a realistic basis, the mere fact that the failure mode has been listed suggests that there is some likelihood that it could occur, and therefore that it should be subjected to further analysis.

(Sometimes it may be prudent to list a wildly unlikely but nonetheless dangerous failure mode in an FMEA, purely to place on record the fact that it was considered and then rejected. In these cases, a comment like "This failure mode is considered too unlikely to justify further analysis" should be recorded in the failure effects column.)

Is the risk tolerable?

One of the most difficult aspects of the management of safety is the extent to which beliefs about what is tolerable vary from individual to individual and from group to group. A wide variety of factors influence these beliefs, by far the most dominant of which is *the degree of control which any individual thinks he or she has over the situation.* People are nearly always prepared to tolerate a higher level of risk when they believe that they are personally in control of the situation than when they believe that the situation is out of their control.

For example, people tolerate much higher levels of risk when driving their own cars than they do as aircraft passengers. (The extent to which this issue governs perceptions of risk is given by the startling statistic that 1 person in 11 000 000 who travels by air between New York and Los Angeles in the USA is likely to be killed while doing so, while 1 person in 14 000 who makes the trip by road is likely to be killed. And yet some people insist on making this trip by road because they believe that they are 'safer'!)

This example illustrates the relationship between the probability of being killed which any one person is prepared to tolerate and the extent to which that person believes he or she is in control. In more general terms, this might vary for a particular individual as shown in Figure 5.2.

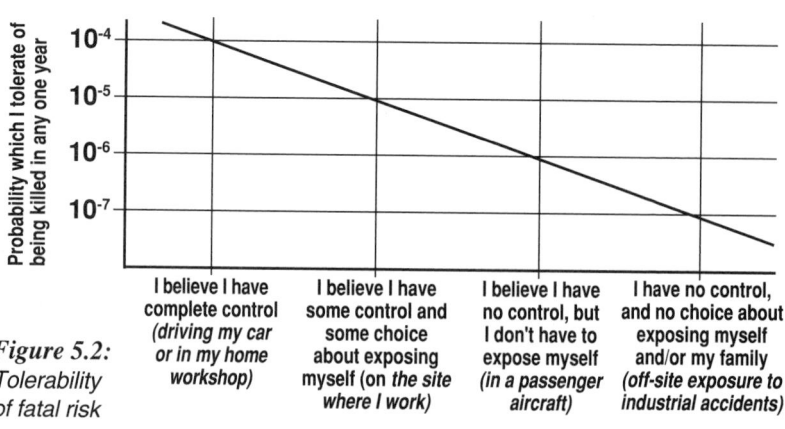

Figure 5.2:
Tolerability
of fatal risk

The figures given in this example are not meant to be prescriptive and they do not necessarily reflect the views of the author – they merely illustrate what one individual might decide that he or she is prepared to tolerate. Note also that they are based on the perspective of one individual going about his or her daily business. This view then has to be translated into a degree of risk for the whole population (all the workers on a site, all the citizens of a town or even the entire population of a country).

In other words, if I tolerate a probability of 1 in 100 000 (10^{-5}) of being killed at work in any one year and I have 1 000 co-workers who all share the same view, then we all tolerate that on average 1 person per year on our site will be killed at work every 100 years – and that person may be me, and it may happen this year.

Bear in mind that any quantification of risk in this fashion can only ever be a rough approximation. In other words, if I say I tolerate a probability of 10^{-5}, it is never more than a ballpark figure. It indicates that I am prepared to tolerate a probability of being killed at work which is roughly 10 times lower than that which I tolerate when I use the roads (about 10^{-4}).

Always bearing in mind that we are dealing with approximations, the next step is to translate the probability which myself and my co-workers are prepared to tolerate that any one of us might be killed by *any* event at work into a tolerable probability for *each single event* (failure mode or multiple failure) which could kill someone.

For example, continuing the logic of the previous example, the probability that any one of my 1 000 co-workers will be killed in any one year is 1 in 100 (assuming that everyone on the site faces roughly the same hazards). Furthermore, if the activities carried out on the site embody (say) 10 000 events which could kill someone, then the average probability that each event could kill one person must be reduced to 10^{-6} in any one year. This means that the probability of an event which is likely to kill ten people must be reduced to 10^{-7}, while the probability of an event which has a 1 in 10 chance of killing one person must be reduced to 10^{-5}.

The techniques by which one moves up and down hierarchies of probability in this fashion are known as probabilistic or quantitative risk assessments. This approach is explored further in Appendix 3. The key points to bear in mind at this stage are that:

- the decision as to what is tolerable should start with the *likely victim*. How one might involve such 'likely victims' in this decision in the industrial context is discussed later in this chapter

- it is possible to link what one person tolerates directly and quantitatively to a tolerable probability of individual failure modes.

Although perceived degree of control usually dominates decisions about the tolerability of risk, it is by no means the only issue. Other factors which help us decide what is tolerable include the following:

• *individual values:* To explore this issue in any depth is well beyond the scope of this book. Suffice it to contrast the views on tolerable risk likely to be held by a mountaineer with those of someone who suffers from vertigo, or those of an underground miner with those of someone who suffers from claustrophobia.

• *industry values:* While every industry nowadays recognizes the need to operate as safely as possible, there is no escaping the fact that some are intrinsically more dangerous than others. Some even compensate for higher levels of risk with higher pay levels. The views of any individual who works in that industry ultimately boil down to his or her perception of whether the intrinsic risks are 'worth it' – in other words, whether the benefit justifies the risk.

• *the effect on 'future generations':* The safety of children – especially unborn children – has an especially powerful effect on peoples' views about what is tolerable. Adults frequently display a surprising and even distressing disregard for their own safety. (Witness how much time has to be spent persuading some people to wear protective clothing.) However, threaten their offspring and their attitude changes completely.

For example, the author worked with one group which had occasion to discuss the properties of a certain chemical. Words like 'toxic' and 'carcinogenic' were treated with indifference, even though most of the members of this group were the people most at risk. However, as soon as it emerged that the chemical was also mutagenic and teratogenic, and the meaning of these words was explained to the group, the chemical was suddenly viewed with much greater respect.

• *knowledge:* perceptions of risk are greatly influenced by how much people know about the asset, the process of which it forms part and the failure mechanisms associated with each failure mode. The more they know, the better their judgement. (Ignorance is often a two-edged sword. In some situations people take the most appalling risks out of sheer ignorance, while in others they wildly exaggerate the risks – also out of ignorance. On the other hand, we need to remind ourselves constantly of the extent to which familiarity can breed contempt.)

A great many other factors also influence perceptions of risk, such as the value placed on human life by different cultural groups, religious values and even factors such as the age and marital status of the individual.

All of these factors mean that it is impossible to specify a standard of tolerability for any risk which is absolute and objective. This suggests that the tolerability of any risk can only be assessed on a basis which is both relative and subjective – 'relative' in the sense that the risk is compared with other risks about which there is a fairly clear consensus, and 'subjective' in the sense that the whole question is ultimately a matter of judgement. But *whose* judgement?

Who should evaluate risks?

The very diversity of the factors discussed above mean that it is simply not possible for any one person – or even one organization – to assess risk in a way which will be universally acceptable. If the assessor is too conservative, people will ignore and may even ridicule the evaluation. If the assessor is too relaxed, he or she might end up being accused of playing with people's lives (if not actually killing them).

This suggests further that a satisfactory evaluation of risk can only be done by a group. As far as possible, this group should represent people who are likely to have a clear understanding of the failure mechanism, the failure effects (especially the nature of any hazards), the likelihood of the failure occurring and what possible measures can be taken to anticipate or prevent it. The group should also include people who have a legitimate view on the tolerability or otherwise of the risks. This means representatives of the likely victims (most often operators or maintainers in the case of direct safety hazards) and management (who are usually held accountable if someone is injured or if an environmental standard is breached).

If it is applied in a properly focused and structured fashion, the collective wisdom of such a group will do much to ensure that the organization does its best to identify and manage all the failure modes that could affect safety or the environment. (The use of such groups is in keeping with the worldwide trend towards laws which say that safety is the responsibility of all employees, not just the responsibility of management.)

Groups of this nature can usually reach consensus quite quickly when dealing with direct safety hazards, because they include the people at risk. Environmental hazards are not quite so simple, because society at large is the 'likely victim' and many of the issues involved are unfamiliar. So any group which is expected to consider whether a failure could breach an environmental standard or regulation must find out beforehand which of these standards and regulations cover the process under review.

Safety and Proactive Maintenance

If a failure could affect safety or the environment, the RCM process stipulates that we must try to prevent it. The above discussion suggests that:

> *For failure modes which have safety or environmental consequences, a proactive task is only worth doing if it reduces the probability of the failure to a tolerably low level*

If a proactive task cannot be found which achieves this objective to the satisfaction of the group performing the analysis, we are dealing with a safety or environmental hazard which cannot be adequately anticipated or prevented. This means that something must be *changed* in order to make the system safe. This 'something' could be the asset itself, a process or an operating procedure. Once-off changes of this sort are classified as 'redesigns', and are usually undertaken with one of two objectives:
- to reduce the probability of the failure occurring to a tolerable level
- to change things so that the failure no longer has safety or environmental consequences.

The question of redesign is discussed in more detail in Chapter 9.

Note that when dealing with safety and environmental issues, RCM does not consider the cost of the failure in financial terms. If it is not safe we have an obligation either to prevent it from failing, or to make it safe. This suggests that the decision process for failure modes which have safety or environmental consequences can be summarized as shown in Figure 5.3:

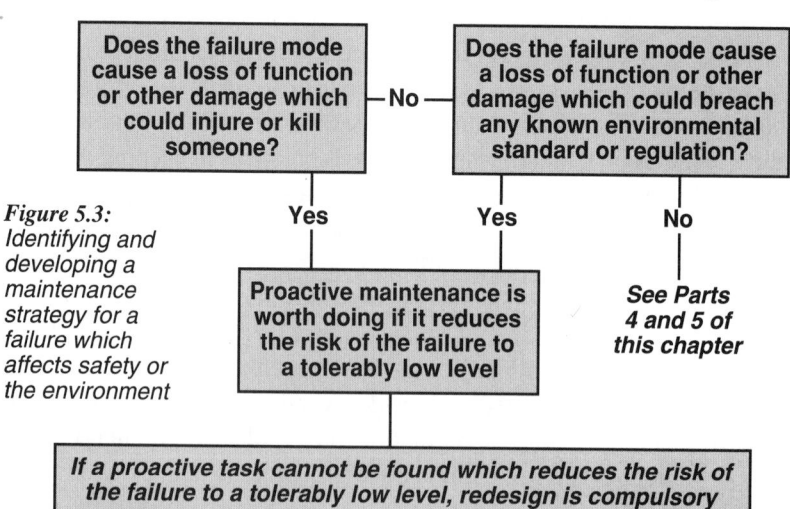

Figure 5.3: Identifying and developing a maintenance strategy for a failure which affects safety or the environment

Does the failure mode cause a loss of function or other damage which could injure or kill someone? — No — Does the failure mode cause a loss of function or other damage which could breach any known environmental standard or regulation?

Yes   Yes   No

Proactive maintenance is worth doing if it reduces the risk of the failure to a tolerably low level

See Parts 4 and 5 of this chapter

If a proactive task cannot be found which reduces the risk of the failure to a tolerably low level, redesign is compulsory

The basis on which we determine the technical feasibility and frequency of different types of proactive task is discussed in Chapters 6 and 7.

RCM and Safety Legislation

A question often arises concerning the relationship between RCM and safety legislation (environmental legislation is dealt with directly). Nowadays, most legislation governing safety merely demands that users are able to demonstrate that they are doing whatever is prudent to ensure that their assets are safe. This has led to rapidly increasing emphasis on the concept of an *audit trail*, which basically requires users of assets to be able to produce documentary evidence that there is a rational, defensible basis for their maintenance programs. In the vast majority of cases, RCM wholly satisfies this type of requirement.

However, some regulations demand that specific tasks should be done on specific types of equipment at specific intervals. If the RCM process suggests a different task and/or a different interval, it is wise to continue doing the task specified by the legislation and to discuss the suggested change with the appropriate regulatory authority.

5.4 Operational Consequences

How Failures Affect Operations

The primary function of most equipment in industry is connected in some way with the need to earn revenue or to support revenue earning activities.

For example, the primary function of most of the assets used in manufacturing is to add value to materials, while customers pay directly for access to telecommunications and transport equipment (buses, trucks, trains or aircraft).

Failures which affect the primary functions of these assets affect the revenue-earning capability of the organization. The magnitude of these effects depends on how heavily the equipment is loaded and the availability of alternatives. However, in nearly all cases the effects are greater – often much greater – than the cost of repairing the failures. This is also true of equipment in service industries such as entertainment, commerce and even banking.

For example, if the lights fail at a ball game, fans tend to want their money back. The same applies if projectors fail at the movies. If the air-conditioning fails in a shop or restaurant, customers walk out. Banks lose business if their ATM's fail.

In general, failures affect operations in four ways:

* *they affect total output.* This occurs when equipment stops working altogether or when it works too slowly. This results either in increased production costs if the plant has to work extra time to catch up, or lost sales if the plant is already fully loaded.

* *they affect product quality.* If a machine can no longer hold manufacturing tolerances or if a failure causes materials to deteriorate, the likely result is either scrap or expensive rework. In a more general sense, "quality" also covers concepts such as the precision of navigation systems, the accuracy of targeting systems and so on.

* *they affect customer service.* Failures affect customer service in many ways, ranging from the late delivery of orders to the late departure of passenger aircraft. Frequent or serious delays sometimes attract heavy penalties, but in most cases they do not result in an *immediate* loss of revenue. However chronic service problems eventually cause customers to lose confidence and take their business elsewhere.

* *increased operating costs in addition to the direct cost of repair.* For instance, the failure might lead to the increased use of energy or it might involve switching to a more expensive alternative process.

In non-profit enterprises such as military undertakings, certain failures can also affect the ability of the organization to fulfil its primary function sometimes with devastating results.

"For want of a nail, a shoe was lost. For want of a shoe, a horse was lost. For want of a horse, a message was lost. For want of a message, a battle was lost. For want of a battle, a war was lost. All for want of a horseshoe nail."

While it may be difficult to cost out the results of losing a war, failures of this sort still have economic implications at a more mundane level. If they occur too often, it may be necessary to keep (say) two horses in order to ensure that one will be available to do the job – or sixty battle tanks instead of fifty – or six aircraft carriers instead of five. Redundancy on this scale can be very expensive indeed.

The severity of these consequences mean that if an evident failure does not pose a threat to safety or the environment, the RCM process focuses next on the operational consequences of failure.

A *failure* has operational consequences if it has a direct adverse effect on operational capability

As we have seen, these consequences tend to be *economic* in nature, so they are usually evaluated in economic terms. However, in certain more extreme cases (such as losing a war), the 'cost' may have to be evaluated on a more qualitative basis.

Avoiding Operational Consequences

The overall economic effect of any failure mode which has operational consequences depends on two factors:

• how much the failure costs each time it occurs, in terms of its effect on operational capability plus repair costs

• how often it happens.

In the previous section of this chapter, we did not pay much attention to how often failures are likely to occur. (Failure rates have little bearing on safety-related failures, because the objective in these cases is to avoid any failures on which to base a rate.) However, if the failure consequences are economic, the total cost *is* affected by how often the consequences are likely to occur. In other words, to assess the economic impact of these failures, we need to assess how much they are likely to cost *over a period of time*.

Consider for example the pump shown in Figure 2.1 and again in Figure 5.4. The pump is controlled by one float switch which activates it when the level in Tank Y drops to 120 000 liters, and another that turns it off when the level in Tank Y reaches 240 000 liters. A low level alarm is located

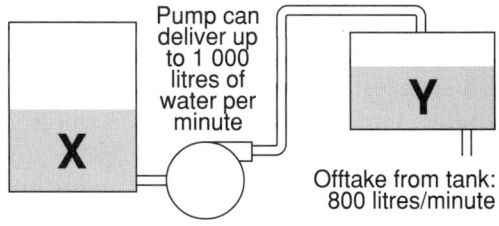

Figure 5.4: Stand-alone pump

just below the 120 000 liter level. If the tank runs dry, the downstream process has to be shut down. This costs the organization using the pump $5 000 per hour.

	FAILURE MODE	FAILURE EFFECT
1	Bearing seizes due to normal wear and tear	Motor trips but no alarm sounds in control room. Level in tank drops until low level alarm sounds at 120 000 litres. Downtime to replace the bearing 4 hours. *(The mean time between occurrences of this failure mode is about 3 years.)*

Figure 5.5: FMEA for bearing failure on the stand-alone pump

Assume that it has already been agreed that one failure mode which can affect this pump is 'Bearing seizes due to normal wear and tear'. For the sake of simplicity, assume that the motor on this pump is equipped with an overload switch, but there is no trip alarm wired to the control room.

This failure mode and its effects might be described on an RCM Information Worksheet as shown in Figure 5.5 above.

Water is drawn out of the tank at a rate of 800 liters per minute, so the tank runs dry 2.5 hours after the low level alarm sounds. It takes 4 hours to replace the bearing, so the downstream process stops for 1.5 hours. So this failure costs:

$$1.5 \times \$5\,000 = \$7\,500$$

in lost production every three years, plus the cost of replacing the bearing.

Assume that it is *technically feasible* to check the bearing for audible noise once a week (the basis upon which we make this kind of judgement is discussed at length in the next chapter). If the bearing is found to be noisy, *the operational consequences of failure can be avoided by ensuring that the tank is full* before starting work on the bearing. This provides five hours of storage so the bearing can now be replaced in four hours without interfering with the downstream process.

Assume also that the pump is located in an unmanned pumping station. It has been agreed that the check should be carried out by a maintenance craftsman, and that the total time needed to do each check is twenty minutes. Assume further that the total cost of employing the craftsman is $24 per hour, in which case it costs $8 to perform each check. If the MTBF of the bearing is 3 years, he will do about 150 checks per failure. In other words, the cost of the checks is:

$$150 \times \$8 = \$1\,200$$

every three years, again plus the cost of replacing the bearing.

In this example, the scheduled task is clearly cost-effective relative to the cost of the operational consequences of the failure plus the cost of repair. This suggests that if a failure has operational consequences, the basis for deciding whether a proactive task is worth doing is economic, as follows:

> ***For failure modes with operational consequences, a proactive task is worth doing if, over a period of time, it costs less than the cost of the operational consequences plus the cost of repairing the failure which it is meant to prevent***

Conversely, if a cost-effective proactive task cannot be found, then *it is not worth doing any scheduled maintenance* to try to anticipate or prevent the failure mode under consideration. In some cases, the most cost-effective option at this point might simply be to decide to live with the failure.

However, if a proactive task cannot be found and the failure consequences are still intolerable, it may be desirable to change the design of the asset (or to change the process) in order to reduce total costs by:

• reducing the frequency (and hence the total cost) of the failure
• reducing or eliminating the consequences of the failure
• making a proactive task cost-effective.

Redesign is discussed in more detail in Chapter 9.

Note that in the case of a failure mode with safety and environmental consequences, the objective is to reduce the probability of the failure to a very low level indeed. In the case of operational consequences, the objective is to reduce the probability (or frequency) to an economically tolerable level. As mentioned at the start of part 3 of this chapter, this frequency is likely to be several orders of magnitude greater than we would tolerate for most safety hazards, so the RCM process assumes that a proactive task which reduces the probability of a safety-related failure to a tolerable level will also deal with the operational consequences of that failure.

To begin with, we again only consider the desirability of making changes *after* we have established whether it is possible to extract the desired performance from the asset as it is currently configured. However, in this case modifications also need to be cost-justified, whereas they were the compulsory default action for failure modes with safety or environmental consequences.

In the light of these comments, the decision process for failures with operational consequences can be summarized as shown in Figure 5.6:

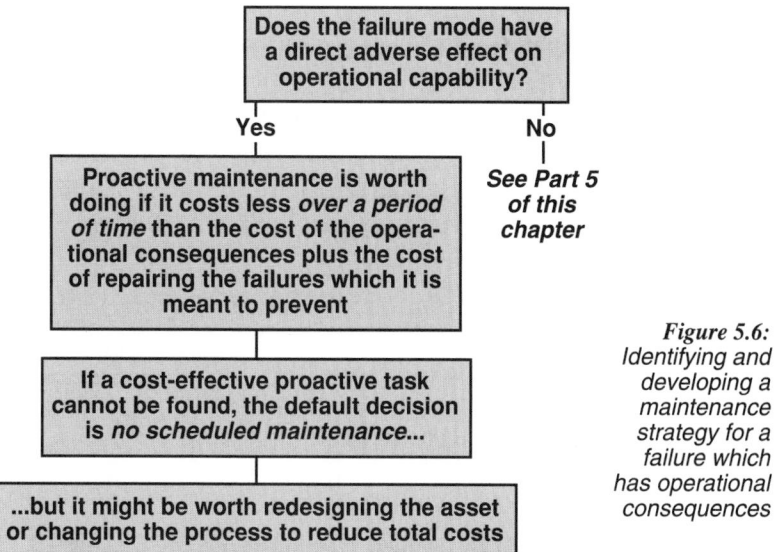

Figure 5.6:
Identifying and developing a maintenance strategy for a failure which has operational consequences

Note that this analysis is carried out for each individual failure mode, and not for the asset as a whole. This is because each proactive task is designed to prevent a specific failure mode, so the economic feasibility of each task can only be compared to the costs of the failure mode which it is meant to prevent. In each case, it is a simple go/no go decision.

In practice, when assessing individual failure modes in this way, it is not always necessary to do a detailed cost-benefit study based on actual downtime costs and MTBF's as shown in the example on Page 106. This is because the economic desirability of proactive tasks is often intuitively obvious when assessing failure modes with operational consequences.

However, whether or not the economic consequences are evaluated formally or intuitively, this aspect of the RCM process must still be applied thoroughly. (In fact, this step is surprisingly often overlooked by people new to the process. Maintenance people in particular have a tendency to implement tasks on the basis of technical feasibility alone, which results in elegant but excessively costly maintenance programs.)

Finally, bear in mind that the operational consequences of any failure are heavily influenced by the context in which the asset is operating. This is yet another reason why care should be taken to ensure that the context is identical before applying a maintenance program developed for one asset to another. The key issues were discussed in Part 3 of Chapter 2.

5.5 Non-operational Consequences

The consequences of an evident failure which has no direct adverse effect on safety, the environment or operational capability are classified as *non-operational*. The only consequences associated with these failures are the direct costs of repair, so these consequences are also *economic*.

Consider for example the pumps shown in Figure 5.7. This set-up is similar to that shown in Figure 5.4, except that there are now two pumps (both identical to the pump in Figure 5.4).

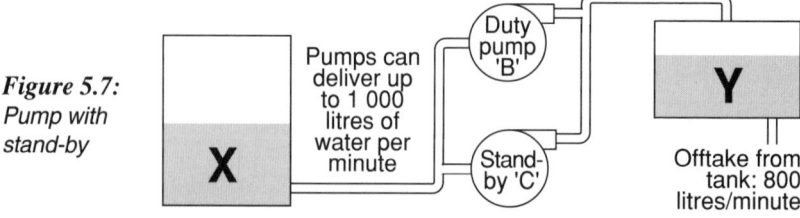

Figure 5.7:
Pump with
stand-by

The duty pump is switched on by one float switch when the level in Tank Y drops to 120 000 liters, and switched off by another when the level reaches 240 000 liters. A third switch is located just below the low level switch of the duty pump, and this switch is designed both to sound an alarm in the control room if the water level reaches it, and to switch on the stand-by pump. If the tank runs dry, the downstream process has to be shut down. This also costs the organization which uses the pump $5 000 per hour.

As before, assume that it has been agreed that one failure mode which can affect the duty pump is 'bearing seized', and that this seizure is caused by normal wear and tear. Assume that the motor on the duty pump is also equipped with an over-load switch, but again there is no trip alarm wired to the control room. This failure mode and its effects might be described on an RCM Information Worksheet as shown in Figure 5.8:

	FAILURE MODE	FAILURE EFFECT
1	Bearing seizes due to normal wear and tear	Motor trips but no alarm sounds in control room. Level in tank drops until low level alarm sounds at 120 000 litres, and stand-by pump is switched on automatically. Time required to replace the bearing 4 hours. *(The mean time between occurrences of this failure is about 3 years.)*

Figure 5.9: FMEA for failure of bearing on duty pump with stand-by

In this example, the standby pump is switched on when the duty pump fails, so the tank does not run dry. So the only cost associated with this failure is:

the cost of replacing the bearing.

Assume however that it is still *technically feasible* to check the bearing for audible noise once a week. If the bearing were found to be noisy, the operators would switch over manually to the standby pump and the bearing would be replaced.

Assume that these pumps are also located in an unmanned pumping station, and that it has again been agreed that the check – which also takes twenty minutes – should be done by a maintenance craftsman at a cost of $8 per check. So once again, he will do about 150 checks per failure. In other words, the cost of the proactive maintenance program per failure is:

150 x $8 = $1 200 plus the cost of replacing the bearing.

In this example, the cost of doing the scheduled task is now much greater than the cost of not doing it. As a result, it is not worth doing the proactive task *even though the pump is technically identical to the pump described in Figure 5.3.* This suggests that it is only worth trying to prevent a failure which has non-operational consequences if, over a period of time, the cost of the preventive task is less than the cost of correcting the failure. If it is not, then scheduled maintenance is not worth doing.

For failure modes with non-operational consequences, a proactive task is worth doing if over a period of time, it costs less than the cost of repairing the failures it is meant to prevent

If a proactive task is not worth doing, then in rare cases a modification might be justified for much the same reasons as those which apply to failures with operational consequences.

Further Points Concerning Non-operational Consequences

Two more points need to be considered when reviewing failures with non-operational consequences, as follows:

- *secondary damage:* Some failure modes cause considerable secondary damage if they are not anticipated or prevented, which adds to the cost of repairing them. A suitable proactive task could make it possible to prevent or anticipate the failure and avoid this damage. However, such a task is only justified if the cost of doing it is less than the cost of repairing the failure and the secondary damage.

   For example, in Figure 5.7 the description of the failure effects suggests that the seizure of the bearing causes no secondary damage. If this is so, then the analysis is valid. However, if the unanticipated failure of the bearing also causes (say) the shaft to shear, then a proactive task which detects imminent bearing failure would enable the operators to shut down the pump before the shaft is damaged. In this case the cost of the unanticipated failure of the bearing is:
   the cost of replacing the bearing and the shaft.
   On the other hand, the cost of the proactive task (per bearing failure) is still:
   $1 200 plus the cost of replacing the bearing.
   Clearly, the task is worth doing if it costs more than $1 200 to replace the shaft. If it costs less than $1 200, then *this task* is still not worth doing.

- *protected functions:* it is only valid to say that a failure will have non-operational consequences because a standby or redundant component is available if it is reasonable to assume that the protective device will be functional when the failure occurs. This of course means that a suitable maintenance program must be applied to the protective device (the standby pump in the example given above). This issue is discussed at length in the next part of this chapter.

   If the consequences of the multiple failure of a protected system are particularly serious, it may be worth trying to prevent the failure of the protected function as well as the protective device in order to reduce the probability of the multiple failure to a tolerable level. (As explained on Page 97, if the multiple failure has safety consequences, it may be wise to assess consequences as if the protection was not present at all, and then to revalidate the protection as part of the task selection process.)

5.6 Hidden Failure Consequences

Hidden Failures and Protective Devices

Chapter 2 mentioned that the growth in the number of ways in which equipment can fail has led to corresponding growth in the variety and severity of failure consequences which fall into the evident categories. It also mentioned that protective devices are being used increasingly in an attempt to eliminate (or at least reduce) these consequences, and explained how these devices work in one of five ways:

* to alert operators to abnormal conditions
* to shut down the equipment in the event of a failure
* to eliminate or relieve abnormal conditions which follow a failure and which might otherwise cause much more serious damage
* to take over from a function which has failed
* to prevent dangerous situations from arising.

In essence, the function of these devices is to ensure that the consequences of the failure of the protected function are much less serious than they would be if there were no protection. So any protective device is in fact part of a system with at least two components:

* the protective device
* the protected function.

For example, Pump C in Figure 5.7 can be regarded as a protective device, because it 'protects' the pumping function if Pump B should fail. Pump B is of course the protected function.

The existence of such a system creates two sets of failure possibilities, depending on whether the protective device is fail-safe or not. We consider the implications of each set in the following paragraphs, starting with devices which are fail-safe.

Fail-safe protective devices

In this context, *fail-safe* means that the failure of the device on its own will become evident to the operating crew under normal circumstances.

> *In the context of this book, a 'fail-safe' device is one whose failure on its own will become evident to the operating crew under normal circumstances*

This means that in a system which includes a fail-safe protective device, there are three failure possibilities in any period, as follows.

The first possibility is that *neither device fails*. In this case everything proceeds normally.

The second possibility is that the *protected function fails before the protective device*. In this case the protective device carries out its intended function and, depending on the nature of the protection, the consequences of failure of the protected function are reduced or eliminated.

The third possibility is that the *protective device fails before the protected function*. This would be evident because if it were not, the device would not be fail-safe in the sense defined above. If normal good practice is followed, the chance of the protected device failing while the protective device is in a failed state can be almost eliminated, either by shutting down the protected function or by providing alternative protection while the failed protective device is being rectified.

For instance, an operator could be asked to keep an eye on a pressure gauge – and his finger by a stop button – while a pressure switch is being replaced.

This means that the consequences of the failure of a fail-safe protective device usually fall into the 'operational' or 'non-operational' categories. This sequence of events is summarized in Figure 5.9.

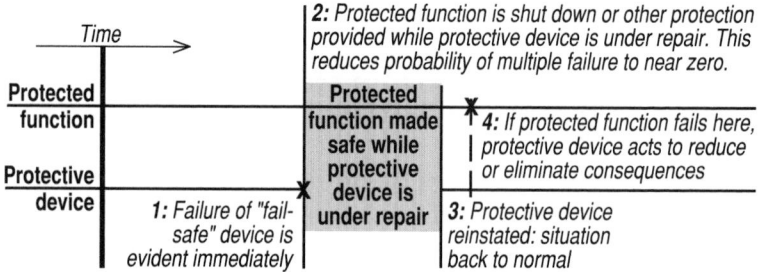

Figure 5.9: *Failure of a "fail-safe" protective device*

Protective devices which are not fail-safe

In a system which contains a protective device which is not fail-safe, the fact that the device is unable to fulfil its intended function is *not* evident under normal circumstances. This creates four failure possibilities in any given period, two of which are the same as those which apply to a fail-safe device. The first is where *neither device fails*, in which case everything proceeds normally as before.

The second possibility is that the *protected function fails at a time when the protective device is still functional*. In this case the protective device also carries out its intended function, so the consequences of the failure of the protected function are again reduced or eliminated altogether.

For instance, consider a pressure relief valve (the protective device) mounted on a pressure vessel (the protected function). If the pressure rises above tolerable limits, the valve relieves and so reduces or eliminates the consequences of the overpressurization. Similarly, if Pump B in Figure 5.7 fails, Pump C takes over.

The third possibility is that the *protective device fails while the protected function* is still working. In this case, the failure has no direct consequences. In fact no-one even knows that the protective device is in a failed state.

For example, if the pressure relief valve was jammed shut, no-one would be aware of the fact as long as the pressure in the vessel remained within normal operating limits. Similarly, if Pump C were to fail somehow while Pump B was still working, no-one would be aware of the fact unless or until Pump B also failed.

The above discussion suggests that hidden functions can be identified by asking the following question:

> **Will the loss of function caused by this failure mode on its own become evident to the operating crew under normal circumstances?**

If the answer to this question is no, the failure mode is hidden. If the answer is yes, it is evident. Note that in this context, 'on its own' means that nothing else has failed. Note also that we assume *at this point in the analysis* that no attempts are being made to check whether the hidden function is still working. This is because such checks are a form of scheduled maintenance, and the whole purpose of the analysis is to find out whether such maintenance is necessary. These two issues are discussed in more detail later in this chapter.

   The fourth possibility during any one cycle is that *the protective device fails, then the protected function fails* while the protective device is in a failed state. This situation is known as a *multiple failure*. (This is a real possibility simply because the failure of the protective device is not evident, and so no-one would be aware of the need to take corrective – or alternative – action to avoid the multiple failure.)

> **A multiple failure only occurs if a protected function fails while the protective device is in a failed state**

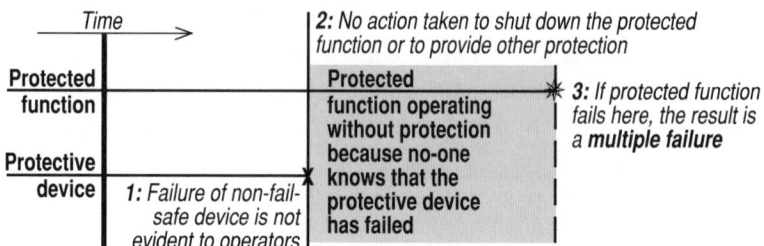

Figure 5.10:
Failure of a protective device whose function is hidden

The sequence of events which leads to a multiple failure is summarized in Figure 5.10.

In the case of the relief valve, if the pressure in the vessel rises excessively while the valve is jammed, the vessel will probably explode (unless someone acts very quickly or unless there is other protection in the system). If Pump B fails while Pump C is in a failed state, the result will be total loss of pumping.

Given that failure prevention is mainly about avoiding the consequences of failure, this example also suggests that when we develop maintenance programs for hidden functions, our objective is actually to prevent – or at least to reduce the probability of – the associated multiple failure.

> ***The objective of a maintenance program for a hidden function is to prevent – or at least to reduce the probability of – the associated multiple failure***

How hard we try to prevent the hidden failure depends on the consequences of the multiple failure.

For example, Pumps B and C might be pumping cooling water to a nuclear reactor. In this case, if the reactor could not be shut down fast enough, the ultimate consequences of the multiple failure could be a melt-down, with catastrophic safety, environmental and operational consequences.

On the other hand, the two pumps might be pumping water into a tank which has enough capacity to supply a downstream process for two hours. In this case, the consequence of the multiple failure would be that production stops after two hours, and then only if neither of the pumps could be repaired before the tank ran dry. Further analysis might suggest that at worst, this multiple failure might cost the organization (say) $2 000 in lost production.

In the first of these examples, the consequences of the multiple failure are very serious indeed, so we would go to great lengths to preserve the integrity of the hidden function. In the second case, the consequences of the multiple failure are purely economic, and how much it costs would influence how hard we would try to prevent the hidden failure.

Further examples of hidden failures and the multiple failures which could follow if they are not detected are:

- *vibration switches:* A vibration switch designed to shut down a large fan might be configured in such a way that its failure is hidden. However, this only matters if the fan vibration rises above tolerable limits (*a second failure*), causing the fan bearings and possibly the fan itself to disintegrate (*the consequences of the multiple failure*).

- *ultimate level switches: Ultimate* level switches are designed to activate an alarm or shut down equipment if a primary level switch fails to operate. In other words, if an ultimate low level switch jams, there are no consequences unless the primary switch also fails (*the second failure*), in which case the vessel or tank would run dry (*the consequences of the multiple failure*).

- *fire hoses:* The failure of a fire hose has no direct consequences. It only matters if there is a fire (*a second failure*), when the failed hose may result in the place burning down and people being killed (*the consequence of the multiple failure*).

Other typical hidden functions include emergency medical equipment, most types of fire detection, fire warning and fire fighting equipment, emergency stop buttons and trip wires, secondary containment structures, pressure and temperature switches, overload or overspeed protection devices, standby plant, redundant structural components, over-current circuit breakers and fuses and emergency power supply systems.

The Required Availability of Hidden Functions

So far, this part of this chapter has defined hidden failures and described the relationship between protective devices and hidden functions. The next question involves a closer look at the performance we require from hidden functions.

One of the most important conclusions which has been drawn so far is that the only direct consequence of a hidden failure is increased exposure to the risk of a multiple failure. Since it is the latter which we most wish to avoid, a key element of the performance required from a hidden function must be connected with the associated multiple failure.

We have seen that where a system is protected by a device which is not fail-safe, a multiple failure only occurs if the protected device fails while the protective device is in a failed state, as illustrated in Figure 5.10.

So the *probability* of a multiple failure in any period must be given by the probability that the protected function will fail while the protective device is in a failed state during the same period. Figure 5.11 shows that this can be calculated as follows:

$$\begin{array}{ccc} \text{Probability of a} \\ \text{multiple failure} \end{array} = \begin{array}{c} \text{Probability of failure of} \\ \text{the protected function} \end{array} \times \begin{array}{c} \text{Average unavailability} \\ \text{of the protective device} \end{array}$$

The tolerable probability of the multiple failure is determined by the users of the system, as discussed in the next part of this chapter and again in Appendix 3. The probability of failure of the protected function is usually a given. So if these two variables are known, the allowed unavailability of the protective device can be expressed as follows:

$$\text{Allowed unavailability of the protective device} = \frac{\text{Probability of a multiple failure}}{\text{Probability of failure of the protected function}}$$

So a crucial element of the performance required from any hidden function is the availability required to reduce the probability of the associated multiple failure to a tolerable level. The above discussion suggests that this availability is determined in the following three stages:

- first establish what probability the organization is prepared to tolerate for the multiple failure

- then determine the probability that the protected function will fail in the period under consideration (this is also known as the demand rate)

- finally, determine what availability the hidden function must achieve to reduce the probability of the multiple failure to the required level.

When calculating the risks associated with protected systems, there is sometimes a tendency to regard the probability of failure of the protected and protective devices as fixed. This leads to the belief that the only way to change the probability of a multiple failure is to change the hardware (in other words, to modify the system), perhaps by adding more protection or by replacing existing components with ones which are thought to be more reliable.

In fact, this belief is incorrect, because *it is usually possible to vary both the probability of failure of the protected function and (especially) the unavailability of the protective device* by adopting suitable maintenance and operating policies. As a result, *it is also possible to reduce the probability of the multiple failure to almost any **desired** level within reason* by adopting such policies. (Zero is of course an unattainable ideal.)

Figure 5.11:
CALCULATING THE *PROBABILITY OF A MULTIPLE FAILURE

The *probability that a *protected function will fail* in any period is the inverse of its mean time between failures, as illustrated in Figure 5.11a below:

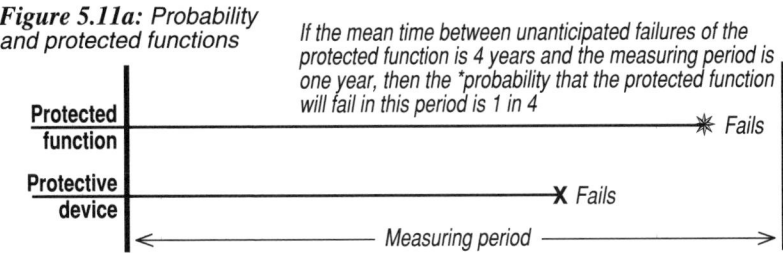

Figure 5.11a: Probability and protected functions

*If the mean time between unanticipated failures of the protected function is 4 years and the measuring period is one year, then the *probability that the protected function will fail in this period is 1 in 4*

The probability that the *protective device will be in a failed state* at any time is given by the percentage of time which it is in a failed state. This is of course measured by its unavailability (also known as *downtime* or *fractional dead time*), as shown in Figure 511b below:

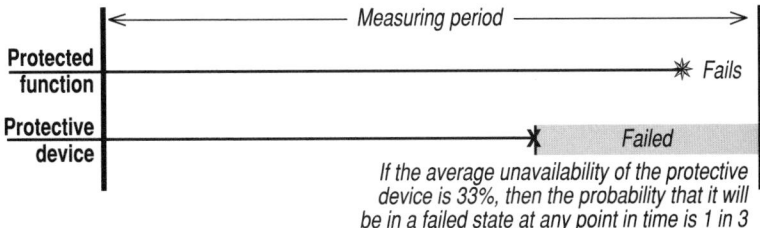

Figure 5.11b:
Probability and protective devices

If the average unavailability of the protective device is 33%, then the probability that it will be in a failed state at any point in time is 1 in 3

The *probability of the multiple failure* is calculated by multiplying the probability of failure of the protected function by the average unavailability of the protective device. For the case described in Figure 5.11(a) and (b) above, the probability of a multiple failure would be as indicated in Figure 511(c) below:

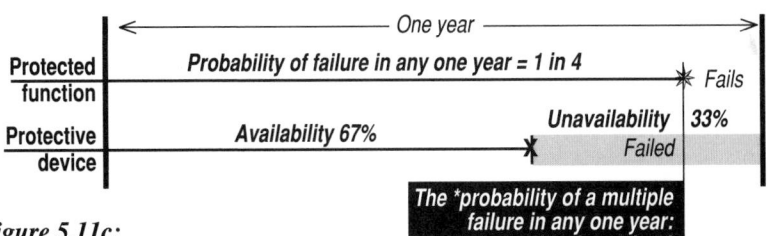

Figure 5.11c:
Probability of a multiple failure

*The *probability of a multiple failure in any one year:
1 in 4 x 1 in 3 = 1 in 12*

* *See note on page 96*

For example, the consequences of both pumps in Figure 5.7 being in a failed state may be such that the users *are prepared to tolerate* a probability of multiple failure of less than 1 in 1 000 in any one year (or 10^{-3}). Assume that it has also been estimated that if the duty pump is suitably maintained, the mean time between unanticipated failures of the duty pump can be increased to ten years, which corresponds to a probability of failure in any one year of one in ten, or 10^{-1}.

So to reduce the probability of the multiple failure to less than 10^{-3}, the unavailability of the standby pump must not be allowed to exceed 10^{-2}, or 1%. In other words it must be maintained in such a way that its availability exceeds 99%. This is illustrated in Figure 5.12 below.

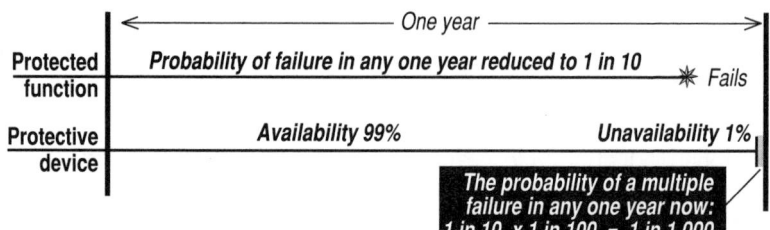

Figure 5.12:
Desired availability of a protected device

In practice, the probability which is considered to be tolerable for any multiple failure depends on its consequences. In the vast majority of cases *the assessment has to be made by the users of the asset.* These consequences vary hugely from system to system, so what is deemed to be tolerable varies equally widely. To illustrate this point, Figure 5.13 suggests four possible such assessments for four different systems:

Failure of Protected Function	Failed State of Protective Device	Multiple Failure	Tolerable Rate of Multiple Failure
Spelling error in interoffice memo or e-mail	Spell-checker in a word-processing program unable to detect errors	Spelling mistake undetected	10 per month?
10kW motor on pump B overloaded	Trip switch jammed in closed position	Motor burns out: $500 to rewind	1 in 50 years?
Duty pump B fails	Standy pump C failed	Total loss of pumping capability: $10 000 in lost production	1 in 1 000 years?
Boiler over-pressurized	Relief valves jammed shut	Boiler blows up: 10 people die	1 in 10 000 000 years?

Figure 5.13: Multiple failure rates

As before, these levels of tolerability are not meant to be prescriptive and do not necessarily reflect the views of the author. They are meant to demonstrate that in any protected system, *someone* must decide what is tolerable before it is possible to decide on the level of protection needed, and that this assessment will differ for different systems.

Part 3 of this chapter suggested that if the multiple failure could affect safety, 'someone' should be a group which includes representatives of the likely victims together with their managers. This is also true of multiple failures which have economic consequences.

For instance, in the case of the spelling error, the 'likely victim' is the author of the correspondence. In most organizations, the consequences are likely to be no more than mild embarrassment (if anyone even notices the error). In the case of the electric motor, the person most likely to be held accountable (in other words, the 'likely victim') will either be the manager responsible for the maintenance budget, or the maintenance manager in person. In the case of loss of pumping, the larger sums involved mean that higher levels of management should become involved in setting tolerability criteria.

Figure 5.13 also suggests that the probabilities which any organization might be prepared to tolerate for failures which have economic consequences tend to decrease as the magnitude of the consequences increases. This further suggests that it should be possible for any organization to develop a schedule of tolerable 'standard' economic risks which could in turn be used to help develop maintenance programs designed to deliver those risks. This might take the form shown in Figure 5.14.

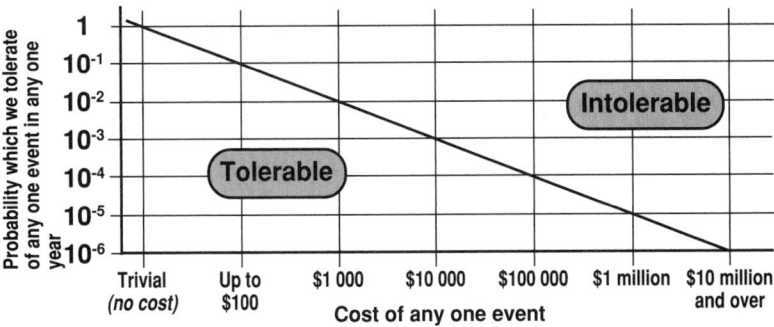

Figure 5.14: Tolerability of economic risk

Yet again, please note that these levels of tolerability are not meant to be prescriptive and are not meant to be any kind of proposed universal standard. The economic risks which any organization is prepared to tolerate are quite literally that organization's business.

Figures 5.2 and 5.14 suggest that it might be possible to produce a schedule of risk which combines safety risks and economic risks in one continuum. How this might be done is discussed in Appendix 3.

In some cases, it may be unnecessary – indeed it is sometimes impossible – to perform a rigorous quantitative analysis of the probability of multiple failure in the manner described above. In such cases, it may be enough to make a judgment about the required availability of the protective device based on a qualitative assessment of the reliability of the protected function and the possible consequences of the multiple failure. This approach is discussed further in Chapter 8. However, if the multiple failure is particularly serious, then a rigorous analysis *should* be performed.

The following paragraphs consider in more detail how it is possible to influence:
- the rate at which protected functions fail
- the availability of protected devices.

Routine Maintenance and Hidden Functions

In a system which incorporates a non-fail-safe protective device, the probability of a multiple failure can be reduced as follows:
• reduce the rate of failure of the *protected* function by:
  * doing some sort of proactive maintenance
  * changing the way in which the protected function is operated
  * changing the design of the protected function.
• increase the availability of the *protective* device by
  * doing some sort of proactive maintenance
  * checking periodically if the protective device has failed
  * modifying the protective device.

Prevent the failure of the protected function
We have seen that the probability of a multiple failure is partly based on the rate of failure of the protected function. This could almost certainly be reduced by improving the maintenance or operation of the protected device, or even (as a last resort) by changing its design.

Specifically, if the failures of a protected function can be anticipated or prevented, the mean time between (unanticipated) failures of this function would be increased. This in turn would reduce the probability of the multiple failure.

For example, one way to prevent the simultaneous failure of Pumps B and C is to try to prevent unanticipated failures of Pump B. By reducing the number of these failures, the mean time between failures of Pump B would be increased and so the probability of the multiple failure would be correspondingly reduced, as shown in Figure 5.12.

However, bear in mind that the reason for installing a protective device is that the protected function is vulnerable to unanticipated failures with serious consequences.

Secondly, if no action is taken to prevent the failure of the protective device, it will inevitably fail at some stage and hence cease to provide any protection. *After this point, the probability of the multiple failure is equal to the probability of the protected function failing on its own.*

This situation must be intolerable, or a protective device would not have been installed to begin with. This suggests that we must at least try to find a practical way of preventing the failure of protective devices which are not fail safe.

Prevent the hidden failure

In order to prevent a multiple failure, we must try to ensure that the hidden function is not in a failed state if and when the protected function fails. If a proactive task could be found which was good enough to ensure 100% availability of the protective device, then a multiple failure is theoretically almost impossible.

For example, if a proactive task could be found which could ensure 100% availability of Pump C while it is in the standby state, then we can be sure that C would always take over if B failed.

(In this case a multiple failure is only possible if the users operate Pump C while B is being repaired or replaced. However, even then the risk of the multiple failure is low, because B should be repaired quickly and so the amount of time the organization is at risk is fairly short. Whether or not the organization is prepared to take the risk of running Pump C while Pump B is down depends on the consequences of the multiple failure and on whether it is possible to arrange other forms of protection, as discussed earlier.)

In practice, it is most unlikely that any proactive task would cause any function, hidden or otherwise, to achieve an availability of 100% indefinitely. What it must do, however, is deliver the availability needed to reduce the probability of the multiple failure to a tolerable level.

For example, assume that a proactive task is found which enables Pump C to achieve an availability of 99%. If the mean time between unanticipated failures of Pump B is 10 years, then the probability of the multiple failure would be 10^{-3} (1 in 1000) in any one year, as discussed earlier.

If the availability of Pump C could be increased to 99.9% then the probability of the multiple failure would be reduced to 10^{-4} (1 in 10 000), and so on.

So for a hidden failure, a proactive task is only worth doing if it secures the availability needed to reduce the probability of the multiple failure to a tolerable level.

For hidden failures, a proactive task is worth doing if it secures the availability needed to reduce the probability of a multiple failure to a tolerable level

The ways in which failures can be prevented are discussed in Chapters 6 and 7. However, these chapters also explain that it is often impossible to find a proactive task which secures the required availability. This applies especially to the type of equipment which suffers from hidden failures. So if we cannot find a way to *prevent* a hidden failure, we must find some other way of improving the availability of the hidden function.

Detect the hidden failure
If it is not possible to find a suitable way of *preventing* a hidden failure, it is still possible to reduce the risk of the multiple failure by checking the hidden function periodically to find out if it is still working. If this check (called a 'failure-finding' task) is carried out at suitable intervals and if the function is rectified as soon as it is found to be faulty, it is still possible to secure high levels of availability. Scheduled failure-finding is discussed in detail in Chapter 8.

Modify the equipment
In a very small number of cases, it is either impossible to find any kind of routine task which secures the desired level of availability, or it is impractical to do it at the required frequency. However, something must still be done to reduce the risk of the multiple failure to a tolerable level, so in these cases, it is usually necessary to 'go back to the drawing board' and reconsider the design.

If the multiple failure could affect safety or the environment, redesign is compulsory. If the multiple failure only has economic consequences, the need for redesign is assessed on economic grounds.

Ways in which redesign can be used to reduce the risk or to change the consequences of a multiple failure are discussed in Chapter 9.

Hidden Functions: The Decision Process

All the points made so far about the development of a maintenance strategy for hidden functions can be summarized as shown in Figure 5.15:

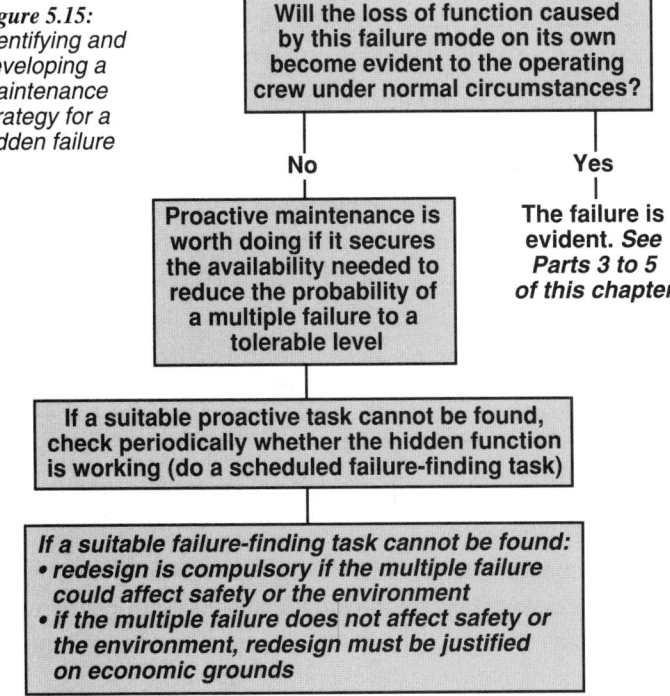

Figure 5.15: Identifying and developing a maintenance strategy for a hidden failure

Will the loss of function caused by this failure mode on its own become evident to the operating crew under normal circumstances?

No   **Yes**

Proactive maintenance is worth doing if it secures the availability needed to reduce the probability of a multiple failure to a tolerable level

The failure is evident. *See Parts 3 to 5 of this chapter*

If a suitable proactive task cannot be found, check periodically whether the hidden function is working (do a scheduled failure-finding task)

If a suitable failure-finding task cannot be found:
• **redesign is compulsory if the multiple failure could affect safety or the environment**
• **if the multiple failure does not affect safety or the environment, redesign must be justified on economic grounds**

Further Points about Hidden Functions

Six issues need special care when asking the first question in Figure 5.15. They are as follows:
• the distinction between functional failures and failure modes
• the question of time
• the primary and secondary functions of protective devices
• what exactly is meant by 'the operating crew'
• what are 'normal circumstances'
• fail-safe' devices.

These are all discussed in more detail in the following paragraphs.

Functional failure and failure mode

At this stage in the RCM process, every failure mode which is reasonably likely to cause each functional failure will already have been identified on the RCM Information Worksheet. This has two key implications:

- firstly, we are *not* asking what failures could occur. All we are trying to establish is whether each failure mode *which has already been identified as a possibility* would be hidden or evident if it did occur.

- secondly, we are *not* asking whether the operating crew can diagnose the failure mode itself. We are asking if the *loss of function* caused by the failure mode will be evident under normal circumstances. (In other words, we are asking if the failure mode has any effects or symptoms which under normal circumstances, would lead the observer to believe that the item is no longer capable of fulfilling its intended function – or at least that something out of the ordinary had occurred.)

For example, consider a motor vehicle which suffers from a blocked fuel line. The average driver (in other words, the average "operator") would not be able to diagnose this failure mode without expert assistance, so there might be a temptation to call this a hidden failure. However, the *loss of the function* caused by this failure mode *is* evident, because the car stops working.

The question of time

There is often a temptation to describe a failure as 'hidden' if a considerable period of time elapses between the moment the failure occurs and the moment it is discovered. In fact, this is not the case. If the loss of function eventually becomes apparent to the operators, and it does so as a direct and inevitable result of this failure *on its own*, then the failure is treated as evident, no matter how much time elapses between the failure in question and its discovery.

For example, a tank fed by Pump A in Figure 5.4 may take weeks to empty, so the failure of this pump might not be apparent as soon as it occurs. This might lead to the temptation to describe the failure as hidden. However, this is not so because the tank runs dry as a direct and inevitable result of the failure of Pump A *on its own*. Therefore the fact that Pump A is in a failed state *will* inevitably become evident to the operating crew.

Conversely, the failure of Pump C in Figure 5.7 will only become evident if Pump B also fails (unless someone makes a point of checking Pump C from time to time.) If pump B were to be operated and maintained in such a way that it is never necessary to switch on Pump C, it is possible that the failure of Pump C *on its own* would never be discovered.

This example demonstrates that time is not an issue when considering hidden failures. We are simply asking whether anyone will eventually become aware of the fact that the failure has occurred *on its own*, and *not* if they will be aware *when* it occurs.

Primary and secondary functions
Thus far we have focused on the primary function of protective devices, which is to be capable of fulfilling the function they are designed to fulfil when called upon to do so. As we have seen, this is usually after the protected function has failed. However, an important secondary function of many of these devices is that they should not work when nothing is wrong.

For instance, the primary function of a pressure switch might be listed as follows:
• to be capable of transmitting a signal when pressure falls below 250 psi
The implied secondary function of this switch is:
• to be incapable of transmitting a signal when pressure is above 250 psi.

The failure of the first function is hidden, but the failure of the second is evident because if it occurs, the switch transmits a spurious shutdown signal and the machine stops. If this is likely to occur in practice, it should be listed as a failure mode of the function which is interrupted (usually the primary function of the machine). As a result, there is usually no need to list the implied second *function* separately, but the failure mode would be listed under the relevant function if it is reasonably likely to occur.

The operating crew
When asking whether a failure is evident, the term *operating crew* refers to anyone who has occasion to observe the equipment or what it is doing at any time in the course of their normal daily activities, and who can be relied upon to report that it has failed.

Failures can be observed by people with many different points of view. They include operators, drivers, quality inspectors, craftsmen, supervisors, and even the tenants of buildings. However, whether any of these people can be relied upon to detect and report a failure depends on four critical elements:

• the observer must be in a position either to detect the failure mode itself or to detect the loss of function caused by the failure mode. This may be a physical location or access to equipment or information (including management information) which will draw attention to the fact that something is wrong.

- the observer must be able to recognize the condition as a failure.

- the observer must understand and accept that it is part of his or her job to report failures.

- the observer must have access to a procedure for reporting failures.

Normal circumstances

Careful analysis often reveals that many of the duties performed by operators are actually maintenance tasks. It is wise to start from a zero base when considering these tasks, because it may transpire that either the tasks or their frequencies need to be radically revised. In other words, when asking if a failure will become evident to the operating crew under 'normal' circumstances, the word *normal* has the following meanings:

- that nothing is being done to *prevent* the failure. If a proactive task is currently successfully preventing the failure, it could be argued that the failure is 'hidden' because it does not occur. However in Chapter 4 it was pointed out that failure modes and effects should be listed and the rest of the RCM process applied as if no proactive tasks are being done, because one of the main purposes of the exercise is to review whether we should be doing any such tasks in the first place.

- that no specific task is being done to *detect* the failure. A surprising number of tasks which already form part of an operator's normal duties are in fact routines designed to check if hidden functions are working.

  For example, pressing a button on a control panel every day to check if all the alarm lights on the panel are working is in fact a failure-finding task.

  We shall see later that failure-finding tasks are covered by the RCM task selection process, so once again it should be assumed at this stage in the analysis that this task is not being done (even though the task is currently genuinely part of the operator's normal duties). This is because the RCM process might reveal a more effective task, or the need to do the same task at a higher or lower frequency.

(Quite apart from the question of maintenance tasks, there is often considerable doubt about what the 'normal' duties of the operating crew actually are. This occurs most often where standard operating procedures are either poorly documented or do not exist. In these cases, the RCM review process does much to help clarify what these duties should be, and can do much to help lay the foundations of a full set of operating procedures. This applies especially to high-technology plants.)

'Fail-safe' devices

It often happens that a protective circuit is said to be fail-safe when it is not. This usually occurs when only part of a circuit is considered instead of the circuit as a whole.

An example is again provided by a pressure switch, this time attached to a hydrostatic bearing. The switch was meant to shut down the machine if the oil pressure in the bearing fell below a certain level. It emerged during discussion that if the electrical signal from the switch to the control panel was interrupted, the machine would shut down, so the failure of the switch was initially judged to be evident.

However, further discussion revealed that a diaphragm inside the switch could deteriorate with age, so the switch could become incapable of sensing changes in the pressure. This failure was hidden, and the maintenance program for the switch was developed accordingly.

To avoid this problem, take care to include the sensors and the actuators in the analysis of any control loop, as well as the electrical circuit itself.

5.7 Conclusion

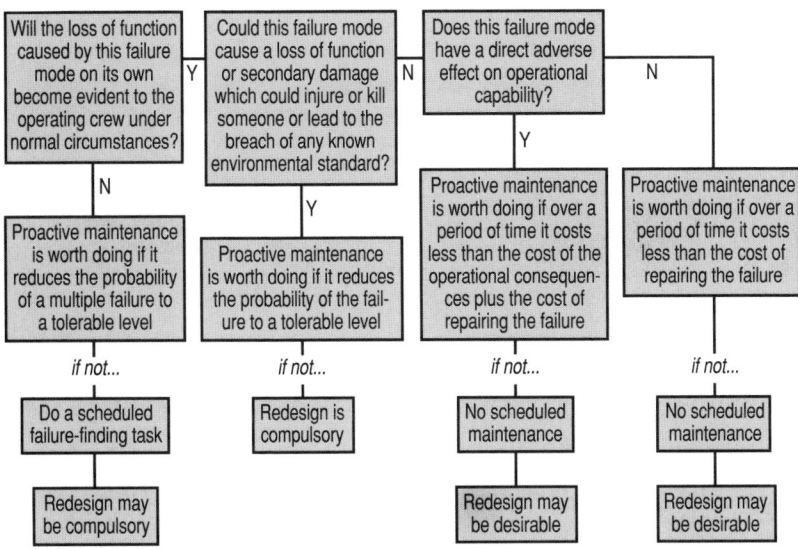

Figure 5.16: *The evaluation of failure consequences*

This chapter has demonstrated how the RCM process provides a comprehensive strategic framework for managing failures. As summarized in Figure 5.16, this framework:

- classifies all failures on the basis of their consequences. In so doing it separates hidden failures from evident failures, and then ranks the consequences of the evident failures in descending order of importance
- provides a basis for deciding whether proactive maintenance is worth doing in each case
- suggests what action should be taken if a suitable proactive task cannot be found.

The different types of proactive tasks and default actions are discussed in the next four chapters, together with an integrated approach to consequence evaluation and task selection.

6 Proactive Maintenance 1: Preventive Tasks

6.1 Technical Feasibility and Proactive Tasks

As mentioned in Chapter 1, the actions which can be taken to deal with failures can be divided into the following two categories:

- *proactive tasks:* these are tasks undertaken before a failure occurs, in order to prevent the item from getting into a failed state. They embrace what is traditionally known as 'predictive' and 'preventive' maintenance, although RCM uses the terms *scheduled restoration, scheduled discard* and *on-condition maintenance*

- *default actions:* these deal with the failed state, and are chosen when it is not possible to identify an effective proactive task. Default actions include *failure-finding, redesign* and *run-to-failure.*

These two categories correspond to the sixth and seventh of the seven questions which make up the basic RCM decision process, as follows:

- *what can be done to predict or prevent each failure?*
- *what if a suitable predictive or preventive task cannot be found?*

Chapters 6 and 7 focus on the sixth question. This deals with the criteria used to decide whether proactive tasks are *technically feasible.* They also look in more detail at how we decide whether specific categories of tasks are worth doing. (Chapters 8 and 9 review default actions.)

The previous chapter explained that a proactive task is worth doing if it reduces the consequences of failure enough to justify the direct and indirect costs of doing the task. It also mentioned that before considering whether a task is worth doing, we must of course determine whether it is *technically feasible.* Technical feasibility is defined as follows:

A task is technically feasible if it is physically possible for the task to reduce, or enable action to be taken to reduce, the consequences of the associated failure mode to an extent that might be acceptable to the owner or user of the asset

Two issues dominate proactive task selection from the technical viewpoint. These are:

• the relationship between the age of the item under consideration and how likely it is to fail

• what happens once a failure has started to occur.

The rest of this chapter considers tasks which could apply when there is a relationship between age (or exposure to stress) and failure. Chapter 7 considers the more difficult cases where there is no such relationship.

6.2 Age and Deterioration

Any physical asset which is required to fulfil a function which brings it into contact with the real world will be subjected to a variety of stresses. These stresses cause the asset to deteriorate by lowering its *resistance to stress.* Eventually this resistance drops to the point at which the asset can no longer deliver the desired performance – in other words, it fails. This process was first illustrated in Figure 4.3, and is shown again in a slightly different form in Figure 6.1.

Exposure to stress is measured in a variety of ways including output, distance travelled, operating cycles, calendar time or running time. These units are all related to time, so it is common to refer to total exposure to stress as the *age* of the item. This connection between stress and time suggests that there should be a direct relationship between the rate of deterioration and the age of the item. If this is so, then it follows that the point at which failure occurs should also depend on the age of the item, as shown in Figure 6.2.

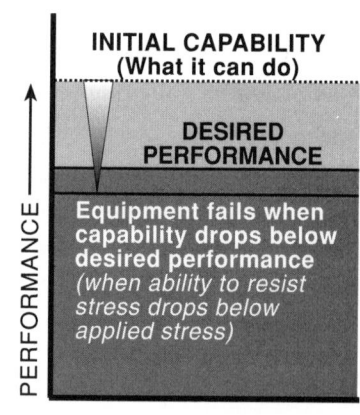

Figure 6.1:
Deterioration to failure

However, Figure 6.2 is based on two key assumptions, as follows:

• deterioration is directly proportional to the applied stress, and

• the stress is applied consistently.

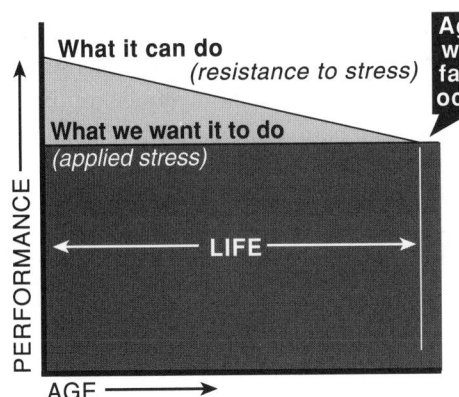

Figure 6.2:
Absolute predictability

If this were true of all assets, we would be able to predict equipment life with great precision. The classical view of preventive maintenance suggests that this can be done – all we need is enough information about failures.

In reality, however, the situation is much less clear cut. This chapter starts looking at the real world by considering a situation where there is a clear relationship between age and failure. Chapter 7 moves on to a more general view of reality.

Age-related Failures

Even parts which seem to be identical vary slightly in their initial resistance to failure. The rate at which this resistance declines with age also varies. Furthermore, no two parts are subject to exactly the same stresses throughout their lives. Even when these variations are quite small, they can have a disproportionate effect on the age at which the part fails. This is illustrated in Figure 6.3, which shows what happens to two components that are put into service with similar resistance to failure.

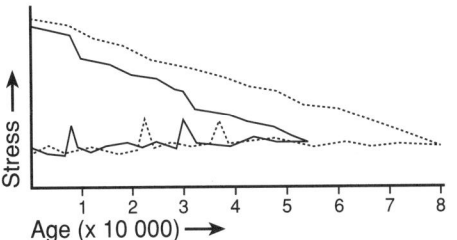

Figure 6.3:
A realistic view of age-related failures

Part B is exposed to a generally higher level of stress throughout its life than part A, so it deteriorates more quickly. Deterioration also accelerates in response to the two stress peaks at 8 000 miles and 30 000 miles. On the other hand, for some reason part A seems to deteriorate at a steady pace despite two stress peaks at 23 000 miles and 37 000 miles. So one component fails at 53 000 miles and the other at 80 000 miles.

This example shows that the failure age of identical parts working under apparently identical conditions varies widely. In practice, although some parts last much longer than others, the failures of a large number of parts which deteriorate in this fashion would tend to congregate around some average life, as shown in Figure 6.4.

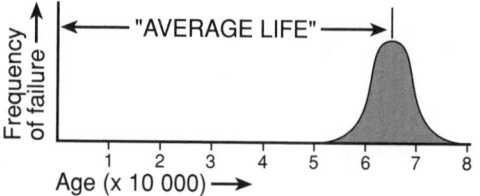

Figure 6.4:
Frequency of failure and "average life"

So even when resistance to failure does decline with age, the point at which failure occurs is often much less predictable than common sense suggests. Chapter 12 explores the quantitative implications of this situation in more depth. It also explains that the failure frequency curve shown in Figure 6.4 can be drawn as a conditional probability of failure curve, as shown in Figure 6.5 below. (The term *useful life* defines the age at which there is a rapid increase in the conditional probability of failure. It is used to distinguish this age from the *average life* shown in Figure 6.4.)

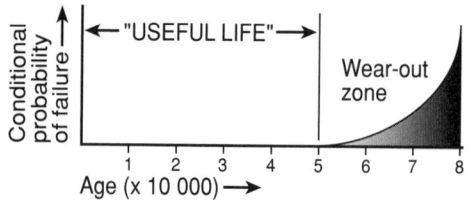

Figure 6.5:
Conditional probability of failure and "useful life"

If large numbers of apparently identical age-related failure modes are analyzed in this fashion, it is not unusual to find a number which occur prematurely. Why this occurs is also discussed in Chapter 12. The result of such premature failures is a conditional probability curve as shown in Figure 6.6. This is the same as failure pattern B in Figure 1.5.

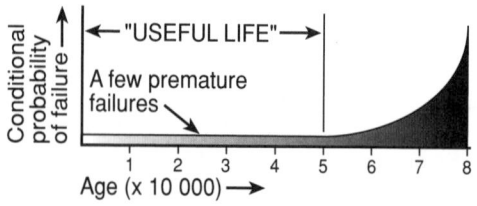

Figure 6.6:
The effect of premature failures

Even this is actually a somewhat simplistic view of age-related failures, because there are in fact three sets of ways in which the probability of failure can increase as an item gets older. These are shown in Figure 6.7.

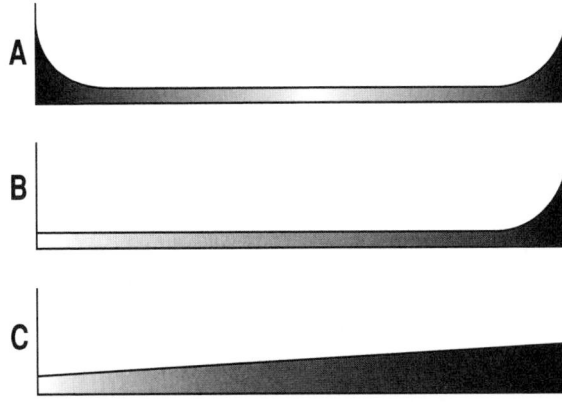

Figure 6.7:
Failures which
are age-related

These patterns were introduced in Chapter 1 and are discussed at much greater length in Chapter 12. The characteristic shared by patterns A and B is that they both display a point at which there is a rapid increase in the conditional probability of failure. Pattern C shows a steady increase in the probability of failure, but no distinct wear-out zone. The next three parts of this chapter consider the implications of these failure patterns from the viewpoint of preventive maintenance.

6.3 Age-Related Failures and Preventive Maintenance

For centuries – certainly since machines have come into widespread use – mankind has tended to believe that most equipment tends to behave as shown in Figures 6.4 to 6.6. In other words, most people still tend to assume that similar items performing a similar duty will perform reliably for a period, perhaps with a small number of random early failures, and then most of the items will 'wear out' at about the same time.

In general, age-related failure patterns apply to items which are very simple, or to complex items which suffer from a dominant failure mode. In practice, they are commonly found under conditions of direct wear (most often where equipment comes into direct contact with the product). They are also associated with fatigue, corrosion, oxidation and evaporation.

*Wear-out characteristics most often occur where
equipment comes into direct contact with the product.
Age-related failures also tend to be associated with
fatigue, oxidation, corrosion and evaporation.*

Examples of points where *equipment comes into contact with the product*
include furnace refractories, pump impellers, valve seats, seals, machine
tooling, screw conveyors, crusher and hopper liners, the inner surfaces
of pipelines, dies and so on.

Fatigue affects items – especially metallic items – which are subjected
to reasonably high-frequency cyclic loads. The rate and extent to which
oxidation and *corrosion* affect any item depend of course on its chemical
composition, the extent to which it is protected and the environment in
which it is operating. *Evaporation* affects solvents and the lighter fractions of petrochemical products.

Under certain circumstances, two preventive options which are available for reducing the incidence of failure modes like these are *scheduled
restoration tasks* and *scheduled discard tasks*. These two categories of
tasks are considered in more detail in the next section of this chapter.

6.4   Scheduled Restoration and Scheduled Discard

Failure modes which conform to Patterns A or B in Figure 6.7 become
more likely to occur after the end of the *useful life* as shown in Figure 6.5.
If an item or component is one of those which survives to the end of this
life, it is possible to remove it from service before it enters the wear-out
zone and take some sort of action either to prevent it from failing, or at
least to reduce the consequences of the failure. Sometimes, this action
entails doing something to restore the initial capability of the item or component that has been removed. If this is done at fixed intervals without
attempting to assess the condition of the item or component concerned
before subjecting it to the restoration process, the action is known as
scheduled restoration. Specifically:

*Scheduled restoration entails restoring the initial capability
of an existing item or component at or before a specified age
limit, regardless of its apparent condition at the time.*

Scheduled restoration tasks also used to be known as *scheduled rework tasks*. They include overhauls or turnarounds that are performed at pre-set intervals in order to prevent specific age-related failure modes.

In the case of some failure modes that are age-related, it is simply impossible to restore anything like the initial capability of the affected item or component once it has reached the end of its useful life. In these cases, initial capability can only be restored by discarding it and replacing it with a new one. In other cases, scheduled restoration of the existing item may be technically possible, but it is much more cost-effective to replace it with a new one. In both cases, if the item or component is replaced with a new one at fixed intervals without attempting to assess the condition of the old one beforehand, the task is known as *scheduled discard*.

> ***Scheduled discard tasks entail discarding an item or component at or before a specified age limit, regardless of its condition at the time***

Note that the terms scheduled restoration and scheduled discard can often be applied to exactly the same task, and which term is appropriate is a function of the level at which the analysis is being performed.

For instance, if a pump impeller wears out at a predictable rate and so is replaced with a new one at fixed intervals, the replacement task could be described as scheduled discard of the *impeller* or scheduled restoration of the *pump*.

This is why we tend to consider scheduled restoration and scheduled discard together. However, the distinction does become important when considering a failure mode which could be prevented by either of the two tasks when considered at the same level of analysis.

For instance, a certain type of electric motor may be known to suffer from failure of the windings after a predictable amount of time in service. In this case, it may be possible to restore initial capability by rewiring the motor (scheduled restoration) or by replacing it with a new one (scheduled discard).

For this reason, the remainder of this section of this chapter considers the common features of scheduled restoration and scheduled discard together, but also takes care to highlight key differences.

Scheduled Restoration and Scheduled Discard Task Frequencies

The frequency with which scheduled restoration and scheduled discard tasks are done is governed by the useful life of the item as shown in Figure 6.5. In other words:

The frequency of scheduled restoration and scheduled discard tasks is governed by the age at which the item or component shows a rapid increase in the conditional probability of failure.

In the case of Pattern C, at least four different restoration intervals need to be analyzed to determine the optimum interval (if one exists at all).

In general, there is a particularly widely held belief that all items 'have a life', and that overhauling the item or installing a new part before this 'life' is reached will automatically make it 'safe'. This is not always true, so RCM takes special care to focus on safety when considering scheduled restoration and scheduled discard tasks.

In fact, RCM recognizes two different types of life-limits when dealing with these tasks. The first apply to tasks meant to avoid failures which have safety consequences, and are called *safe-life* limits. Those which are intended to prevent failures that do not have safety consequences are called *economic-life* limits.

Safe-life limits
Safe-life limits only apply to failures that have safety or environmental consequences, so the associated tasks must reduce the probability of a failure occurring before the end of the life limit to a tolerable level. (A method of deciding what is tolerable was discussed in Chapter 5 part 3 and Appendix 3 of this book. Probabilities as low as 10^{-6} and sometimes even 10^{-9} are sometimes used in this context). This means that safe-life limits cannot apply to items that conform to pattern A, because infant mortality means that a significant number of items must fail prematurely. In fact, they cannot apply to *any* failure mode where there is a significant probability of the failure occurring when the item enters service.

Ideally, safe-life limits should be determined before the item is put into service. It should be tested in a simulated operating environment to determine what life is actually achieved, and a conservative fraction of this life used as the safe-life limit. This is illustrated in Figure 6.8.

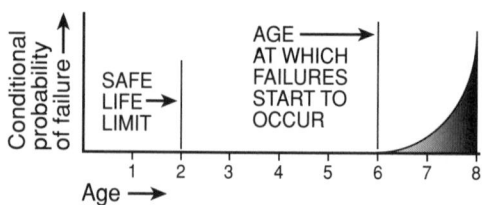

Figure 6.8:
Safe-life limits

There is never a perfect correlation between a test environment and the operating environment. Testing a long-lived part to failure is also costly and obviously takes a long time, so there is usually not enough test data for survival curves to be drawn with confidence. In these cases safe-life limits are sometimes established by dividing the average by an arbitrary factor as large as three or four. This implies that the conditional probability of failure at the life limit should essentially be zero.

Economic-life limits
Operating experience sometimes suggests that scheduled restoration or scheduled discard is desirable on economic grounds. The associated life-limit is known as an *economic-life limit*. It is usually equal to the useful life, rather than a fraction of this life. The economics of scheduled restoration and scheduled discard are discussed in more detail later in this section of this chapter.

The Technical Feasibility of Scheduled Restoration

The above comments indicate that for a scheduled restoration task to be technically feasible, the first criteria which must be satisfied are that

• there must be a point at which there is an increase in the conditional probability of failure (in other words, the item must have a 'useful life')
• we must be reasonably sure what the useful life is.

Secondly, most of the items must survive to this age. If too many items fail before reaching it, the net result is an increase in unanticipated failures. Not only could this have unacceptable consequences, but it means that the associated restoration tasks are done out of sequence. This in turn disrupts the entire schedule planning process. (If the failure mode has safety or environmental consequences, the probability of a failure occurring before the safe-life limit must be reduced to a vanishingly low level - effectively zero - as discussed above.)

Finally, scheduled restoration must restore the initial capability of the asset, or at least something close enough to the initial capability to ensure that the item continues to be able to fulfil its intended function for a period which should ideally be equal to the original life limit.

For example, no-one in their right mind would try to overhaul a domestic light bulb, simply because it is not possible to restore it to its original condition (regardless of the economics of the matter). On the other hand, it could be argued that retreading a truck tire restores the tread to something approaching its original condition.

These points lead to the following general conclusions about the technical feasibility of scheduled restoration:

Scheduled restoration tasks are technically feasible if:
- *there is an identifiable age at which the item shows a rapid increase in the conditional probability of failure*
- *most of the items survive to that age (all of the items if the failure has safety or environmental consequences)*
- *they restore the original resistance to failure of the item.*

The Technical Feasibility of Scheduled Discard Tasks

The above comments indicate that scheduled discard tasks are technically feasible under the following circumstances:

Scheduled discard tasks are technically feasible if:
- *there is an identifiable age at which the item shows a rapid increase in the conditional probability of failure*
- *most of the items survive to that age (all of the items if the failure has safety or environmental consequences).*

There is usually no need to ask if the task will restore the initial capability because the item is replaced with a new one.

The Effectiveness of Scheduled Restoration and Scheduled Discard

Even if it is technically feasible, scheduled restoration and scheduled discard might still not be worth doing because other tasks may deal with the failure consequences even more effectively, as explained in Chapter 7.

If a more effective task cannot be found, there is often a temptation to select scheduled restoration or scheduled discard purely on the grounds of technical feasibility. An age limit applied to an item which behaves as shown in Figure 6.6 means that some items will receive attention before they need it while others might fail early, but the net effect may be an overall reduction in the number of unanticipated failures. However even then scheduled restoration or discard might not be worth doing, because, as mentioned earlier, a reduction in the number of failures is not sufficient if the failure has *safety* or *environmental* consequences. This is so because to be *worth doing*, the task should reduce the probability of failures which have these consequences to a vanishingly low level (effectively zero).

On the other hand, if the consequences are economic, we need to be sure that over a period of time, the cost of doing the scheduled restoration task or scheduled discard task will be less than the cost of allowing the failure to occur. In other words, the only justification for an economic life limit is cost-effectiveness. This is so because scheduled restoration increases the number of jobs passing through the workshop, while scheduled discard increases the consumption of the items or components which are subject to discard. Why this is so is illustrated by the example at the foot of this page.

When considering failures which have operational consequences, bear in mind also that scheduled restoration and scheduled discard may themselves also affect operations. However, in most cases, this effect is likely to be less than the consequences of the failure because:

- the task would normally be done when it is likely to have the least effect on operations (usually during a so-called production window)

- the task is likely to take less time than it would to repair the failure because it is possible to plan more thoroughly for the scheduled task.

Figure 6.9 shows an age-related failure mode where the useful life is 12 months, while the average life is 18 months. In a period of 3 years, the failure occurs *twice* if no preventive maintenance is done, while the preventive task would be done *three* times. In other words, the preventive task has to be done 50% more often than the corrective task which would have to be performed if the failure was allowed to occur on its own.

If each failure costs (say) $2 000 in lost production and repair, failures would cost $4 000 over a three year period. If the total cost of each preventive task is (say) $1 100, these tasks would cost $3 300 over the same period. So in this case, the task *is* cost-effective.

On the other hand, if the average life was 24 months and all other figures remained the same, failures only occur 1.5 times every three years, and would cost $3 000 over this period. The scheduled task still costs $3 300 over the same period, so it would *not* be cost-effective.

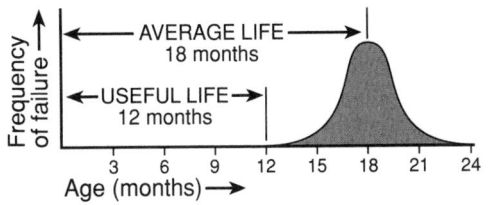

Figure 6.9:
"Useful life" and
"average life"

If there are no operational consequences, scheduled restoration and scheduled discard are only justified if they cost substantially less than the cost of repair (which may be the case if the failure causes extensive secondary damage).

All this means that in general, an economic life-limit is *worth applying* if it avoids or reduces the operational consequences of an unanticipated failure, and/or if the failure which it prevents causes significant secondary damage. Clearly, we must know the failure pattern before we can assess the cost effectiveness of scheduled discard tasks.

For new assets, this means that a failure mode which has major economic consequences should also be put into an age-exploration program to find out if a life limit is applicable. However, there is seldom enough evidence to include scheduled restoration or scheduled discard in an initial scheduled maintenance program. In practice, the frequency of these tasks can only be determined satisfactorily on the basis of reliable historical data. This is seldom available when assets first go into service, so it is usually impossible to specify scheduled restoration or scheduled discard in prior-to-service maintenance programs. (For example scheduled restoration tasks were only assigned to seven components in the initial program developed for the Douglas DC 10). However, items subject to very expensive failure modes should be put into age exploration programs as soon as possible to find out if they would benefit from these tasks.

6.5 Failures which are Not Age-related

One of the most challenging developments in modern maintenance management has been the discovery that very few failure modes actually conform to any of the failure patterns shown in Figure 6.7. As discussed in the following paragraphs, this is due primarily to a combination of variations in applied stress and increasing complexity.

Variable stress
Contrary to the assumptions listed in part 2 of this chapter, deterioration is not always proportional to the applied stress, and stress is not always applied consistently. For instance, part 3 of Chapter 4 mentioned that many failures are caused by increases in applied stress, which are caused in turn by incorrect operation, incorrect assembly or external damage.

Examples of such increases in stress given in Chapter 4 included operating errors (starting up a machine too quickly, accidentally putting it into reverse while it is going forward, feeding material into a process too quickly) assembly errors (over-torquing bolts, misfitting parts) and external damage (lightning, the 'thousand-year flood', and so on).

In all of these cases, there is little or no re-lationship between how long the asset has been in service and the likelihood of the failure occurring, This is shown in Figure 6.10, which is basically the same as Figure 4.4 with a time dimension added. (Ideally, 'preventing' failures of this sort should be

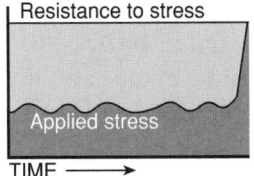

Figure 6.10

a matter of preventing whatever causes the increase in stress levels, rather than a matter of doing anything to the asset.)

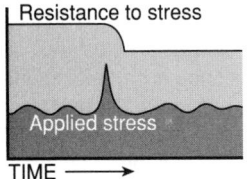

Figure 6.11

In Figure 6.11, the stress peak perma-nently reduces resistance to failure, but does not actually cause the item to fail *(an earth-quake cracks a structure but does not cause it to fall down)*. The reduced failure resis-tance makes the part vulnerable to the next peak, which may or may not occur before the part is replaced for another reason.

In Figure 6.12, the stress peak only tem-porarily reduces failure resistance *(a ther-moplastic material that softens when tem-perature rises and hardens when it drops)*.

Finally in Figure 6.13 a stress peak acce-lerates the decline of failure resistance and eventually greatly shortens the life of the

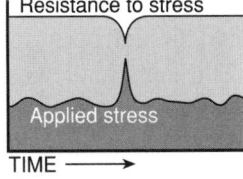

Figure 6.12

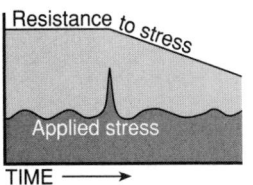

Figure 6.13

component. When this happens, the cause and effect relationship can be very difficult to establish, because the failure could occur months or even years after the stress peak.

This often happens if a part is damaged during installation (which might happen if a ball-bear-ing is misaligned), if it is damaged prior to instal-lation (the bearing is dropped on the floor in the parts store) or if it is mistreated in service (dirt gets into the bearing). In these cases, failure prevention is ideally a matter of ensuring that maintenance and in-stallation work is done correctly and that parts are looked after properly in storage.

In all four of these examples, when the items enter service it is not possible to predict when the failures will occur. For this reason, such failures are described as 'random'.

Complexity

The failure processes depicted in Figure 6.7 apply to fairly simple mechanisms. In the case of complex items, the situation becomes even less predictable. Items are made more complex to improve their performance (by incorporating new or additional technology or by automation) or to make them safer (using protective devices).

For example, Nowlan and Heap[1978] cite developments in the field of civil aviation. In the 1930's, an air trip was a slow, somewhat risky affair, undertaken in reasonably favorable weather conditions in an aircraft with a range of a few hundred miles and space for about twenty passengers. The aircraft had one or two reciprocating engines, fixed landing gear, fixed pitch propellers and no wing flaps.

Today an air trip is much faster and very much safer. It is undertaken in almost any weather conditions in an aircraft with a range of thousands of miles and space for hundreds of passengers. The aircraft has several jet engines, anti-icing equipment, retractable landing gear, moveable high-lift devices, pressure and temperature control systems for the cabin, extensive navigation and communications equipment, complex instrumentation and complex ancillary support systems.

In other words, better performance and greater safety are achieved at the cost of greater complexity. This is true in most branches of industry.

Greater complexity means balancing the lightness and compactness needed for high performance, with the size and mass needed for durability. This combination of complexity and compromise:

• increases the number of components which can fail, and also increases the number of interfaces or connections between components. This in turn increases the number and variety of failures which can occur.

For example, a great many mechanical failures involve welds or bolts, while a significant proportion of electrical and electronic failures involve the connections between components. The more such connections there are, the more such failures there will be.

• reduces the margin between the initial capability of each component and the desired performance (in other words, the 'can' is closer to the 'want'), which reduces scope for deterioration before failure occurs.

These two developments in turn suggest that complex items are more likely to suffer from random failures than simple items.

Figure 6.14:
Failures which
are not age-
related

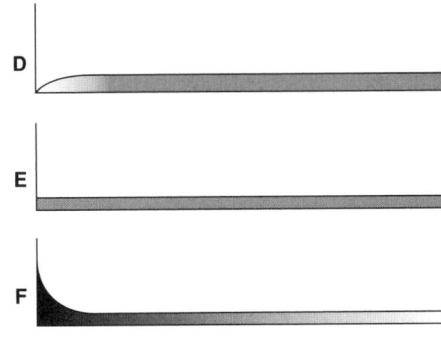

Patterns D, E and F
The combination of variable
stress and erratic response to
stress coupled with the increasing complexity mean that in practice, a
high and rising proportion of failure modes conform to the failure patterns
shown in Figure 6.14. The most important characteristic of patterns D, E
and F is that after the initial period, there is little or no relationship
between reliability and operating age. In the case of such failure modes, age
limits do little or nothing to reduce the probability of failure.

(In fact, scheduled overhauls can actually *increase* overall failure rates
by introducing infant mortality into otherwise stable systems. This is
borne out by the high and rising number of nasty accidents around the
world which have occurred either while maintenance is under way or
immediately after a maintenance intervention. It is also borne out by the
machine operator who says that "every time maintenance works on it over
the weekend, it takes us until Wednesday to get it going again".)

From the maintenance management viewpoint, the main conclusion to
be drawn from these failure patterns is that the idea of a wear-out age
simply does not apply to random failures, so the idea of fixed interval
replacement or overhaul prior to such an age cannot apply.

As mentioned in Chapter 1, an intuitive awareness of these facts has led
some people to abandon the idea of preventive maintenance altogether.
Although this can be the right thing to do for failures with minor consequences, when the failure consequences are serious, *something* must be
done to prevent the failures or at least to avoid the consequences.

The continuing need to prevent certain types of failure, and the growing inability of classical techniques to do so, are behind the growth of new
types of failure management. Foremost among these are the techniques
known as predictive or on-condition maintenance. These techniques are
discussed at length in the next chapter.

7 Proactive Maintenance 2: Predictive Tasks

7.1 Potential Failures and On-condition Maintenance

The previous chapter explained that there is often little or no relationship between how long an asset has been in service and how likely is to fail. However, although many failure modes are not age-related, most of them give some sort of warning that they are in the process of occurring or are about to occur. If evidence can be found that something is in the final stages of failure, it may be possible to take action to prevent it from failing completely and/or to avoid the consequences.

Figure 7.1 illustrates what happens in the final stages of failure. It is called the *P-F curve,* because it shows how a failure starts, deteriorates to the point at which it can be detected (point 'P') and then, if it is not detected and corrected, continues to deteriorate – usually at an accelerating rate – until it reaches the point of functional failure ('F').

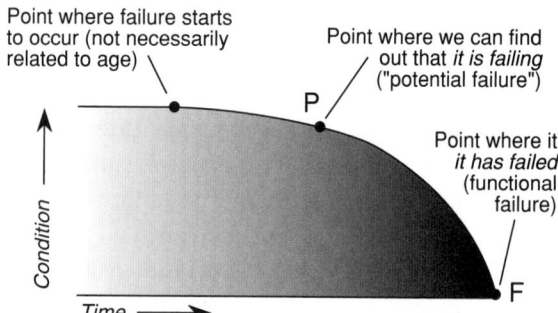

Figure 7.1:
The P-F curve

The point in the failure process at which it is possible to detect whether the failure is occurring or is about to occur is known as a *potential failure.*

> *A potential failure is an identifiable condition*
> *which indicates that a functional failure is either*
> *about to occur or in the process of occurring*

In practice, there are thousands of ways of finding out if failures are in the process of occurring.

Examples of potential failures include hot spots showing deterioration of furnace refractories or electrical insulation, vibrations indicating imminent bearing failure, cracks showing metal fatigue, particles in gearbox oil showing imminent gear failure, excessive tread wear on tires, etc.

If a potential failure is detected between point P and point F in Figure 7.1, it may be possible to take action to prevent or to avoid the consequences of the functional failure. (Whether or not it is possible to take meaningful action depends on how quickly the failure occurs, as discussed in part 2 of this chapter.) Tasks designed to detect potential failures are known as *on-condition tasks*.

> ***On-condition tasks entail checking for potential failures, so that action can be taken to prevent the functional failure or to avoid the consequences of the functional failure***

On-condition tasks are so called because the items which are inspected are left in service *on the condition* that they continue to meet specified performance standards. This is also known as *predictive* maintenance (because we are trying to predict whether – and possibly when – the item is going to fail on the basis of its present behavior) or *condition-based* maintenance (because the need for corrective or consequence-avoiding action is based on as assessment of the condition of the item.)

7.2 The P-F Interval

In addition to the potential failure itself, we need to consider the amount of time (or the number of stress cycles) which elapse between the point at which a potential failure occurs – in other words, the point at which it becomes *detectable* – and the point where it deteriorates into a functional failure. As shown in Figure 7.2, this interval is known as the *P-F interval*.

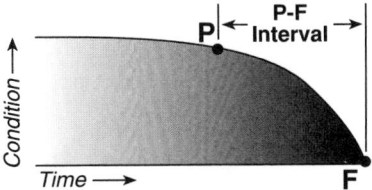

> ***The P-F interval is the interval between the occurrence of a potential failure and its decay into a functional failure***

Figure 7.2: The P-F interval

The P-F interval tells us how often on-condition tasks must be done. If we want to detect the potential failure before it becomes a functional failure, the interval between checks must be less than the P-F interval.

> ### On-condition tasks must be carried out at intervals less than the P-F interval

The P-F interval is also known as the *warning period*, the *lead time to failure* or the *failure development period*. It can be measured in any units which provide an indication of exposure to stress (running time, units of output, stop-start cycles etc), but for practical reasons, it is most often measured in terms of elapsed time. For different failure modes, it varies from fractions of a second to several decades.

Note that if an on-condition task is done at intervals which are longer than the P-F interval, there is a chance that we will miss the failure altogether. On the other hand, if we do the task at too small a percentage of the P-F interval, we will waste resources on the checking process.

For instance, if the P-F interval for a given failure mode is two weeks, the failure will be detected if the item is checked once a week. Conversely, if it is checked once a month, it is possible to miss the whole failure process. On the other hand, if the P-F interval is three months it is a waste of effort to check the item every day.

In practice it is usually sufficient to select a task frequency equal to half the P-F interval. This ensures that the inspection will detect the potential failure before the functional failure occurs, while (in most cases) providing a reasonable amount of time to do something about it. This leads to the concept of the *nett P-F interval*.

The Nett P-F Interval

The nett P-F interval is the minimum interval likely to elapse between the *discovery* of a potential failure and the occurrence of the functional failure. This is illustrated in Figures 7.3 and 7.4, which both show a failure with a P-F interval of nine months.

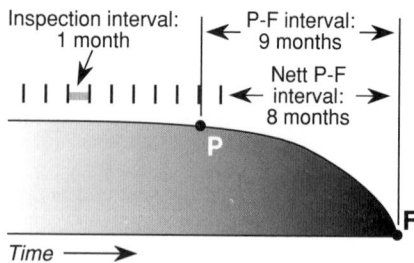

Figure 7.3:
Nett P-F interval (1)

Figure 7.3 shows that if the item is inspected monthly, the nett P-F interval is 8 months. On the other hand, if it is inspected at six monthly intervals as shown in Figure 7.4, the nett P-F interval is 3 months. So in the first case the minimum amount of time available to do something about the failure is five months longer than in the second, but the inspection task has to be done six times more often.

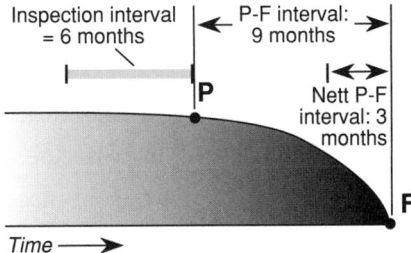

Figure 7.4:
Nett P-F interval (2)

The nett P-F interval governs the amount of time *available* to take whatever action is needed to reduce or eliminate the consequences of the failure. Depending on the operating context of the asset, warning of incipient failure enables the users of an asset to reduce or avoid consequences in a number of ways, as follows

- *downtime:* corrective action can be planned at a time which does not disrupt operations. The opportunity to plan the corrective action properly also means that it is likely to be done more quickly.

  For example, if an electrical component is found to be overheating before it burns out, it may be possible to replace it when the machine is normally idle. Note that in this case, the failure of the component is not 'prevented' – it might be doomed whatever happens – but the operational consequences of the failure are avoided.

- *repair costs:* users may be able to take action to eliminate the secondary damage which would be caused by unanticipated failures. This would reduce the downtime and the repair costs associated with the failure.

  For instance, a timely warning might enable users to switch a machine off before (say) a collapsing bearing allows a rotor to touch a stator.

- *safety:* warning of failure provides time either to shut down a plant before the situation becomes dangerous, or to move people who might otherwise be injured out of harm's way.

  For instance, if a crack in a wall is discovered in good time, it may be possible to shore up the foundations and so prevent the wall from deteriorating so much that it falls down. It is highly likely that we would have to vacate the premises while this work is done, but at least we avoid the safety consequences which would arise if the wall fell down.

For an on-condition task to be technically feasible, the nett P-F interval must be *longer* than the time required to take action to avoid or reduce the consequences of the failure. If the nett P-F interval is too short for any sensible action to be taken, then the on-condition task is clearly not technically feasible.

In practice, the time required varies widely. In some cases it may be a matter of hours (say until the end of an operating cycle or the end of a shift) or even minutes (to shut down a machine or evacuate a building). In other cases it can be weeks or even months (say until a major shutdown).

In general, longer P-F intervals are desirable for two reasons:

• it is possible to do whatever is necessary to avoid the consequences of the failure (including planning the corrective action) in a more considered and hence more controlled fashion.

• fewer on-condition inspections are required.

This explains why so much energy is being devoted to finding potential failure conditions and associated on-condition techniques which give the longest possible P-F intervals. However, note that it is possible to make use of very short P-F intervals in certain cases.

For example, failures which affect the balance of large fans cause serious problems very quickly, so on-line vibration sensors are used to shut the fans down when such failures occur. In this case, the P-F interval is very short, so monitoring is continuous. Note also that once again, the monitoring device is being used *to avoid the consequences* of the failure.

P-F Interval Consistency

The P-F curves illustrated so far in this chapter indicate that the P-F interval for any given failure is constant. In fact, this is not the case – some actually vary over a quite considerable range of values, as shown in Figure 7.5.

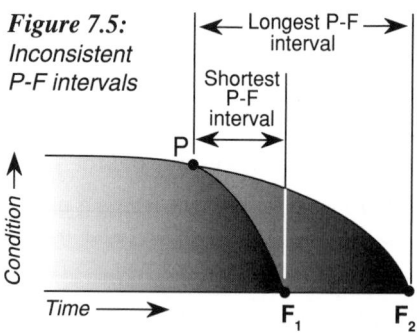

Figure 7.5: Inconsistent P-F intervals

For example, when discussing the P-F interval associated with a change in noise levels, someone might say: "This thing rattles away for anything from two weeks to three months before it collapses." In another case, tests might show that anything from six months to five years elapses from the moment a crack becomes detectable at a particular point in a structure until the moment the structure fails.

Clearly, in these cases a task interval should be selected which is substantially less than the shortest of the likely P-F intervals. In this way, we can always be reasonably certain of detecting the potential failure before it becomes a functional failure. If the nett P-F interval associated with this minimum interval is long enough for suitable action to be taken to deal with the consequences of the failure, then the on-condition task is technically feasible.

On the other hand, if the P-F interval is wildly inconsistent – as some of them can be – then it is not possible to establish a meaningful task interval, and the task in question should again be abandoned in favor of some other way of dealing with the failure.

7.3 Technical Feasibility of On-condition Tasks

In the light of the above discussion, the criteria which any on-condition task must satisfy to be technically feasible can be summarized as follows:

Scheduled on-condition tasks are technically feasible if:

- *it is possible to define a clear potential failure condition*
- *the P-F interval is reasonably consistent*
- *it is practical to monitor the item at intervals less than the P-F interval*
- *the nett P-F interval is long enough to be of some use (in other words, long enough for action to be taken to reduce or eliminate the consequences of the functional failure).*

7.4 Categories of On-condition Techniques

The four major categories of on-condition techniques are as follows:

- *condition monitoring* techniques, which involve the use of specialized equipment to monitor the condition of other equipment
- techniques based on variations in *product quality*
- *primary effects monitoring* techniques, which entail the intelligent use of existing gauges and process monitoring equipment
- inspection techniques based on the *human senses.*

These are each reviewed in the following paragraphs.

Condition monitoring

The most sensitive on-condition maintenance techniques usually involve the use of some type of equipment to detect potential failures. In other words, equipment is used to monitor the condition of other equipment. These techniques are known as *condition monitoring* to distinguish them from other types of on-condition maintenance.

Condition monitoring embraces several hundred different techniques, so a detailed study of the subject is well beyond the scope of this chapter. However, Appendix 4 provides a brief summary of about 100 of the better known techniques. All of these techniques are designed to detect failure *effects* (or more precisely, potential failure effects, such as changes in vibration characteristics, changes in temperature, particles in lubricating oil, leaks, and so on). They are classified accordingly in Appendix 4 under the following headings:

• dynamic effects
• particle effects
• chemical effects
• physical effects
• temperature effects
• electrical effects.

These techniques can be seen as highly sensitive versions of the human senses. Many of them are now very sensitive indeed, and a few give several months (if not several years) warning of failure. However, a major limitation of nearly every condition monitoring device is that it monitors only one condition. For instance, a vibration analyzer only monitors vibration and cannot detect chemicals or temperature changes. So greater sensitivity is bought at the price of the versatility inherent in the human senses.

The P-F intervals associated with different monitoring techniques vary from a few minutes to several months. Different techniques also pinpoint failures with different degrees of precision. Both of these factors must be considered when assessing the *feasibility* of any technique.

In general, condition monitoring techniques can be spectacularly effective when they are appropriate, but when they are inappropriate they can be a very expensive and sometimes bitterly disappointing waste of time. As a result, the criteria for assessing whether on-condition tasks are technically feasible and worth doing should be applied especially rigorously to condition monitoring techniques.

Product quality variation

In some industries, an important source of data about potential failures is the quality management function. Often the emergence of a defect in an article produced by a machine is directly related to a failure mode in the machine itself. Many of these defects emerge gradually, and so provide timely evidence of potential failures. If the data gathering and evaluation procedures exist already, it costs very little to use them to provide warning of equipment failure

One popular technique which can often be used in this way is Statistical Process Control (SPC). SPC entails measuring some attribute of a product such as a dimension, filling level or packing weight, and using the measurements to draw conclusions about the stability of the process.

Figure 2.6 in Chapter 2 showed how such measurements might appear for a process which is in control and in specification. Figures 3.4 and 3.5 in Chapter 3 showed two ways in which a process could be out of control and out of specification (in other words, failed). In a great many cases, the transition from being in control to failed takes place gradually. SPC charts frequently track this transition.

For instance, Figure 7.6 overleaf shows a typical SPC chart on which the readings are in control to start with. A failure mode occurs which causes the measurements to start drifting in one direction.

For example, as a grinding wheel wears, the diameter of successive workpieces increases until the wheel is adjusted or replaced.

In zone 2 in Figure 7.6 the process is out of control but still within specification. (Oakland[1991] describes how it is possible to identify very gradual shifts of this sort using a 'cusum chart'.) This shift in the mean is a clearly identifiable condition which indicates that a functional failure is about to occur. In other words, it is a potential failure. If nothing is done to rectify the situation the process eventually begins to produce out-of-spec products, as shown in zone 3 in Figure 7.6.

This example describes only one of many ways in which SPC can be used to measure and manage process variability. A full description of all the techniques is well beyond the scope of this book. However, the key point to note at this stage is that if deviations on charts like these can be related directly to specific failure modes, then the charts are sources of on-condition data which can make a valuable contribution to overall pro-active maintenance efforts.

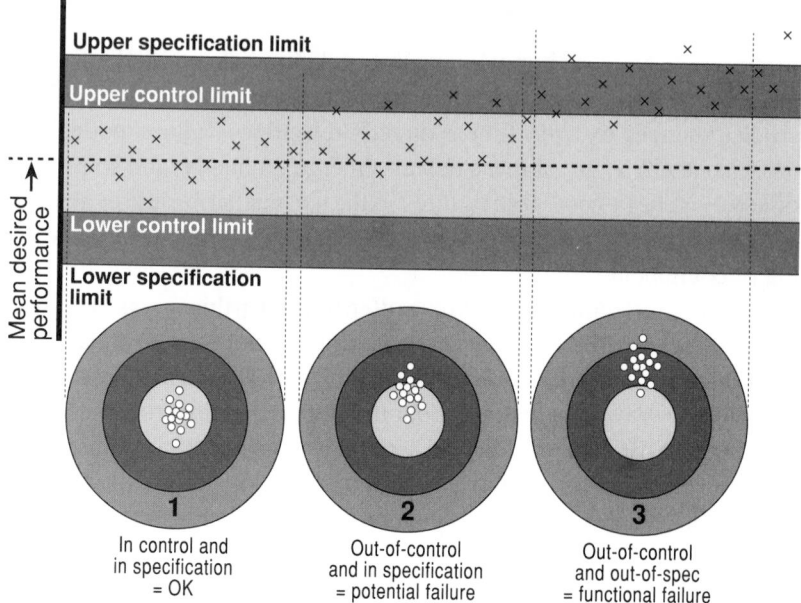

Figure 7.6: On-condition maintenance and SPC

Primary effects monitoring

Primary effects (speed, flow rate, pressure, temperature, power, current, etc) are yet another source of information about equipment condition. The effects can be monitored by a person reading a gauge and perhaps recording the reading manually, by a computer as part of a process control system, or even by a traditional chart recorder.

The records of these effects or their derivatives are compared with reference information, and so provide evidence of a potential failure. However, in the case of the first option in particular, take care to ensure that:

• the person taking the reading knows what the reading should be when all is well, what reading corresponds to a potential failure and what corresponds to functional failure

• the readings are taken at a frequency which is less than the P-F interval (in other words, the frequency should be less than the time it takes the pointer on the dial to move from the potential failure level to the functional failure level when the failure mode in question is occurring)

• that the gauge itself is maintained in such a way that it is sufficiently accurate for this purpose.

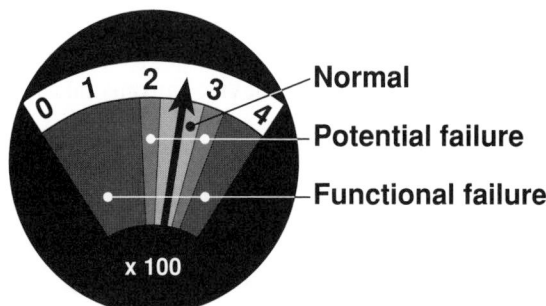

Normal

Potential failure

Functional failure

Figure 7.7:
Using gauges
for on-condition
maintenance

The process of taking readings can be greatly simplified if gauges are marked up (or even colored) as shown in Figure 7.7. In this case, all the operator – or anyone else – needs to do is look at the gauge and report if the pointer is in the potential failure (yellow?) zone, or take more drastic action if it is in the functional failure (red?) zone. However the gauge must still be monitored at intervals which are less than the P-F interval.

(For obvious reasons, this suggestion only applies to gauges which are measuring a steady state. Also take care to ensure that gauges marked up in this way are not taken off and remounted in the wrong place.)

The human senses

Perhaps the best known on-condition inspection techniques are those based on the human senses (look, listen, feel and smell). The two main disadvantages of using these senses to detect potential failures are that:

• by the time it is possible to detect most failures using the human senses, the process of deterioration is already quite far advanced. This means that the P-F intervals are usually short, so the checks must be done more frequently than most and response has to be rapid

• the process is subjective, so it is difficult to develop precise inspection criteria, and the observations depend very much on the experience and even the state of mind of the observer.

However, the advantages of using these senses are as follows:

• the average human being is highly versatile and can detect a wide variety of failure conditions, whereas any one condition monitoring technique can only be used to monitor one type of potential failure

• it can be very cost-effective if the monitoring is done by people who are at or near the assets anyway in the course of their normal duties

- a human is able to exercise judgement about the severity of the potential failure and hence about the most appropriate action to be taken, whereas a condition monitoring device can only take readings and send a signal.

Selecting the Right Category

Many failure modes are preceded by more than one – often several – different potential failures, so more than one category of on-condition task might be appropriate. Each of these will have a different P-F interval, and each will require different types and levels of skill.

For example, consider a ball bearing whose failure is described as 'bearing seizes due to normal wear and tear'. Figure 7.8 shows how this failure could be preceded by a variety of potential failures, each of which could be detected by a different on-condition task.

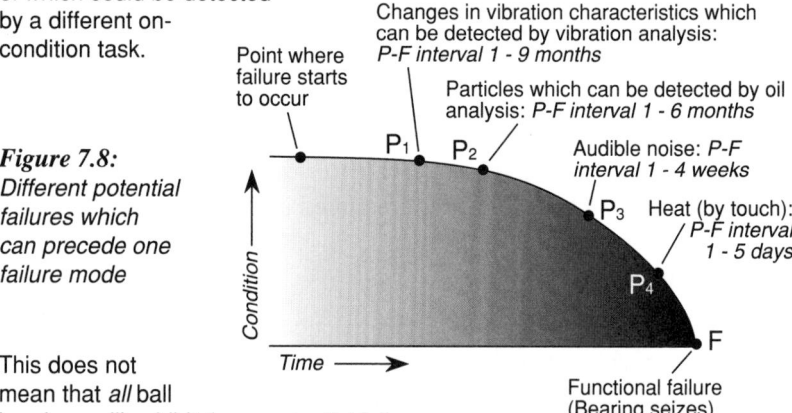

Figure 7.8:
Different potential failures which can precede one failure mode

This does not mean that *all* ball bearings will exhibit these potential failures, nor will they necessarily have the same P-F intervals.

The extent to which any technique is technically feasible and worth doing depends very much on the operating context of the bearing. For instance:

- the bearing may be buried so deep in the machine that it is impossible to monitor its vibration characteristics

- it is only possible to detect particles in the oil if the bearing is operating in a totally enclosed oil-lubricated system

- background noise levels may be so high that it is impossible to detect the noise made by a failing bearing

- it may not be possible to reach the bearing housing to feel how hot it is.

This means that no one single category of tasks will always be more cost-effective than any other. It is important to bear this in mind, because there is a tendency in some quarters to present condition monitoring in particular as 'the answer' to all our maintenance problems.

In fact, if RCM is correctly applied to typical modern, complex industrial systems, it is not unusual to find that condition monitoring as defined in this part of this chapter is technically feasible for no more than 20% of failure modes, and worth doing in less than half these cases. (All four categories of on-condition maintenance together are usually suitable for about 25 - 35% of failure modes.) This is not meant to imply that condition monitoring should not be used – where it is good it is very, very good – but that we must also remember to develop suitable strategies for managing the other 90% of our failure modes. In other words, condition monitoring is only part of the answer – and a fairly small part at that.

So to avoid unnecessary bias in task selection, we need to:

• consider *all* the warnings which are reasonably likely to precede each failure mode, together with the *full* range of on-condition tasks which could be used to detect those warnings.

• apply the RCM task selection criteria rigorously to determine which (if any) of the tasks is likely to be the most cost-effective way of anticipating the failure mode under consideration.

As with so much else in maintenance, the 'right' choice ultimately depends on the operating context of the asset.

7.5 On-condition Tasks: Some of the Pitfalls

When considering the technical feasibility of on-condition maintenance, two issues need special care. They concern the distinction between potential and functional failures, and the distinction between potential failure and age. These issues are discussed in more detail below.

Potential and functional failures
In practice, confusion often arises over the distinction between potential and functional failures. This happens because certain conditions can correctly be regarded as potential failures in one context and as functional failures in another. This is especially common in the case of leaks.

For example, a minor leak in a flanged joint on a pipeline might be regarded as a potential failure if the pipeline is carrying water. In this case, the on-condition task would be 'Check pipe joints for leaks'. The task frequency is based on the amount of time it takes for an 'acceptable' minor leak to become an 'unacceptable' major leak, and suitable corrective action would be initiated whenever a minor leak was discovered.

However, if the same pipeline was carrying a toxic substance like cyanide, any leak at all would be regarded as a functional failure. In this case it is not feasible to ask anyone to check for leaks, so some other method would need to be found to manage the failure. This would almost certainly entail some sort of modification.

This example re-emphasizes how important it is to agree what is meant by a functional failure *before* considering what should be done to prevent it.

The P-F interval and operating age

When applying these principles for the first time, people often have difficulty in distinguishing between the 'life' of a component and the P-F interval. This leads them to base on-condition task frequencies on the real or imagined 'life' of the item. If it exists at all, this life is usually many times greater than the P-F interval, so the task achieves little or nothing. In reality, we measure the life of a component forwards from the moment it enters service. The P-F interval is measured back from the functional failure, so the two concepts are often completely unrelated. The distinction is important because failures which are not related to age (in other words, random failures) are as likely to be preceded by a warning as those which are not.

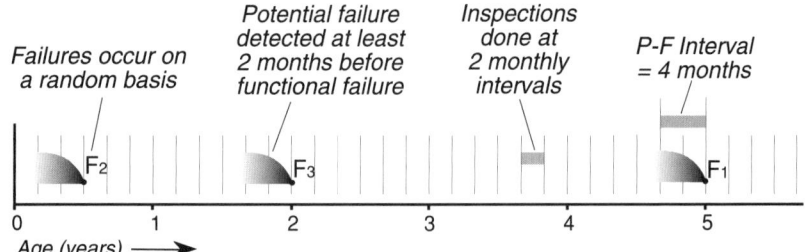

Figure 7.9: Random failures and the P-F interval

For example, Figure 7.9 depicts a component which conforms to a random failure pattern (pattern E). One of the components failed after five years, a second after six months and a third after two years. In each case, the functional failure was preceded by a potential failure with a P-F interval of four months.

Figure 7.9 shows that in order to detect the potential failure, we need to do an inspection task every 2 months. Because the failures occur on a random basis, we don't know when the next one is going to happen, so the cycle of inspections must begin as soon as the item is put into service. In other words, the timing of the *inspections* has nothing to do with the age or life of the component.

However, this does not mean that on-condition tasks apply *only* to items which fail on a random basis. They can also be applied to items which suffer age-related failures, as discussed later in this chapter.

7.6 Linear and Non-linear P-F Curves

Part 1 of this chapter explained that the final stages of deterioration can be described by the P-F curve. In this part of this chapter, we consider this curve in more detail, starting with a look at non-linear P-F curves and then going on to consider linear P-F curves.

The final stages of deterioration

Figure 7.1 on Page 144 suggests that deterioration usually accelerates in the final stages. To see why this is so, let us consider in more detail what happens when a ball bearing fails due to 'normal wear and tear'.

Figure 7.10 overleaf illustrates a typical vertically-loaded ball bearing which is rotating clockwise. The most heavily and frequently loaded part of the bearing will be the bottom of the outer race. As the bearing rotates, the inner surface of the outer race moves up and down as each ball passes over it. These cyclic movements are tiny, but they are sufficient to cause subsurface fatigue cracks which develop as shown in Figure 7.10.

Figure 7.10 also explains how these cracks eventually give rise to detectable symptoms of deterioration. These are of course potential failures, and the associated P-F intervals are shown in Figure 7.8 on page 154. This example raises several further points about potential failures, as follows:

• in the example, the deterioration process accelerates. This suggests that if a quantitative technique such as vibration analysis is used to detect potential failures, we cannot predict when failure will occur by drawing a straight line based on just two observations.

This in turn leads to the notion that after an initial deviation is observed, additional vibration readings should be taken at progressively shorter intervals until some further point is reached at which action should be taken. In practice, this can only be done if the P-F interval is long enough to allow time for the additional readings. It also does not escape the fact that the initial readings need to be taken at a frequency which is known to be less than the P-F interval.

(In fact, if the shape of the P-F curve is fairly well known and the P-F interval is reasonably consistent, it should not be necessary to take additional readings after the first sign of deviation is discovered. This suggests that the process of deterioration should only be tracked by taking additional readings if the P-F curve is poorly understood or if the P-F interval is highly inconsistent.)

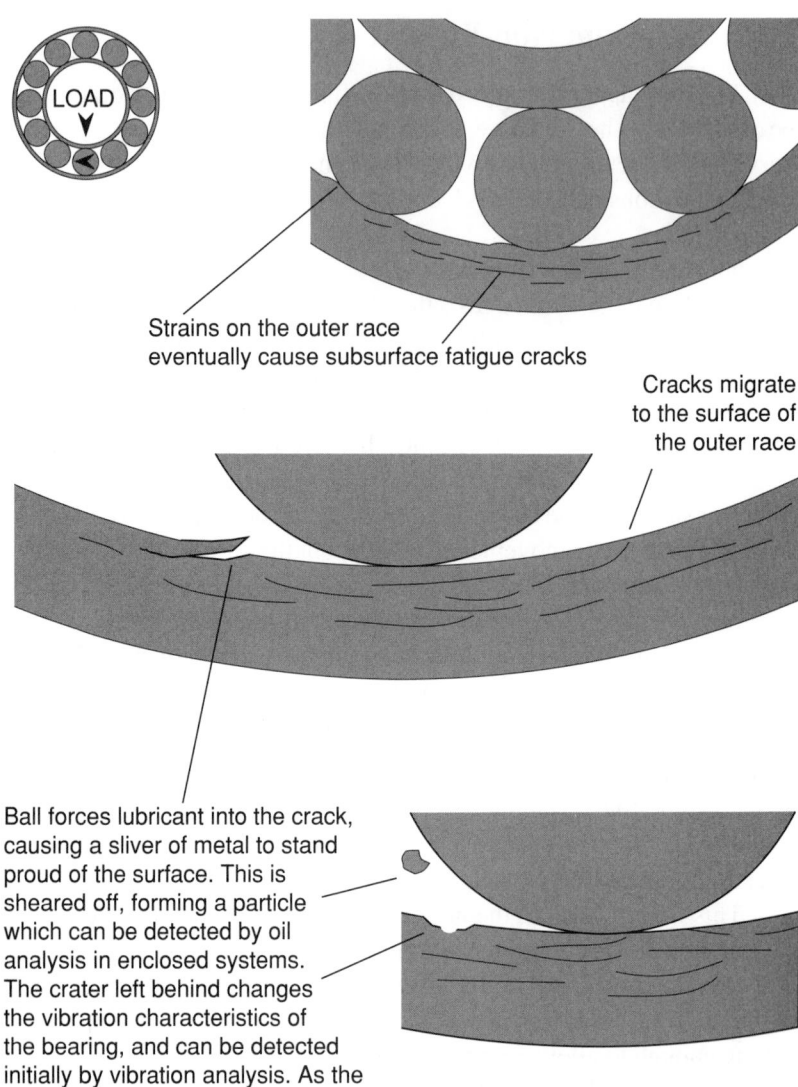

Strains on the outer race
eventually cause subsurface fatigue cracks

Cracks migrate
to the surface of
the outer race

Ball forces lubricant into the crack, causing a sliver of metal to stand proud of the surface. This is sheared off, forming a particle which can be detected by oil analysis in enclosed systems. The crater left behind changes the vibration characteristics of the bearing, and can be detected initially by vibration analysis. As the balls pass over the crater, they make it bigger. Soon the balls themselves get damaged because they are no longer rolling on a smooth surface. At some point, the bearing becomes audibly noisy, and then starts getting hotter. Deterioration continues at an accelerating pace until the balls eventually disintegrate and the bearing seizes.

Figure 7.10:
How a rolling element bearing fails due to 'normal wear and tear'

- different failure modes can often exhibit similar symptoms.

  For example, the symptoms described in Figure 7.10 are based on failure due to normal wear and tear. Very similar symptoms would be exhibited in the final stages of the failure of a bearing where the failure process has been initiated by dirt, lack of lubrication or brinelling.

  In practice, the precise root cause of many failures can only be identified using sophisticated instruments. For instance, it might be possible to determine the root cause of the failure of a bearing by using a ferrograph to separate particles from the lubricating oil and examining the particles under an electron microscope.

  However, if two different failures have the same symptoms and if the P-F interval is broadly similar for each set of symptoms – as it probably would be in the case of the bearing examples – the distinction between root causes is irrelevant from the failure *detection* viewpoint. (The distinction does of course become relevant if we are seeking to *eliminate* the root cause of the failure.)

- failure only becomes detectable when the fatigue cracks migrate to the surface and the surface starts breaking up. The point at which this happens in the life of any one bearing depends on the speed of rotation of the bearing, the magnitude of the load, the extent to which the outer race itself rotates, whether the bearing surface is damaged prior to or during installation, how hot the bearing gets in service, the alignment of the shaft relative to the housing, the materials used to manufacture the bearing, how well it was made, etc. Effectively this combination of variables makes it impossible to predict how many operating cycles must elapse before the cracks reach the surface, and hence when the bearing will start exhibiting the symptoms mentioned in Figure 7.10. (For those interested in pursuing this subject further, chaos theory – in particular the 'butterfly effect' – shows how tiny differences between the initial conditions which apply to any dynamic system lead to dramatic differences after the passage of time. This may explain why minute variations between the initial conditions of two rolling element bearings can lead to huge differences between the ages at which they fail. See Gleick[1987])

Deterioration accelerates in the final stages of most failures. For instance, deterioration is likely to accelerate when bolts start to loosen, when filter elements get blinded, when V-belts slacken and start slipping, when electrical contactors overheat, when seals start to fail, when rotors become unbalanced and so on. But it does not accelerate in *every* case.

Linear P-F curves

If an item deteriorates in a more or less linear fashion over its entire life, it stands to reason that the final stages of deterioration will also be more or less linear. A close look at Figures 6.2 and 6.3 suggests that this is likely to be true of age-related failures.

For example, consider tire wear. The surface of a tire is likely to wear in a more or less linear fashion until the tread depth reaches the legal minimum. If this minimum is (say) 2 mm, it is possible to specify a depth of tread greater than 2 mm which provides adequate warning that functional failure is imminent. This is of course the potential failure level.

If the potential failure is set at (say) 3 mm, then the P-F interval is the distance the tire could be expected to travel while its tread depth wears down from 3 mm to 2 mm, as illustrated in Figure 7.11.

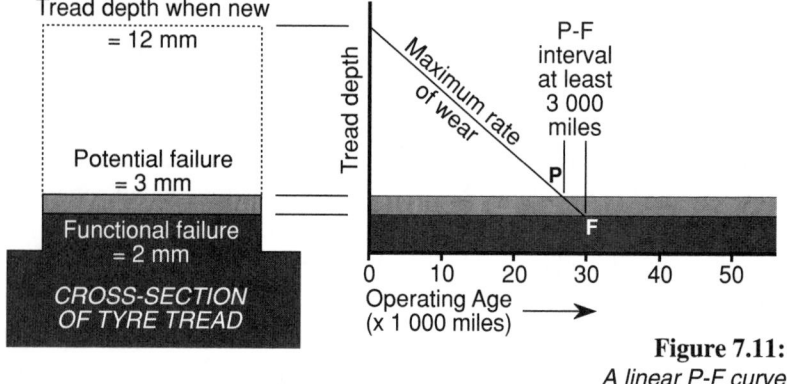

Figure 7.11:
A linear P-F curve

Figure 7.11 also suggests that if the tire enters service with a tread depth of (say) 12 mm, it should be possible to predict the P-F interval based on the total distance usually covered before the tire has to be retreaded. For instance, if the tires last *at least* 30 000 miles before they have to be retreaded, it is reasonable to conclude that the tread wears at a maximum rate of 1 mm for every 3 000 miles travelled. This amounts to a P-F interval of 3 000 miles. The associated on-condition task would call for the driver to:

> 'Check tread depth every 1 500 miles and report tires whose tread depth is less than 3 mm.'

Not only will this task ensure that wear is detected before it exceeds the legal limit, but it also allows plenty of time – 1 500 miles in this case – for the vehicle operators to plan to remove the tire before it reaches the limit.

In general, linear deterioration between 'P' and 'F' is only likely to be encountered where the failure mechanisms are intrinsically age-related (except in the case of fatigue, which is a somewhat more complex case. This failure process is discussed in more detail in later.)

Note that the P-F interval and the associated task frequency can only be deduced in this way if deterioration is linear. As we have seen, the P-F interval cannot be determined in this way if deterioration accelerates between 'P' and 'F'.

A further point about linear failures concerns the point at which one should start to look for potential failures.

For example, Figure 7.11 suggests that it would be a waste of time to measure the overall depth of the tire tread at ten or twenty thousand miles, because we know that it only approaches the potential failure point at 30 000 miles. So perhaps we should only start measuring the tread depth of each tire after it has passed the point where we know tread depth will be approaching 3 mm – in other words, when the tire has been in service for more than (say) 25 000 miles.

However, if we want to ensure that this checking regime is adopted in practice, consider how the checks for a 4-wheeled truck would have to be planned if the actual history of a set of tires is as follows:

Item	Distance travelled by truck and by each tire (miles)			
Truck	140 000	141 500	143 000	144 500
Left front tire	39 000	40 500	42 000	1 000*
Right front tire	21 000	22 500	24 000	25 500
Left rear tire	12 000	500†	2 000	3 500
Right rear tire	24 000	25 500	27 000	28 500

* *Tread depth of L/F tire dropped below 3 mm and tire replaced at depot*
† *6 inch nail caused tire to blow out at 13 000 miles - replaced with new spare*

If we are seriously going to try to ensure that the driver only checks each tire after it passes 25 000 miles in service, we have to devise a system which tells him to:
• start checking the L/F tire only when the truck reached 126 000 miles
• check the L/F and R/R tires when the truck reached 141 500 miles
• and again at 143 000 miles
• but check the R/R and R/F tires at 144 500 miles, but not the L/F tire.

Clearly this is nonsense, because the cost of administering such a planning system would be far greater than the cost of asking the driver to check the tread depth of every tire on the vehicle every 1 500 miles. In other words, in this example the cost of fine-tuning the planning system would be far greater than the cost of doing the tasks. So we would simply ask the driver to check the tread depth of every tire at 1 500 mile intervals, rather than direct his attention to specific tires.

However, if the process of deterioration is linear and the task itself is very expensive, then it might be worth ensuring that we only start checking for potential failures when it is really necessary.

For instance, if an on-condition task entails shutting down and opening up a large turbine to check the turbine discs for cracks, and *we are certain that deterioration only becomes detectable after the turbine has been in service for a certain length of time (in other words, the failure is age-related)*, then we should only start taking the turbine out of service to check for the cracks after it has passed the age at which there is a reasonable likelihood that detectable cracks will start to emerge. Thereafter, the frequency of checking is based on the rate at which a detectable crack is likely to deteriorate into a failure.

For the record, the age at which cracks are likely to start becoming detectable is known as the *crack initiation life*, whereas the time (or number of stress cycles) which elapse from the moment a crack becomes detectable until it grows so large that the item fails is known as the *crack propagation life*.

In cases like these, the cost of doing the task would be much greater than the cost of the associated planning systems, so it is worth ensuring that we only start doing the tasks when it is really necessary. However, if it is felt that this fine-tuning is worthwhile, bear in mind that the planning process has to employ two completely different timeframes, as follows:

- the first time-frame is used to decide *when we should start* doing the on-condition tasks. This is the *operating age* at which potential failures are likely to start becoming detectable.

- the second time-frame governs *how often* we should do the tasks after this age has been reached. This time-frame is of course the *P-F interval*.

For example, it might be felt that the turbine disc is unlikely to develop any detectable cracks until it has been in service for at least 50 000 hours, but that it takes a minimum of ten thousand hours for a detectable crack to deteriorate into disc failure. This suggests that we don't need to start checking for cracks until the item has been in service for 50 000 hours, *but thereafter it must be checked at intervals of less than ten thousand hours.*

Planning with this degree of sophistication requires a very detailed understanding of the failure mode under consideration, together with highly sophisticated planning systems. In practice, few failure modes are this well understood. When they are, even fewer organizations possess planning systems which can switch from one time frame to another as described above, so this issue needs to be approached with care.

In closing this discussion, it must be stressed that all the curves – P-F and age-related – which have been drawn in this part of this chapter have been drawn for *one failure mode at a time*.

For instance, in the example concerning tires, the failure process was 'normal' wear. Different failure modes (such as flat spots worn on the tires due to emergency braking or damage to the carcass caused by hitting kerbs) would lead to different conclusions because both the technical characteristics and the consequences of these failure modes are different.

It is one matter to speculate on the nature of P-F curves in general, but it is quite another to determine the magnitude of the P-F interval in practice. This issue is considered in the next section of this chapter.

7.7 How to Determine the P-F Interval

It is usually a fairly simple matter to determine the P-F interval for age-related failure modes whose final stages of deterioration are linear. It is done by applying logic similar to that used in the tire example above. On the other hand, the P-F interval can be surprisingly difficult to determine in the case of random failures where deterioration accelerates. The main problem with random failures is that we don't know when the next one is going to occur, so we don't know when the next failure mode is going to start on its way down the P-F curve. So if we don't even know where the P-F curve is going to start, how can we go about finding out how long it is? The following paragraphs review five possibilities, only the fourth and fifth of which have any merit.

Continuous observation
In theory, it is possible to determine the P-F interval by continuously observing an item which is in service until a potential failure occurs, noting when that happens, and then continuing to observe the item until it fails completely. (Note that we cannot chart a full P-F curve by observing the item intermittently, because when we eventually discovered that it was failing we still wouldn't know precisely when the failure process started. What is more, if the P-F interval is shorter than the intermittent observation period we might miss the P-F curve altogether, in which case we would have to start all over again with a new item.)

Clearly this approach is impractical, firstly because continuous observation is very expensive – especially if we were to try to establish every P-F interval in this way. Secondly, waiting until the functional failure occurs means that the item actually has to fail. This might end up with us saying to the boss after (say) the compressor blew up: "Oh, we knew it was failing, but we just wanted to see how long it would take before it finally went so that we could determine the P-F interval!"

Start with a short interval and gradually extend it
The impracticality of the above approach leads some people to suggest that P-F intervals can be established by starting the checks at some quite short but arbitrary interval (say 10 days), and then waiting until "we find out what the interval should be", perhaps by gradually extending the interval. Unfortunately, this is again the point at which the functional failure occurs, so we would still end up blowing up the compressor.

This approach is of course potentially very dangerous, because there is also no guarantee that the initial arbitrary interval, no matter how short, will be shorter than the P-F interval to begin with (unless serious consideration is given to the failure process itself).

Arbitrary intervals
The difficulties associated with the two approaches described above lead some people to suggest – quite seriously – that some arbitrary 'reasonably short' interval should be selected for *all* on-condition tasks. This arbitrary approach is the least satisfactory (and the most dangerous) way to set on-condition task frequencies, because there is again no guarantee that the 'reasonably short' arbitrary interval will be shorter than the P-F interval. On the other hand, the true P-F interval may be much longer than the arbitrary interval, in which case the task ends up being done much more often than necessary.

For instance, if a daily task really only needs to be done once a month, that task is costing *thirty times* as much as it should.

Research
The best way to establish a precise P-F interval is to simulate the failure in such a way that there are no serious consequences when it eventually does occur. For example this is done when aircraft components are tested to failure on the ground rather than in the air. This not only provides data about the life of the components, as discussed in Chapter 6, but it also enables the observers to study at leisure how failures develop and how quickly this happens. However, laboratory testing is expensive and it takes time to yield results, even when it is accelerated. So it is only worth doing in cases where a fairly large number of components are at risk – such as an aircraft fleet – and the failures have very serious consequences.

A rational approach
The above paragraphs indicate that in most cases, it is either impossible, impractical or too expensive to try to determine P-F intervals on an empirical basis. On the other hand, it is even more unwise simply to take a shot in the dark. Despite these problems, P-F intervals can still be estimated with surprising accuracy on the basis of judgement and experience.

The first trick is to ask the right question. It is essential that anyone who is trying to determine a P-F interval understands that we are asking *how quickly the item fails.* In other words, we are asking how much time (or how many stress cycles) elapse from the moment the potential failure becomes detectable until the moment it reaches the functionally failed state. We are *not* asking how often it fails or how long it lasts.

The second trick is to ask the right people – people who have an intimate knowledge of the asset, the ways in which it fails and the symptoms of each failure. For most equipment, this usually means the people who operate it, the craftsmen who maintain it and their first-line supervisors. If the detection process requires specialized instruments such as condition monitoring equipment, then appropriate specialists should also take part in the analysis.

In practice, the author has found that an effective way to crystallize thinking about P-F intervals is to provide a number of mental 'coat-hooks' on which people can hang their thoughts. For instance, one could ask: "do you think that the P-F interval is likely to be of the order of days, weeks or months?" If the answer is (say) weeks, the next step is to ask: "One, two, four or eight weeks?"

If everyone in the group achieves consensus, then the P-F interval has been established and the analysts go on to consider other task selection criteria such as the consistency of the P-F interval and whether the nett interval is long enough to avoid the failure consequences.

If the group cannot achieve consensus, then it is not possible to provide a positive answer to the question "what is the P-F interval?". When this happens, the associated on-condition task must be abandoned as a way of detecting the failure mode under consideration, and the failure must be dealt with in some other way.

The third trick is to concentrate on one failure mode at a time. In other words, if the failure mode is wear, then the analysts should concentrate on the characteristics of wear, and should not discuss (say) corrosion or fatigue (unless the symptoms of the other failure modes are almost identical and the rate of deterioration is also very similar).

Finally, it must be clearly understood by everyone taking part in such an analysis that the objective is to arrive at an on-condition task interval which is less than the P-F interval, but not so much less that resources will be squandered on the checking process.

The effectiveness of such a group is redoubled if management expresses an appreciation of the fact that it is made up of human beings, and that humans are not infallible. However, the analysts must also be aware that if the failure has safety consequences, the price of getting it badly wrong could (literally) be fatal for themselves or their colleagues, so they need to take special care in this area.

7.8 When On-condition Tasks are Worth Doing

On-condition tasks must satisfy the following criteria to be worth doing:

• if a failure is *hidden*, it has no direct consequences. So an on-condition task intended to prevent a hidden failure should reduce the risk of the multiple failure to an acceptably low level. In practice, because the function is hidden, many of the potential failures which normally affect evident functions would also be hidden. What is more, much of this type of equipment suffers from random failures with very short or non-existent P-F intervals, so it is fairly unusual to find an on-condition task which is technically feasible and worth doing for a hidden function. But this does not mean that one should not be sought.

• if the failure has *safety* or *environmental* consequences, an on-condition task is only worth doing if it can be relied on to give enough warning of the failure to ensure that action can be taken in time to avoid the safety or environmental consequences.

• if the failure does not involve safety, the task must be cost-effective, so over a period of time, the cost of doing the on-condition task must be less than the cost of not doing it. The question of cost-effectiveness applies to failures with operational and non-operational consequences, as follows:

  - *Operational* consequences are usually expensive, so an on-condition task which reduces the rate at which the operational consequences occur is likely to be cost-effective. This is because the cost of inspection is usually low. This was illustrated in the example on pages 104 and 105.

  - The only cost of a functional failure which has *non-operational* consequences is the cost of repair. Sometimes this is almost the same as the cost of correcting the potential failure which precedes it. In such cases, even though an on-condition task may be technically feasible, it would not be cost-effective, because over a period of time, the cost of the inspections plus the cost of correcting the potential failures would be greater than the cost of repairing the functional failure (see pages 107 and 108.) However, an on-condition task may be justified if the functional failure costs a lot more to repair than the potential failure, especially if the former causes secondary damage.

7.9   Selecting Proactive Tasks

It is seldom difficult to decide whether a proactive task is *technically feasible*. The characteristics of the failure govern this decision, and they are usually clear enough to make the decision a simple yes/no affair.

Deciding whether they are *worth doing* usually needs more judgement. For instance, Figure 7.8 indicates that it may be technically feasible for two or more tasks of the *same* category to prevent the same failure mode. They may even be so closely matched in terms of cost-effectiveness that which one is chosen becomes a matter of personal preference.

The situation is complicated further when tasks from two *different* categories are both technically feasible for the same failure mode.

For example, most countries nowadays specify a minimum legal tread depth for tires (usually about 2 mm). Tires which are worn below this depth must either be replaced or retreaded. In practice, truck tires – especially tires on similar vehicles in a single fleet working the same routes – show a fairly close relationship between age and failure. Retreading restores nearly all the original failure resistance, so the tires could be scheduled for restoration after they have covered a set distance. This means that all the tires in the truck fleet would be retreaded after they had covered the specified mileage, whether or not they needed it.

Figure 6.4 in chapter 6, repeated below as Figure 7.12, could have been drawn for just such a fleet. This shows that in terms of normal wear, all tires last between 30 000 and 50 000 miles. If a scheduled restoration policy were to be adopted on the basis of this information, there is a rapid increase in the conditional probability of this failure mode at 30 000 miles and none of these failures occur before this age, so all of the tires would be retreaded at 30 000 miles. However, if this policy were adopted many tires would be retreaded long before it was really necessary. In some cases, tires which could have lasted as much as 50 000 miles would be retreaded at 30 000 miles, so they could lose up to 20 000 miles of useful life.

On the other hand, as discussed in part 6 of this chapter, it is possible to define a potential failure condition for tires related to tread depth. Checking tread depth is quick and easy, so it is a simple matter to check the tires every 1 500 miles and to arrange for them to be retreaded only when they need it. In this way, the tires would average 40 000 miles between retreads (due to normal wear) without endangering the drivers, instead of 30 000 miles if the scheduled restoration task is done as described above – an increase in useful tire life of 33%. So in this case on-condition tasks are much more cost-effective than scheduled restoration.

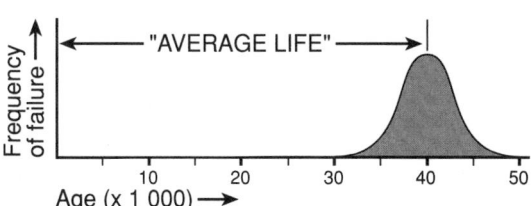

Figure 7.12:
Failure of tyres due to normal wear in a hypothetical truck fleet

This example suggests the following basic order of preference for selecting proactive tasks:

On-condition tasks
On-condition tasks are considered first in the task selection process, for the following reasons:

- they can nearly always be performed without moving the asset from its installed position and usually while it is in operation, so they seldom interfere with the production process. They are also easy to organize.
- they identify *specific* potential failure conditions so corrective action can be clearly defined before work starts. This reduces the amount of repair work to be done, and enables it to be done more quickly.
- by identifying equipment on the point of potential failure, they enable it to realize almost all of its useful life (as illustrated by the tire example).

Scheduled restoration and scheduled discard tasks
If a suitable on-condition task cannot be found for a particular failure mode, the next choice is scheduled restoration or scheduled discard. If they satisfy the technical feasibility and worth doing criteria described in Chapter 6, they may significantly reduce the consequences of the failures at which they are directed. However, these two categories of tasks also have significant disadvantages, as follows:

- they can only be done when items are stopped and (usually) sent to the workshop, so the tasks nearly always affect operations in some way
- the age limits apply to all items, so many items or components which might have survived to higher ages will be removed
- restoration tasks involve shop work, so they generate a much higher workload than on-condition tasks.

These disadvantages mean that when both categories are technically feasible, on-condition tasks are nearly always more cost-effective than scheduled restoration or discard, so the former are considered first.

As mentioned in Chapter 6, scheduled restoration and scheduled discard are usually considered together because they have so much in common. When they are encountered in practice, it is usually obvious whether the failure mode concerned should be dealt with by scheduled discard or scheduled restoration. However, in the case of some failure modes, both categories of tasks can satisfy the criteria for technical feasibility. In these cases, the most cost-effective of the two should be selected.

In general, however, scheduled restoration is usually considered before scheduled discard because it is inherently more conservative to restore things instead of throwing them away.

Combinations of tasks

For a very small number of failure modes which have safety or environmental consequences, a task cannot be found which *on its own* reduces the risk of failure to an acceptably low level, and a suitable modification does not readily suggest itself.

In these cases, it is sometimes possible to find a combination of tasks (usually from two different task categories, such as an on-condition task and a scheduled discard task), which reduces the risk of the failure to an acceptable level. Each task is carried out at the frequency appropriate for that task. However, it must be stressed that situations in which this is necessary are very rare, and care should be taken not to employ such tasks on a 'belt and braces' basis.

The task selection process

The task selection process is summarised in Figure 7.13. This basic order of preference is valid for the majority of failure modes, but it does not apply in every single case. If a lower order task is clearly going to be a more cost-effective method of managing a failure than a higher order task, then the lower order task should be selected.

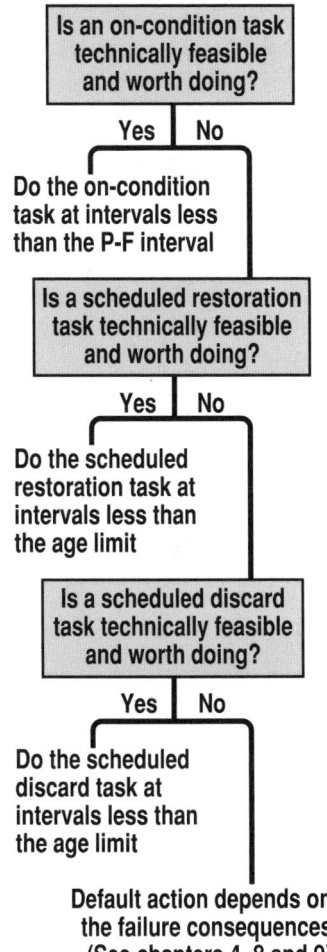

Figure 7.13:
The task selection process

8 Default Actions 1: Failure-finding Tasks

8.1 Default Actions

Previous chapters have mentioned that if a proactive task cannot be found which is both technically feasible and worth doing for any failure mode, then the default action which must be taken is governed by the consequences of the failure, as follows:

- if a proactive task cannot be found which reduces the risk of the multiple failure associated with a *hidden function* to a tolerably low level, then a periodic *failure-finding task* must be performed. If a suitable failure-finding task cannot be found, then the secondary default decision is that the item may have to be redesigned.

- if a proactive task cannot be found which reduces the risk of a failure which could affect *safety* or *the environment* to a tolerably low level, ***the item must be redesigned or the process must be changed.***

- if a proactive task cannot be found which costs less over a period of time than a failure which has *operational* consequences, the initial default decision is ***no scheduled maintenance***. (If this occurs and the operational consequences are still unacceptable, then the secondary default decision is again redesign).

- if a proactive task cannot be found which costs less over a period of time than a failure which has *non-operational* consequences, the initial default decision is ***no scheduled maintenance***, and if the repair costs are too high, the secondary default decision is once again redesign.

The location of the default actions in the RCM decision framework is shown in Figure 8.1 opposite. At this point, we are answering the seventh of the seven questions which make up the RCM decision process:

• *what should be done if a suitable proactive task cannot be found?*

This chapter considers failure-finding. Chapter 9 deals with redesign and run-to failure, and also considers routine tasks which *fall outside the RCM decision framework* such as walk-around checks.

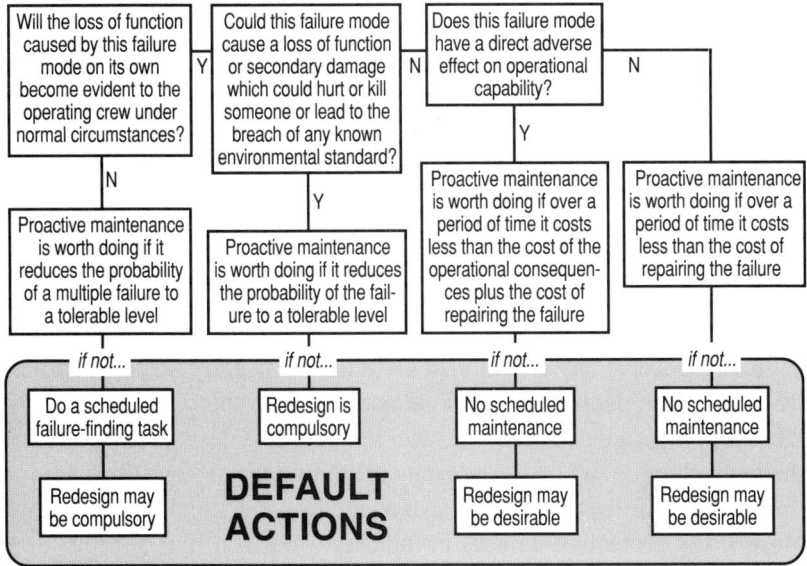

Figure 8.1: Default actions

8.2 Failure-finding

Why bother?

Much of what has been written to date on the subject of maintenance strategy refers to three – and only three – types of maintenance: predictive, preventive and corrective. Predictive tasks entail checking if something is failing. Preventive maintenance means overhauling items or replacing components at fixed intervals. Corrective maintenance means fixing things either when they are found to be failing or when they have failed.

However, there is a whole family of maintenance tasks which falls into none of the above categories. For example, when we periodically activate a fire alarm, we are not checking if it is failing. We are not overhauling or replacing it, nor are we repairing it.

We are simply checking if it still works.

Tasks designed to check whether something still works are known as *failure-finding tasks* or *functional checks*. (In order to rhyme with the other three families of tasks, the author and his colleagues also call them *detective* tasks because they are used to *detect* whether something has failed.)

Failure-finding applies only to hidden or unrevealed failures. Hidden failures in turn only affect protective devices.

If RCM is correctly applied to almost any modern, complex industrial system, it is not unusual to find that up to 40% of failure modes fall into the hidden category. Furthermore, up to 80% of these failure modes require failure-finding, so *up to one third of the tasks generated by comprehensive, correctly applied maintenance strategy development programs are failure-finding tasks.*

A more troubling finding is that at the time of writing, many existing maintenance programs provide for fewer than one third of protective devices to receive any attention at all (and then usually at inappropriate intervals). The people who operate and maintain the plant covered by these programs are aware that another third of these devices exist but pay them no attention, while it is not unusual to find that no-one even knows that the final third exist. This lack of awareness and attention means that most of the protective devices in industry – our last line of protection when things go wrong – are maintained poorly or not at all.

This situation is completely untenable.

If industry is serious about safety and environmental integrity, then the whole question of failure-finding needs to be given top priority as a matter of urgency. As more and more maintenance professionals become aware of the importance of this neglected area of maintenance, it is likely to become a bigger maintenance strategy issue in the next decade than predictive maintenance has been in the last ten years. The rest of this chapter explores this issue in some detail.

Multiple failures and failure-finding

A multiple failure occurs if a protected function fails while a protective device is in a failed state. This phenomenon was illustrated in Figure 5.10 on page 114. Figure 5.11 on page 117 showed that the probability of a multiple failure can be calculated as follows:

$$\text{Probability of a} \atop \text{multiple failure} = \text{Probability of failure of} \atop \text{the protected function} \times \text{Average unavailability} \atop \text{of the protective device} \quad \dots 1$$

This led to the conclusion that the probability of a multiple failure can be reduced by reducing the unavailability of the protective device – in other words, by increasing its availability. Chapter 5 went on to explain that the best way to do this is to prevent the protective device from getting into a failed state by applying some sort of proactive maintenance.

Chapters 6 and 7 described how to decide whether any sort of proactive maintenance is technically feasible and worth doing. However, when the criteria described in these two chapters are applied to hidden functions, it transpires that fewer than 10% of these functions are susceptible to any form of predictive or preventive maintenance.

Nonetheless, although proactive maintenance is often inappropriate, it is still essential to do something to reduce the probability of the multiple failure to the required level. This can be done by checking periodically whether the hidden function is still working.

For example, we cannot *prevent* the failure of a brake light bulb. So if there is no warning circuit to show that a bulb has failed, the only way to reduce the possibility that a burnt-out bulb will fail to warn other drivers of our intentions is to check if it is still working and replace it if it has failed.

Such checks are known as failure-finding tasks.

> ***Scheduled failure-finding entails checking***
> ***a hidden function at regular intervals to***
> ***find out whether it has failed***

This chapter looks at key technical aspects of failure-finding, describes how to determine failure-finding intervals, defines the formal technical feasibility criteria for failure-finding and considers what should be done if a suitable failure-finding task cannot be found.

Technical aspects of failure finding

The objective of failure-finding is to satisfy ourselves that a protective device will provide the required protection if it is called upon to do so. In other words, we are not checking whether the device looks OK – we are checking whether *it still works* as it should. (This is why failure-finding tasks are also known as *functional checks*.) The following paragraphs consider some of the key issues in this area.

Check the entire protective system
A failure-finding task must be sure of detecting all the failure modes which are reasonably likely to cause the protective device to fail. This is especially true of complex devices such as electrical circuits. In these cases, the function of the entire system should be checked *from sensor to actuator*. Ideally, this should be done by simulating the conditions the circuit should respond to, and checking if the actuator gives the right response.

For example, a pressure switch may be designed to shut down a machine if the lubricating oil pressure drops below a certain level. Wherever possible switches like this should be checked by dropping the oil pressure to the required level and checking whether the machine shuts down.

Similarly, a fire detection circuit should be checked from smoke detector to fire alarm by blowing smoke at the detector and checking if the alarm sounds.

Do not disturb

Dismantling anything always creates the possibility that it will be put back together incorrectly. If this happens to a hidden function, the fact that it is hidden means that no-one will know it has been left in a failed state until the next check (or until it is needed). For this reason, we should always look for ways of checking the functions of protective devices without disconnecting or otherwise disturbing them.

This having been said, some devices simply have to be dismantled or removed altogether to check if they are working properly. In these cases, great care must be taken to do the task in such a way that the devices will still work when they are returned to service. (The mathematical implications of the fact that a failure-finding task might induce a failure are considered later in this chapter.)

It must be physically possible to check the function

In a very small but still significant number of cases, it is impossible to carry out a failure-finding tasks of any sort. These are:

• where it is impossible to gain access to the protective device in order to check it (this is almost always a result of thoughtless design).

• when the function of the device cannot be checked without destroying it (as in the case of fusible devices and rupture discs). In most such cases, other technologies are available (such as circuit breakers instead of fuses). However, in one or two cases our only options are to find some other way of managing the risks associated with untestable protection until something better comes along, or to abandon the processes concerned.

Minimize risk while the task is being done

It should be possible to carry out a failure-finding task without significantly increasing the risk of the multiple failure.

An example of a borderline task is overspeeding something in order to check whether the overspeed protection mechanism works.

If a protective device has to be disabled in order to carry out a failure-finding task, or if such a device is checked and found to be in a failed state, then alternative protection should be provided or the protected function should be shut down until the original protection is restored. This issue is discussed in more detail later.

Failure-finding should not be carried out on systems where it is called for but would simply be too dangerous, (If society is serious about safety, it is debatable whether such systems should be allowed to exist at all.)

The frequency must be practical
It must be practical to do the failure-finding task at the required intervals. However, before we can decide whether a required interval is practical, we need to determine what interval is actually 'required'. This issue is considered next.

8.3 Failure-finding Task Intervals

This section of this chapter describes how to determine the frequency of failure-finding tasks. It will start by explaining that this frequency depends on two variables – the desired availability and the frequency of failure of the protective device. It goes on to look at how we establish the 'desired' availability, and then examines different methods which can be used to establish failure-finding intervals under different circumstances.

Failure-finding intervals, availability and reliability

We have seen that predictive and preventive maintenance task intervals are each based on just one variable (P-F interval and useful life respectively). The following paragraphs will show that not one but *two* variables – availability and reliability – are used to set failure-finding intervals.

Figure 8.2 shows a situation in which ten motorbikes have been in service for four years. This means that the total service life of the fleet of bikes in this period is:
10 bikes x 4 years = 40 years.
The brake light on each motorbike has been checked once a year for four years. (This example assumes that no attempt is made to check the lights between the annual checks.) Over the four year period, the lights have been found to be in a failed state on four occasions, as shown in Figure 8.2. So the mean time between failures (MTBF) of the brake lights is:
40 years in service ÷ 4 failures = 10 years.

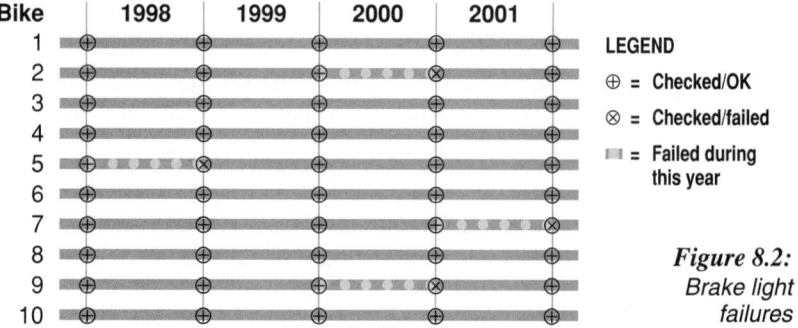

Figure 8.2:
Brake light
failures

In this case, the failure-finding interval of one year is equal to 10% of the MTBF of ten years. However, we don't know exactly when each failed light ceased to function. One might have failed the day after the last check, another the day before the current check, and the rest at some time in between. All we know for sure is that each of the four lights failed some time during the year preceding the check. So in the absence of any better information, we assume that *on average*, each failed light failed half way through the year. In other words, on average, each of the failed lights was out of service for half a year. This means that over the four year period, our failed lights were in a failed state for a total of:

4 failed lights x 0.5 years each in a failed state = 2 years.

So on the basis of the above information, it seems that we can expect an average unavailability from our brake lights of:

2 years in a failed state ÷ 40 years in service = 5%.

This corresponds to an availability of 95%.

The above example suggests that there is a linear correlation between the unavailability (5%), the failure-finding interval (1 year) and the reliability of the protective device as given by its MTBF (10 years), as follows:

Unavailability = 0.5 x failure-finding interval ÷ MTBF of the protective device

It can be shown that this linear relationship is valid for all unavailabilities of less than 5%, provided that the protective device conforms to an exponential survival distribution (failure pattern E or random failure). (See Cox & Tait[1991], Pp 283 - 284 or Andrews & Moss[1993], Pp 110 - 112)

Excluding task time and repair time
Note that the 'unavailability' of the protective device does not include any unavailability incurred while the failure-finding task is being carried out, nor does it include any unavailability caused by the need to repair the device if it is found to be failed. This is so for two reasons:

- the unavailability required to carry out the failure-finding task and to effect any repairs is likely to be very small indeed relative to the unrevealed unavailability between tasks, to the extent that it will usually be negligible on purely mathematical grounds

- both the failure-finding task and any repairs which might be needed should be carried out under tightly controlled conditions. These conditions should greatly reduce - if not completely eliminate - the chance of a multiple failure while the intervention is under way. This entails either shutting down the protected system or arranging alternative protection until the system has been fully restored. If this is done properly, the unavailability resulting from the (controlled) intervention can be ignored in any assessments of the probability of a multiple failure.

In the RCM decision process, the latter point is covered by the criteria for assessing whether a failure-finding task is worth doing. If there is a significant increase in the likelihood of a multiple failure *while the task is under way*, the answer to the question "Does the task reduce the probability of a multiple failure to a tolerable level" will be 'no', and the RCM decision process defaults to the secondary default actions discussed later.

Calculating FFI using availability and reliability only
If we use the abbreviation 'FFI' to describe the failure-finding interval and 'M_{TIVE}' to describe the MTBF of the protective device, the above unavailability equation can be rearranged to give the following formula:

$$FFI = 2 \times \text{unavailability} \times M_{TIVE} \qquad \text{....... } 2$$

This tells us that in order to determine the failure-finding interval for a single protective device, we need to know its *mean time between failures* and the *desired availability* of the device (from which we can determine the unavailability to be used in the formula).

For instance, assume that the riders of our motorbikes decide they are not satisfied with an availability of 95%, and would prefer to see it increased to 99%. The associated unavailability is 1%. If the MTBF of the brake lights stays unchanged at four years, checking interval needs to be changed from once a year to:

$$FFI = 2 \times 1\% \times 4 \text{ years} = 2\% \text{ of } 48 \text{ months} \approx 1 \text{ month.}$$

In other words, based on their availability expectations and the existing failure data, the bikers need to check whether their brake lights are working once a month. If they want an availability of 99.9%, they need to check about twice a week.

(Strictly speaking, the above calculations are only valid if the brake lights on all the bikes are used about the same number of times each week. If there is a wide variation, both the MTBF and the failure-finding interval should be calculated in terms of distance travelled, or even more precisely, in terms of the number of times the brakes – and hence the brake lights – are used. However, the key point to note at this stage is the connection between the checking interval, the desired availability and the MTBF).

For people who are uncomfortable with mathematical formulae, formula (2) above can be used to develop a simple table, as follows

Availability we require for the hidden function	99.99%	99.95%	99.9%	99.5%	99%	98%	95%
Failure-finding interval (as a % of the MTBF)	0.02%	0.1%	0.2%	1%	2%	4%	10%

Figure 8.3: Failure-finding intervals, availability and reliability

Required availability

Having established the relationship between availability, reliability and failure-finding intervals, the next issue to consider is how we decide what availability we require. Part 6 of Chapter 5 explained that this can be done in three stages, as follows:

1: first ask what probability the organization is prepared to tolerate for the *multiple* failure which could occur if the hidden function was not working when called upon to do so

2: then determine the probability that the *protected* function will fail in the period under consideration

3: finally determine what availability the *hidden* function must achieve to reduce the probability of the multiple failure to the desired level

In addition to carrying out these three steps, we need to find out the mean time between failures of the hidden function. Once this has been done, we are in a position to look at Figure 8.3 and select the task frequency which corresponds to the level of availability established in step 3. This process is illustrated in the following example:

Figure 8.4 summarizes the duty/standby pump example in Chapter 5, where:

• in step 1 above, the users decided that they wanted the probability of the multiple failure to be less than 1 in 1000 in any one year

• in step 2 they established that the rate of unanticipated failures of the duty pump could be reduced to an average of 1 in 10 years

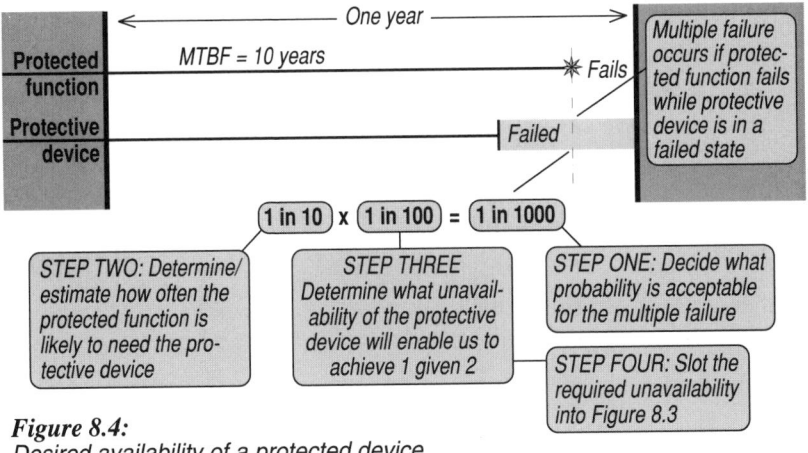

Figure 8.4:
Desired availability of a protected device

- this meant that the unavailability of the standby pump must not exceed 1%, so the availability of this pump has to be 99% or better (step 3).

Figure 8.3 suggests that to achieve an availability of 99% for the standby pump, someone would need to carry out a failure-finding task (in other words, check that it is fully functional) at an interval of 2% of its mean time between failures. Records might show that the standby pump has a mean time between failures of 8 years (or about 400 weeks), so the failure-finding task frequency should be:

2% of 400 weeks = 8 weeks = 2 months.

Rigorous Methods for Calculating FFI

The above example suggests that it is possible to develop a single formula for determining failure-finding intervals which incorporates all the variables considered so far. In fact, this can be done by combining equations (1) and (2) above, as explained in the following paragraphs. Let us begin by defining a few key terms:

- a probability of a multiple failure of 1 in 1 000 000 in any one year implies a *mean time between multiple failures* of 1 000 000 years. Let us call this M_{MF}. If this is so, then the probability of a multiple failure occurring in any one year is $1/M_{MF}$. (See again note on page 96)

- we have seen that if the demand rate of the protected function is (say) once in 200 years, this corresponds to a probability of failure for the protected function of 1 in 200 in any one year, or a *mean time between failures of the protected function* of 200 years. Let us call this M_{TED}, so the probability of failure of the protected function in any one year will be $1/M_{TED}$. This is also known as the *demand rate*.

- as before, M_{TIVE} is the *mean time between failures of the protective device* and FFI is the *failure-finding task interval.*

- U_{TIVE} is the *allowed unavailability of the protective device.*

If we substitute the above expressions, equation (1) becomes:

$$1/M_{MF} = (1/M_{TED}) \times U_{TIVE} \qquad \text{.... 3}$$

This can be rearranged as follows:

$$U_{TIVE} = M_{TED} \div M_{MF} \qquad \text{.... 4}$$

Equation (2) above states that:

$$FFI = 2 \times U_{TIVE} \times M_{TIVE} \qquad \text{.... 2}$$

So substituting U_{TIVE} from equation 4 into equation 2 gives:

$$FFI = \frac{2 \times M_{TIVE} \times M_{TED}}{M_{MF}} \qquad \text{.... 5}$$

This formula allows a failure-finding interval to be determined in a single step, as follows:

If we apply this formula to the figures used in the duty/standby pump system mentioned above, M_{MF} is 1000 years, M_{TIVE} is 8 years and M_{TED} is 10 years, so:

$$FFI = \frac{2 \times 8 \times 10}{1000} \approx 2 \text{ months}$$

Multiple failure modes in a single protective device
Throughout this chapter, all the failure possibilities which could cause each protective device to fail have been grouped together as one single failure mode ('standby pump fails'). The vast majority of protective devices can be treated in this way, because all the failure modes which could cause a protective device to cease to function are checked when the function of the device as a whole is checked.

However, it is sometimes appropriate to carry out a detailed FMEA of the device in order to identify individual failure modes which might *on their own* cause the device to be unable to provide the required protection. This is usually done in two sets of circumstances:

- when some of the failure modes are known to be susceptible to pro-active maintenance, but others are neither predictable nor preventable. In these cases, the appropriate on-condition or scheduled restoration/discard task should be applied to the failure modes which qualify, and failure-finding tasks applied to the *remainder* of the failure modes

- when the protective device is new and the only failure data which are available (from data banks, component suppliers or wherever) apply to parts of the device but not to the device as a whole.

In these cases, equation (5) above can be modified to accommodate the MTBF of each component of the device.

When the failure-finding task can cause the failure
A major practical problem which affects the whole question of failure-finding is that the task itself can cause the very failure which it is supposed to detect. This usually happens in one of two ways:

- the task stresses the system in such a way that it eventually causes it to fail (as might be the case when a switch is tested, where the mere act of switching imposes stresses on the mechanism of the switch)

- if the system needs to be disturbed to do the task, there is always a chance that the person doing it will leave the system in a failed state.

In both cases, the device will be in a failed state from the moment the test is completed. If *p* is the probability that it will be left in such a state after a test, then p (as a decimal) will be its unavailability caused by the testing process. If M_{OTHER} is the mean time between *failures caused by phenomena other than the test*, it can be shown (for a single system) that:

$$ FFI = \frac{2 \times M_{OTHER}}{(1 - p)} \times \left(\frac{M_{TED} - p}{M_{MF}} \right) \qquad \text{...... } 6 $$

In this formula, the expression (1 - p) can be ignored if p is less than 0.05.
    If the act of switching is the *only* cause of failure (in other words, there is no M_{OTHER}) *and* if the failure conforms to an exponential survival distribution, the probability of a multiple failure is the demand rate (in years) multiplied by the number of cycles between failures of the protective device.

For example, if the demand rate is 40 years and the switch lasts an average of 600 000 cycles, then the probability of a multiple failure is:
    1 in (40 x 600 000) = 1 in 24 000 000 years.

This is so because if the failure is caused *only* by switching, then the act of operating the switch to check if it has failed will simultaneously:

- enable you to find out if the last operation of the switch caused it to fail

- stress the switch and so create the possibility that it will fail as a result of the check.

So under this unique set of circumstances (random failure caused *solely* by operating the item), a failure-finding task which involves operating the item to check whether it has failed will have no effect at all on the probability of a multiple failure, regardless of how often the task is done. In other words, the answer to the question "Is it practical to do the task at the required intervals?" is 'no', because there is no suitable interval. So in this case, if the organization wants the probability of a multiple failure of the switch described above to be less than 1 in 100 000 000 years, the only way they can achieve this is by reducing the demand rate on the switch, and/or by installing either more switches or a more reliable switch.

All of this indicates that failure rates which are given as a number of operations should be treated with great caution, for the following reasons:

- they seldom indicate whether the failure under consideration is hidden or evident

- they do not indicate whether the underlying failure-pattern is age-related, in which case some form of scheduled restoration or scheduled discard might be appropriate, or whether it is random

- despite the previous comment, a failure mode caused *solely* by the operation of a switch *is* likely to be age-related. If this is so, then it is equally likely that a preventive task could be identified which reduces the probability of a multiple failure to the required level.

This suggests that as a rule, important switches – especially big circuit breakers – should not be treated as single failure modes. Rather, they should be subjected to a detailed FMEA, and the most appropriate maintenance policy developed for each failure mode.

Sources of Data for FFI Calculations

Most modern industrial undertakings possess several thousand protected systems, most of which incorporate hidden functions. The multiple failures associated with many of these systems will be serious enough to necessitate using one of the rigorous approaches to failure-finding.

If accurate data about the probability of failure of the protected function and the mean time between failures of the hidden function are available, the calculations can be performed quite quickly. If this information is not available – and very often it is not – then it is necessary to estimate what these variables are likely to be *in the context under consideration*. In rare cases, it might be possible to obtain data from one of the following:

- the manufacturers of the equipment
- commercial data banks
- other users of similar equipment.

More often, however, the estimates have to be based on the knowledge and experience of the people who know the most about the equipment. In many cases these are operators and maintenance craftsmen. (When using data from external sources, take special note of the operating context of the items for which the data was gathered compared to the context in which your equipment is operating.)

Once a failure-finding task frequency has been established and the tasks are being done on a regular basis, it becomes possible to verify the assumptions used to determine the frequency quite rapidly. However, this does require the keeping of absolutely meticulous records, not only about when each failure-finding task is done, but also about:

- whether or not the hidden function is found to be functional each time the task is done
- how often the protected function fails (this can often be inferred from the number of times the protected function makes use of the protective device – for instance, from the number of times a pressure relief valve actually has to relieve the pressure in the system).

On the basis of this information the actual mean time between failures can be calculated and, if necessary, the task frequency revised accordingly.

Failure modes where the MTBF and/or the associated failure patterns are completely unknown – and a satisfactory guess cannot be made – should be put into an age-exploration program right away to establish the true picture. If the situation is such that the uncertainty cannot be tolerated while the data are being gathered – in other words, if the consequences of guessing wrong are simply too serious for the organization (or in some cases, society as a whole) to accept – then every effort should be made to change the consequences. This in turn will nearly always necessitate some form of redesign.

An Informal Approach to Setting Failure-finding Intervals

Not every hidden function is important enough to warrant the time and effort needed to do a full rigorous analysis. This applies mainly to multiple failures which do not affect safety or the environment. It could also apply to multiple failures which could affect safety but where the protected function is inherently very reliable and the threat to safety is marginal.

In these cases, it may be sufficient to take a general view of the entire protected system in its operating context, and go straight to a decision on a desired level of availability for the hidden function. This decision is then used in conjunction with the MTBF of the hidden failure to set a task interval, using the table in Figure 8.3. (Some organizations even go so far as to use an availability of 95% for all hidden functions where the associated multiple failure cannot affect safety or the environment. However, general policies of this nature can be dangerous so they should only be used by people who have extensive experience with this type of analysis.)

Once again, if adequate records about hidden failures are not available – and they seldom will be – it will be necessary to guess at the MTBF's to begin with. But again these records should be compiled as quickly as possible to validate the initial estimates.

Other Methods of Calculating Failure-finding Intervals

The range of techniques for setting failure-finding intervals described so far in this chapter is by no means exhaustive. Many additional variants have been developed by the Aladon network of RCM specialists. These include formulae for:
• voting systems
• multiple, independent, fully redundant systems
• deriving cost-optimized intervals for systems where the multiple failures do not affect safety or the environment.
As this book is only intended to provide an introduction to this subject, these formulae are not included in this chapter.

The Practicality of Task Intervals

The methods described so far for calculating failure-finding intervals sometimes produce very short or very long intervals. In some cases, these intervals might be too long or too short, as follows:

• a very short failure-finding task interval has two main implications:
  - sometimes the interval is simply far too short to be practical. Examples would be failure-finding tasks which call for major items of plant to be shut down every few days
  - the task could cause habituation (which might happen if a fire alarm is tested too often).
  In these cases, the proposed task is rejected and we move on to the next stage of the RCM decision-making process, as discussed later.

- we also encounter very long intervals – sometimes as long as a hundred years or more. Here the process is clearly suggesting that we really need not worry about doing the task at all. In these cases the proposed 'task' should be stated as follows: *'the risk/reliability profile is such that failure-finding is felt to be unnecessary'.*

- in rare cases, task intervals emerge which are significantly longer than the demand rate (M_{TED}). It makes little sense to carry out a failure-finding task at intervals (FFI) which are longer than the system is effectively testing itself (M_{TED}), so in these cases, the answer to the question "Is it practical to do the task at the required interval?" will be 'no'. However, bear in mind that if a failure-finding task is not done on a protected system, (and if M_{TIVE} is more than 4 or 5 times greater than M_{TED}, which is usually the case), it can be shown that:

$$M_{MF} = M_{TED} + M_{TIVE}$$

If this value of M_{MF} is too low to be acceptable, then the protection is inadequate and the system will almost certainly have to be redesigned, as discussed in the next chapter.

8.4 The Technical Feasibility of Failure-finding

The issues discussed in Parts 2 and 3 of this chapter mean that for a failure-finding task to be technically feasible, it should be possible to do the task at all, it should be possible to do it without increasing the risk of the multiple failure, and it should be practical to do the task at the required interval.

Failure-finding is technically feasible if
- *it is possible to do the task*
- *the task does not increase the risk of a multiple failure*
- *it is practical to do the task at the required interval.*

The objective of a failure-finding task is to reduce the probability of the multiple failure associated with the hidden function to a tolerable level. It is only worth doing if it achieves this objective.

Failure-finding is worth doing if it reduces the probability of the associated multiple failure to a tolerable level

Failure-finding is a Default Action!

Bear in mind that successful proactive maintenance prevents things from failing, whereas failure-finding accepts that they will spend some time – albeit not very much – in a failed state. This means that proactive maintenance is inherently more conservative (in other words, safer) than failure-finding, so the latter should only be specified if a more effective proactive task cannot be found. For this reason, it is wise to avoid RCM decision diagrams which put failure-finding ahead of proactive maintenance in the task selection process.

What if Failure-finding is Not Suitable?

If it transpires that a failure-finding task is not technically feasible or worth doing, we have exhausted all the possibilities which might enable us to extract the required performance from the existing asset. Where this leaves us is once again governed by the consequences of the multiple failure, as follows:

• if a suitable failure-finding task cannot be found and the multiple failure could affect safety or the environment, something must be changed in order to make the situation safe. In other words, redesign is compulsory

• if a failure-finding task cannot be found and the multiple failure does not affect safety or the environment, then it is acceptable to take no action, but redesign may be justified if the multiple failure has very expensive consequences.

This decision process is summarized in Figure 8.5. (This diagram is a fuller description of this aspect of the process than the two boxes at the foot of the left hand column in Figure 8.1):

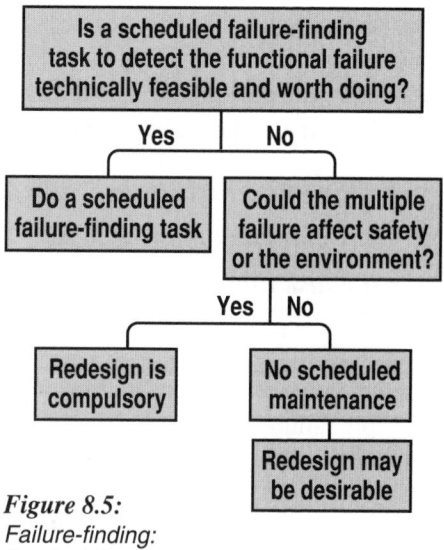

Figure 8.5:
Failure-finding:
the decision process

9 Other Default Actions

Three default actions are shown at the foot of Figure 8.1. The first of these – failure-finding – was covered in Chapter 8. This chapter focuses on *no scheduled maintenance* and *redesign*. It also briefly reviews the role of *walk-around checks*.

9.1 No Scheduled Maintenance

We have seen that failure-finding is the initial default action if a suitable proactive task cannot be found for a *hidden failure*. However, if a suitable failure-finding task cannot be found in turn, then redesign is the compulsory secondary default action if the multiple failure has safety or environmental consequences. We have also seen that if an evident failure has *safety* or *environmental consequences* and a suitable preventive task cannot be found, something must also be changed to make the situation safe.

However, if the failure is evident and it does not affect safety or the environment, or if it is hidden and the multiple failure does not affect safety or the environment, then the initial default decision is to do *no scheduled maintenance*. In these cases, the items are left in service until a functional failure occurs, at which point they are repaired or replaced. In other words, 'no scheduled maintenance' is only valid if:

- a suitable scheduled task cannot be found for a hidden function, and the associated multiple failure does not have safety or environmental consequences

- a cost-effective preventive task cannot be found for failures which have operational or non-operational consequences.

Note that if a suitable preventive task cannot be found for a failure under either of these circumstances, it simply means that we do not carry out *scheduled maintenance* on that component in its present form. It does not mean that we simply forget about it. As we see in the next section of this chapter, there may be circumstances under which it is worth changing the design of the component to reduce overall costs.

9.2 Redesign

The question of equipment design has arisen again and again as we have traced the steps which must be followed to develop a successful maintenance program. In this part of this chapter, we consider two general issues which affect the relationship between design and maintenance, and then consider the part played by redesign in the task selection process.

The term 'redesign' is used in its broadest sense in this chapter. Firstly, it refers to any change to the specification of any item of equipment. This means any action which should result in a change to a drawing or a parts list. It includes *changing the specification of a component, adding a new item, replacing an entire machine* with one of a different make or type, or *relocating a machine*. It also means any other once-off change to a *process* or *procedure* which affects the operation of the plant. It even covers *training* as a method of dealing with a specific failure mode (which can be seen as 'redesigning' the capability of the person being trained.)

Design and Maintenance

Changing anything is expensive. It involves the cost of developing the new idea (designing a new machine, drawing up a new operating procedure), the cost of turning the idea into reality (making a new part, buying a new machine, compiling a new training program) and the cost of implementing the change (installing the part, conducting the training program). Further indirect costs are incurred if equipment or people have to be taken out of service while the change is being implemented. There is also the risk that a change will fail to eliminate or even alleviate the problem it is meant to solve. In some cases, it may even create more problems.

As a result, the whole question of modifications should be approached with great caution. Two issues need particular attention:
• what do we consider first - design or maintenance?
• the relationship between inherent reliability and desired performance.

Which comes first - redesign or maintenance?
Reliability, design and maintenance are inextricably linked. This can lead to a temptation to start reviewing the design of existing equipment before considering its maintenance requirements. In fact, the RCM process considers maintenance first for two reasons.

Most modifications take from six months to three years from conception to commissioning, depending on the cost and complexity of the new design. On the other hand, the maintenance person who is on duty *today* has to maintain the equipment as it exists *today*, not what should be there or what might be there some time in the future. So today's realities must be dealt with before tomorrow's design changes.

Secondly, most organizations are faced with many more apparently desirable design improvement opportunities than are physically or economically feasible. By focusing on failure consequences, RCM does much to help us to develop a rational set of priorities for these projects, especially because it separates those which are essential from those that are merely desirable. Clearly, such priorities can only be established after the review has been carried out.

Inherent reliability vs desired performance
Among other things, Part 2 of Chapter 2 stressed that the inherent reliability of any asset is established by its design and by how it is made, and that maintenance cannot yield reliability beyond that inherent in the design. This led to two important conclusions.

Firstly, if the inherent reliability or built-in capability of an asset is greater than the desired performance, maintenance can help achieve the desired performance. Most equipment *is* adequately specified, designed and built, so it is usually possible to develop a satisfactory maintenance program, as described in previous chapters. In other words, in most cases, RCM helps us to extract the desired performance from the asset as it is currently configured.

On the other hand, if desired performance exceeds inherent reliability, then no amount of *maintenance* can deliver the desired performance. In these cases 'better' maintenance cannot solve the problem, so we need to look beyond maintenance for the solutions. Options include:
• modifying the equipment
• changing operating procedures
• lowering our expectations and deciding to live with the problem.
This reminds us that maintenance is not *always* the answer to chronic reliability problems. It also reminds us that we must establish as soon and as precisely as possible what *we want each piece of equipment to do* in its operating context before we can starting talking sensibly about the appropriateness of its design or its maintenance requirements.

Redesign as the Default Action

Figure 8.1 shows that redesign appears at the bottom of all four columns of the decision diagram. In the case of failures which have safety or environmental consequences, it is the compulsory default action, and in the other three cases, it 'may be desirable'. In this part of this chapter, we consider each case in more detail, starting with the safety case.

Safety or environmental consequences
If a failure could affect safety or the environment and no preventive task or combination of tasks can be found which reduces the risk of the failure to a tolerable level, something must be changed, simply because we are dealing with a safety or environmental hazard which cannot be adequately prevented. In these cases, redesign is usually undertaken with one of two objectives:

- to reduce the probability of the failure mode occurring to a level which is tolerable. This is usually done by replacing the affected component with one which is stronger or more reliable.

- to change the item or the process in such a way that the failure no longer has safety or environmental consequences. This is most often done by installing one or more of the five types of protective devices which were categorized as follows in Chapter 2:
  - to alert operators to abnormal conditions
  - to shut down the equipment in the event of a failure
  - to eliminate or relieve abnormal conditions which follow a failure and which might otherwise cause more serious damage
  - to take over from a function which has failed
  - to prevent dangerous situations from arising.
  Remember that if such a device is added, its maintenance requirements must also be analyzed. Safety or environmental consequences can also be reduced by eliminating hazardous materials from a process, or even by abandoning a dangerous process altogether.

As mentioned in Chapter 5, when dealing with safety or the environment, RCM does not raise the question of economics. If the level of risk associated with any failure is regarded as intolerable, we are obliged either to prevent the failure, or to make the process safe. The alternative is to accept conditions that are known to be unsafe or environmentally unsound. This is no longer acceptable in most industries.

Hidden failures

In the case of hidden failures, the risk of a multiple failure can be reduced by modifying the equipment in one of four ways:

* ***make the hidden function evident by adding another device:*** Certain hidden functions can be made evident by adding another device which draws the attention of the operator to the failure of the hidden function.

For example, a battery used to power a smoke detector is a classical hidden function if no additional protection is provided. However, a warning light is fitted to most such detectors in such a way that the light goes out if the battery fails. In this way the additional protection makes the function of the battery evident. (Note that the light only tells us about the condition of the battery, not about the ability of the detector to detect smoke.)

Special care is needed in this area, because extra functions installed for this purpose also tend to be hidden. If too many layers of protection are added, it becomes increasingly difficult – if not impossible – to define sensible failure-finding tasks. A much more effective approach is to substitute an evident function for the hidden function, as explained in the next paragraph.

* ***substitute an evident function for the hidden function***: In most cases this means substituting a genuinely fail-safe protective device for one which is not fail-safe. This is surprisingly difficult to do in practice, but if it is done, the need for a failure-finding task falls away at once.

For example, one commonly used way to warn the driver of a vehicle that his brake lights have failed is to install a warning light which is switched on if the brake lights fail. (In many cases, this light is also switched on for a short while when the ignition is switched on. However, so are all the other lights on the dashboard. Under these circumstances one missing warning light is likely to be overlooked, so its function is effectively hidden.)

The system might also be configured in such a way that its full function can only be tested by disabling a brake light and seeing if the warning light comes on. This is a clumsy and invasive task which is likely to cause more problems than it solves, so it is likely to be dismissed on the grounds of impracticality. The multiple failures associated with this system could have serious safety consequences, so it is necessary to reconsider the design.

One way to eliminate this problem is to make the function of the brake lights *and* of the warning system evident. This can be done by substituting fibre-optic cables for the warning light, and mounting the cables so that the driver looks through them at the brake lights every time he uses the brakes. (In fact, he sees a pinprick of light at the end of each cable.) In this situation, it is apparent to the driver if either a brake light or a cable fails. In other words, the function of this protective device is now evident, so failure-finding is no longer necessary.

- *substitute a more reliable (but still hidden) device for the existing hidden function:* Figure 8.3 suggests that a more reliable hidden function (in other words, one which has a higher mean time between failures) will enable the organization to achieve one of three objectives:

  - to reduce the probability of the multiple failure without changing the failure-finding task intervals. This increases the level of protection
  - to increase the interval between tasks without changing the probability of the multiple failure. This reduces resource requirements
  - to reduce the probability of the multiple failure *and* increase the task intervals, giving increased protection with less effort.

- *duplicate the hidden function:* If it is not possible to find a single protective device which has a high enough MTBF to give the desired level of protection, it is still possible to achieve any of the above three objectives by duplicating (or even triplicating) the hidden function.

Let us return to the example of a duty pump with a standby. It was explained on page 179 that if the users want the probability of a multiple failure to be less than 1 in 1000, and the unanticipated failure rate of the duty pump is reduced to 1 in 10 years, then the availability of the standby pump has to be 99% or better. This led to the conclusion that a failure-finding task should be done on the standby pump every 2 months in order to achieve an availability of 99% (based on an MTBF for this pump of 8 years).

However, now let us assume that someone has decided that the probability of a multiple failure in this system should not exceed 1 in 100 000 (or 10^{-5}), rather than 1 in 1 000. If the mean time between unanticipated failures of the duty pump (M_{TED}) is unchanged at 10 years, applying formula 4 in Chapter 8 shows that the unavailability (U_{TIVE}) of the standby pump should not exceed:

$$U_{TIVE} = M_{TED}/M_{MF} = 10/100\,000 = 10^{-4}$$

So the unavailability of the standby pump must now not exceed 10^{-4} (0.01%). If the MTBF of the standby pump is unchanged at 8 years, applying formula 2 from Chapter 8 yields the following:

$$FFI = 2 \times 10^{-4} \times 8 \text{ years} = 14 \text{ hours}$$

Activating a standby pump this often is plainly impractical, so more thought has to be given to the design of this system.

In fact, Figure 9.1 opposite shows that if we were to add a second standby pump, and ensure that the availability of each standby pump on its own exceeds 99% (corresponding to an unavailability of 1%, or 10^{-2}), the probability of the multiple failure would be:

$$10^{-1} \times 10^{-2} \times 10^{-2} = 10^{-5}$$

or 1 in 100 000. Figure 8.3 suggests that this can be achieved by doing a failure-finding task on each standby pump at the original frequency of once every 8 weeks. In other words, a much higher level of protection is achieved without changing the task interval.

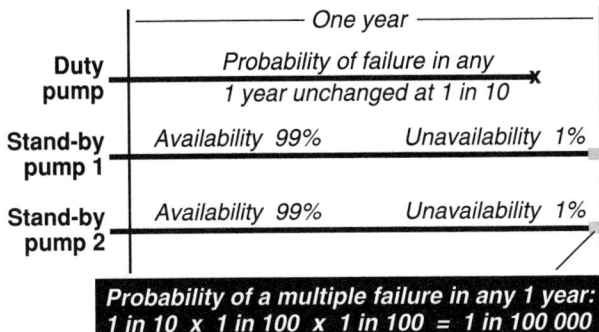

Figure 9.1:
*The effect of
duplicating a
hidden function*

Operational and non-operational consequences
If a technically feasible preventive task cannot be found which is worth doing for failures with operational or non-operational consequences, the immediate default decision is no scheduled maintenance. However, it may still be desirable to modify the equipment to reduce total costs. To achieve this, the plant could be modified to:

• reduce the number of times the failure occurs, or possibly eliminate it altogether, again by making the component stronger or more reliable

• reduce or eliminate the consequences of the failure (for example, by providing a standby capability)

• make a preventive task cost-effective (for instance, by making a component more accessible).

Note that in this case the failure consequences are purely economic so modifications must be cost-justified, whereas they were the compulsory default action if the failure had safety or environmental consequences.

There is no one way to determine whether a modification will be cost-effective. Each case is governed by a different set of variables, which include a before-and-after assessment of maintenance and operating costs, the remaining technologically useful life of the asset, the likelihood that the modification will work, the number of other projects competing for the capital resources of the company and so on.

A detailed cost-benefit study which takes all these factors into account can be very time-consuming, so it is helpful to know beforehand whether this effort is likely to be worthwhile. To help make a quick preliminary assessment, Nowlan & Heap[1978] developed the decision diagram shown in Figure 9.2.

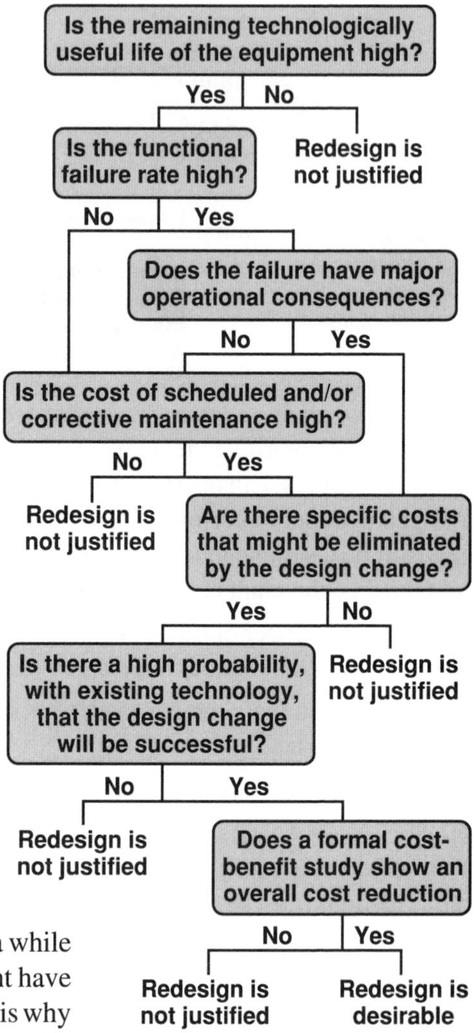

Figure 9.2:
Decision diagram
for a preliminary
assessment
of a proposed
modification

Is the remaining technologically useful life of the equipment high?

Yes | No

Is the functional failure rate high?   **Redesign is not justified**

No | Yes

Does the failure have major operational consequences?

No | Yes

Is the cost of scheduled and/or corrective maintenance high?

No | Yes

Redesign is not justified   **Are there specific costs that might be eliminated by the design change?**

Yes | No

Is there a high probability, with existing technology, that the design change will be successful?   **Redesign is not justified**

No | Yes

Redesign is not justified   **Does a formal cost-benefit study show an overall cost reduction**

No | Yes

Redesign is not justified   **Redesign is desirable**

No matter how reliable, all assets are eventually superseded by new technology. So the first question to ask is whether the asset under consideration is going to be rendered obsolete in the near future. If it is, then it is clearly not worth modifying it. On the other hand, if it is going to be around for a while longer, the modification might have a chance to pay for itself. This is why the first question in Figure 9.2 asks:

Is the remaining technologically useful life of the equipment high?

Some organizations demand that modifications should pay for themselves within a specified period – say, two years. This effectively sets the operational horizon of the equipment at two years. This type of policy reduces the number of projects initiated on the basis of projected cost-benefits and ensures that only projects which will pay for themselves quickly are submitted for approval. So if the answer to the first question in Figure 9.2 is 'no', redesign is probably not justified.

For example Figure 9.3 shows a stainless steel hopper which is periodically blocked by lumps. So far, the RCM process has revealed that this failure mode costs £400 in lost production every time it occurs, and that it cannot be prevented by maintenance. It has been suggested that one way to eliminate the failure mode might be to install a stainless steel grid above the hopper outlet at a cost of £6 000.

If the hopper were due to be superseded within two years, it is highly unlikely that this modification would be worth doing, especially in view of the fact that several months would elapse before it could be commissioned. On the other hand, if the hopper were to remain in service for several more years, the modification would be worth further consideration.

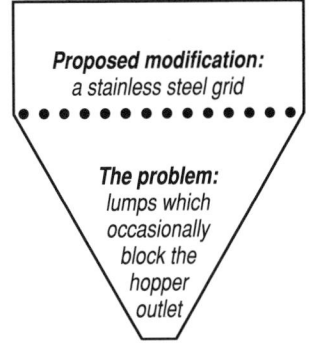

Figure 9.3:
A stainless steel hopper

If the answer to the first question is 'yes', the next question to consider is whether the failure is happening often enough to be a real problem:

Is the functional failure rate high?

This question eliminates items which fail so seldom that the cost of redesign would probably be greater than the benefits to be derived from it (unless of course a preventive task is the reason for a low failure rate. This is why a 'no' answer to this question does not immediately abort the modification – the maintenance task itself might be so expensive that the modification is still justified.)

For example if the blockage in the hopper occurred once every two or three years, no-one would pay much attention to it. If it occurred once a month, it would be worth investigating further.

If the failure rate is high, we start considering the economic implications of the failure:

Does the failure involve major operational consequences?

If the answer is yes, then the question of redesign should be taken further. A 'no' to this question means that the failure only has a minor effect on operating costs, but we must still consider the maintenance costs associated with the failure by asking:

Is the cost of scheduled and/or corrective maintenance high?

Note that this question is approached from two directions. As we have seen, we may get a 'no' answer to the failure rate question only because a very costly preventive task is preventing functional failures.

A 'no' answer to the question of operational consequences means that failures might not be affecting operating capability, but they may result in excessive repair costs. So a 'yes' answer to either of these two questions brings us to the design change itself:

Are there specific costs which might be eliminated by the design change?

This question refers to the operational consequences and the direct costs of proactive and/or corrective maintenance. However, if these costs are not related to a specific design feature, it is unlikely that the problem will be solved by a design change. So a 'no' answer to this question means that it may be necessary to live with the economic consequences of the failure. On the other hand, if the problem can be pinned down to a specific cost element, then the economic potential of redesign is high.

In the case of the hopper, it is hoped that the grid would prevent the lumps from reaching the hopper outlet, and so eliminate the cost of £400 per blockage.

But will the new design work? In other words:

Is there a high probability, with existing technology, that the modification will be successful?

Although a particular design change might be very desirable economically, there is a chance that it will not have the desired effect. A change directed at one failure mode may reveal other failure modes, requiring several attempts to solve the problem. Any design change which entails adding hardware also adds more failure possibilities – maybe too many.

So if a cold-blooded assessment of the proposed change indicates a low probability of success, the change is unlikely to be economically viable.

For instance, in the case of the hopper we would need to be sure that lumps would not simply accumulate on the grid and coagulate into a possibly much more costly problem in the long term.

Any proposed design change which makes it this far deserves a detailed cost-benefit study:

Does an economic trade-off study show an expected cost saving?

Such a study compares the expected reduction in costs over the remaining useful life of the equipment with the costs of carrying out the modification. To be on the safe side, the expected benefit should be regarded as the projected saving if the first attempt at improvement is successful, multiplied by the probability of success at the first try. Alternatively it might be considered that the design change will always be successful, but only some of the savings will be achieved.

If we are certain that the modification to the hopper will work, a discounted cash flow analysis on the figures provided for the hopper (at a discount rate of 10%) shows that the modification will pay for itself
* in five years if the blockage occurs four times per year,
* in seven years if it occurs three times per year and
* in more than ten years if it occurs twice per year.

This type of justification is not necessary, of course, if the reliability characteristics of an item are the subject of contractual warranties or if the changes are needed for reasons other than cost (such as safety).

9.3 Walk-around Checks

Walk-around checks serve two purposes. The first is to spot accidental damage. These checks may include a few specific on-condition tasks for the sake of convenience, but damage in general can occur at any time and is not related to any definable level of failure resistance.

As a result, there is no basis for defining an explicit potential failure condition or a predictable P-F interval. Similarly, the checks are not based on the failure characteristics of any particular item, but are intended to spot unforeseen exceptions in failure behavior.

Walk-around checks are also meant to spot problems due to ignorance or negligence, such as hazardous materials or foreign objects left lying around, spillage, and other items of a housekeeping nature. They also give managers an opportunity to ensure that general standards of maintenance are satisfactory, and can be used to check whether maintenance routines are being done correctly. Again, there are rarely any explicit potential failure conditions and no predictable P-F interval.

Some organizations distinguish between formal scheduled tasks and walk-around checks on the pretext that one is mainly technical and the other predominantly managerial, so they are sometimes done by different people. In fact it does not matter who does them, as long as both are done frequently and thoroughly enough to ensure a reasonable degree of protection from the consequences of the failures concerned.

10   The RCM Decision Diagram

10.1 Integrating Consequences and Tasks

Chapters 5 to 9 have provided a detailed explanation of the criteria used to answer the last three of the seven questions which make up the RCM process. These questions are:

- *in what way does each failure matter?*
- *what can be done to prevent each failure?*
- *what should be done if a suitable preventive task cannot be found?*

This chapter summarizes the most important of these criteria. It also describes the RCM Decision Diagram, which integrates all the decision processes into a single strategic framework. This framework is shown in Figure 10.1 overleaf, and is applied to each of the failure modes listed on the RCM Information Worksheet.

Finally, this chapter describes the RCM Decision Worksheet, which is the second of the two key working documents used in the application of RCM (the Information Worksheet shown in Figure 4.13 being the first).

10.2 The RCM Decision Process

The RCM Decision Worksheet is illustrated in Figure 10.2 opposite. The rest of this chapter demonstrates how this worksheet is used to record the answers to the questions in the Decision Diagram, and in the light of these answers, to record:

- what routine maintenance (if any) is to be done, how often it is to be done and by whom
- which failures are serious enough to warrant redesign
- cases where a deliberate decision has been made to let failures happen.

RCM II
DECISION
WORKSHEET
© 1990 ALADON LTD

| SYSTEM | System Nº | Facilitator: | Date | Sheet Nº |
| SUB-SYSTEM | Sub-system Nº | Auditor: | Date | of |

Information reference			Consequence evaluation							Default action			Proposed task	Initial interval	Can be done by
F	FF	FM	H	S	E	O	H1 S1 O1 N1	H2 S2 O1 N2	H3 S3 O3 N3	H4	H5	S4			

Figure 10.2: The RCM Decision Worksheet

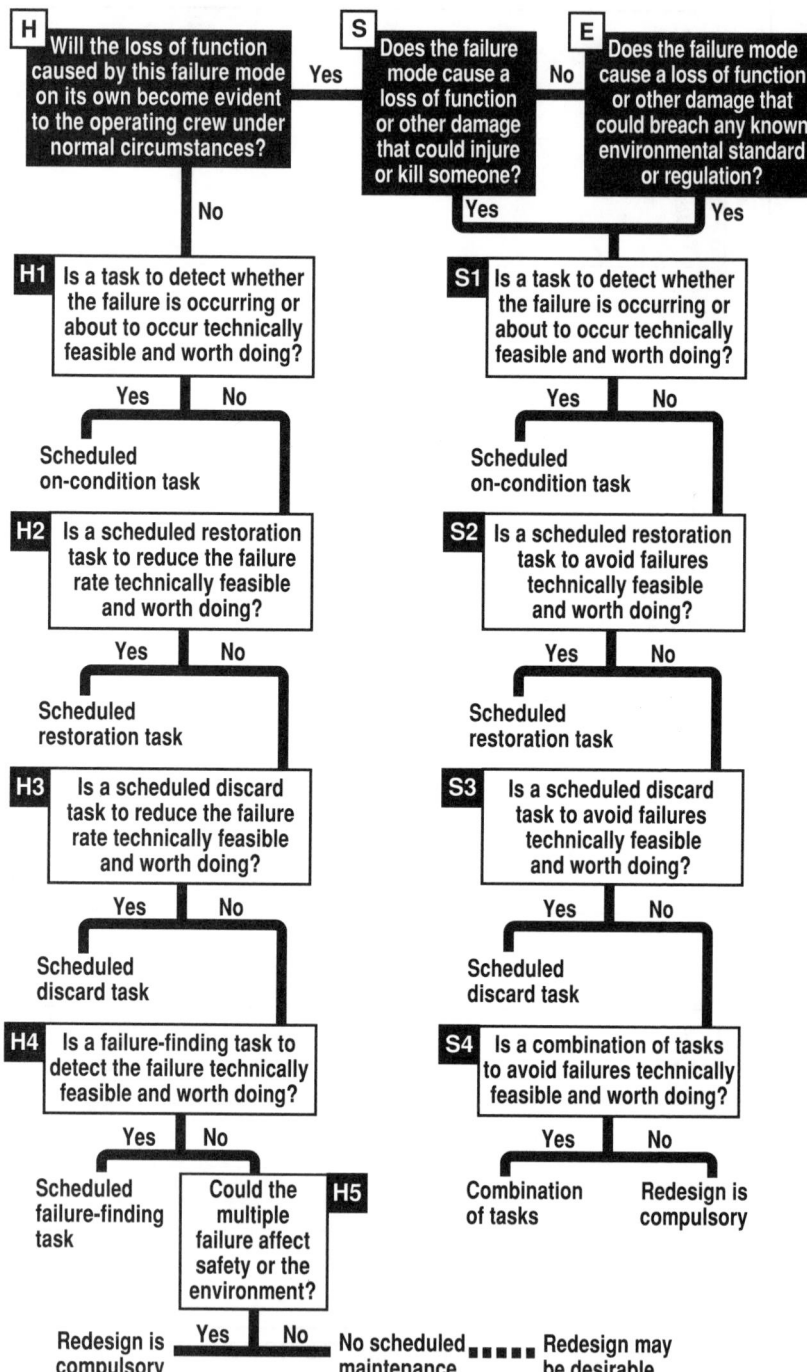

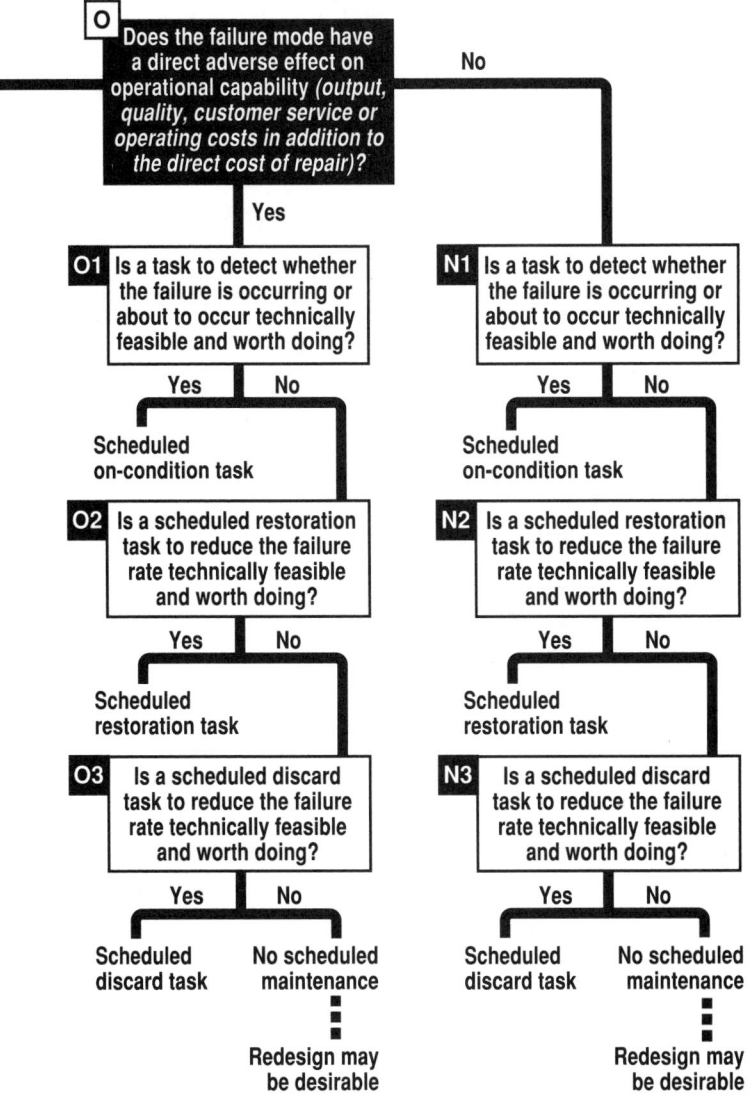

O
Does the failure mode have a direct adverse effect on operational capability *(output, quality, customer service or operating costs in addition to the direct cost of repair)?*

No

Yes

O1 Is a task to detect whether the failure is occurring or about to occur technically feasible and worth doing?

Yes   No

Scheduled on-condition task

O2 Is a scheduled restoration task to reduce the failure rate technically feasible and worth doing?

Yes   No

Scheduled restoration task

O3 Is a scheduled discard task to reduce the failure rate technically feasible and worth doing?

Yes   No

Scheduled discard task   No scheduled maintenance

Redesign may be desirable

N1 Is a task to detect whether the failure is occurring or about to occur technically feasible and worth doing?

Yes   No

Scheduled on-condition task

N2 Is a scheduled restoration task to reduce the failure rate technically feasible and worth doing?

Yes   No

Scheduled restoration task

N3 Is a scheduled discard task to reduce the failure rate technically feasible and worth doing?

Yes   No

Scheduled discard task   No scheduled maintenance

Redesign may be desirable

Figure 10.1:
THE RCM II DECISION DIAGRAM
© 1991 **Aladon Ltd**

The decision worksheet is divided into sixteen columns. The columns headed F, FF and FM identify the failure mode under consideration. They are used to cross-refer the information and decision worksheets, as shown in figure 10.3 below:

RCM II INFORMATION WORKSHEET © 1990 ALADON LTD	SYSTEM *Cooling Water Pumping System*		
	SUB-SYSTEM		

	FUNCTION		FUNCTIONAL FAILURE (Loss of Function)		FAILURE MODE (Cause of Failure)	
1	To transfer water from tank X to tank Y at not less than 800 liters/minute	A	Unable to transfer any water at all	1	Bearing seized due to normal wear and tear	

RCM II DECISION WORKSHEET © 1990 ALADON LTD	SYSTEM *Cooling Water*	
	SUB-SYSTEM	

Information reference			Consequence evaluation					H1 S1 O1 N1	H2 S2 O2 N2	H3 S3 O3 N3	Default action					
F	FF	FM	H	S	E	O					H4	H5	S4			
1	A	1														

Figure 10.3: Cross-referring the information and decision worksheets

The headings on the next ten columns refer to the questions on the RCM Decision Diagram in Figure 10.1, as follows:

- the columns headed H, S, E, O and N are used to record the answers to the questions concerning the consequences of each failure mode
- the next three columns (headed H1, H2, H3 etc) record whether a pro-active task has been selected, and if so, what type of task
- if it becomes necessary to answer any of the default questions, the columns headed H4 and H5, or S4 are used to record the answers.

The last three columns record the task which has been selected (if any), the frequency with which it is to be done and who has been selected to do it. The 'proposed task' column is also used to record the cases where re-design is required or it has been decided that the failure mode does not need scheduled maintenance.

In the following paragraphs, each of these four sections of the decision worksheet is reviewed in the context of the associated questions on the Decision Diagram.

Failure Consequences

The precise meanings of questions H, S, E and O in Figure 10.1 are discussed at length in Chapter 5. These questions are asked for each failure mode, and the answers recorded on the decision worksheet on the basis shown in Figure 10.4 below.

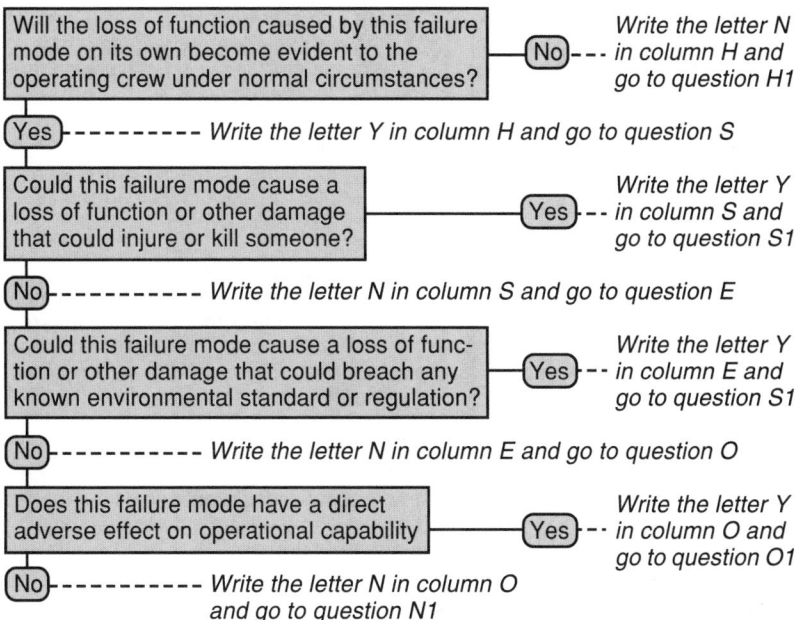

Figure 10.4: Using the decision worksheet to record failure consequences

Figure 10.5 shows how the answers to these questions are recorded on the decision worksheet. Note that:

• each failure mode is dealt with in terms of one category of consequences only. So if it is classified as having environmental consequences, we do not also evaluate its operational consequences (at least when performing the first analysis of any asset). This means that if for instance a 'Y' is recorded in column E, nothing is recorded in column O.

• once the consequences of the failure mode have been categorized, the next step is to seek a suitable preventive task. Figure 7.5 also summarizes the criteria used to decide whether such tasks are *worth doing*.

Information reference			Consequence evaluation				
F	**FF**	**FM**	**H**	**S**	**E**	**O**	
3	A	1	N				*A hidden failure:* To be worth doing, any preventive task must reduce the risk of a multiple failure to an acceptable level
5	B	2	Y	Y			*Safety consequences:* To be worth doing, any preventive task must reduce the risk of this failure on its own to a tolerable level
2	C	4	Y	N	Y		*Environmental consequences:* To be worth doing, any preventive task must reduce the risk of this failure on its own to a tolerable level
1	A	5	Y	N	N	Y	*Operational consequences:* To be worth doing, *over a period of time* any preventive task must cost less than the cost of the operational consequences plus the cost of repair of the failure which it is meant to prevent
1	B	3	Y	N	N	N	*Non-operational consequences:* To be worth doing, *over a period of time* any preventive task must cost less than the cost of repairing the failure which it is meant to prevent

Figure 10.5:
Failure consequences - a summary

Proactive Tasks

The eighth to tenth columns on the decision worksheet are used to record whether a proactive task has been selected, as follows:

- the column headed H1/S1/O1/N1 is used to record whether a suitable on-condition task could be found to anticipate the failure mode in time to avoid, eliminate or minimise the consequences

- the column headed H2/S2/O2/N2 is used to record whether a suitable scheduled restoration task could be found to prevent the failures

- the column headed H3/S3/O3/N3 is used to record whether a suitable scheduled discard task could be found to prevent the failures.

In each case, a task is only suitable if it is worth doing *and* technically feasible. Chapters 6 and 7 explained in detail how to establish whether a task is technically feasible,. These criteria are summarized in Figure 10.6. In essence, for a task to be technically feasible and worth doing it must be possible to provide a positive answer to *all* of the questions shown in Figure 10.6 which apply to that category of tasks, *and* the task must fulfil the 'worth doing' criteria in Figure 10.5. If the answer to any of these questions is 'no' or unknown, then that task as a whole is rejected. If all of the questions can be answered positively, then a Y is recorded in the appropriate column.

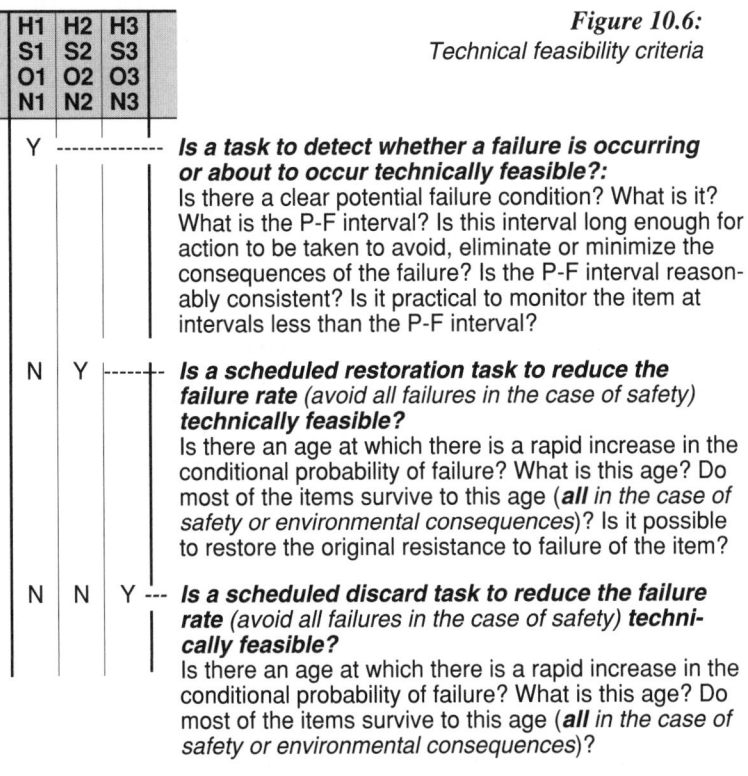

Figure 10.6:
Technical feasibility criteria

H1	H2	H3
S1	S2	S3
O1	O2	O3
N1	N2	N3

Y -------------- **Is a task to detect whether a failure is occurring or about to occur technically feasible?:**
Is there a clear potential failure condition? What is it? What is the P-F interval? Is this interval long enough for action to be taken to avoid, eliminate or minimize the consequences of the failure? Is the P-F interval reasonably consistent? Is it practical to monitor the item at intervals less than the P-F interval?

N   Y ------+- **Is a scheduled restoration task to reduce the failure rate** *(avoid all failures in the case of safety)* **technically feasible?**
Is there an age at which there is a rapid increase in the conditional probability of failure? What is this age? Do most of the items survive to this age (*all in the case of safety or environmental consequences*)? Is it possible to restore the original resistance to failure of the item?

N   N   Y --- **Is a scheduled discard task to reduce the failure rate** *(avoid all failures in the case of safety)* **technically feasible?**
Is there an age at which there is a rapid increase in the conditional probability of failure? What is this age? Do most of the items survive to this age (*all in the case of safety or environmental consequences*)?

If a task is selected, a description of the task and the frequency with which it must be done are recorded as explained later in this chapter, and the analysts move on to the next failure mode. However, as mentioned in Chapter 7, bear in mind that if it seems that a lower order task may be more cost-effective than a higher order task, then the lower order task should also be considered and the more effective of the two chosen.

The Default Questions

The columns headed H4, H5 and S4 on the decision worksheet are used to record the answers to the three default questions. The basis on which these questions are answered is summarized in Figure 10.7. (Note that the default questions are only asked if the answers to the previous three questions are all 'no'.)

Information reference			Consequence evaluation				H1 S1 O1 N1	H2 S2 O2 N2	H3 S3 O3 N3	Default action		
F	FF	FM	H	S	E	O				H4	H5	S4

Figure 10.7:
The default questions

| 3 | A | 1 | N | | | | N | N | N | Y | | |

---------- **Is a failure-finding task technically feasible and worth doing?**
Record yes if it is possible to do the task *and* it is practical to do it at the required frequency *and* it reduces the risk of the multiple failure to an acceptable level.

| 4 | B | 4 | N | | | | N | N | N | N | Y | |
| 4 | C | 2 | N | | | | N | N | N | N | N | |

Could the multiple failure affect safety or the environment?
(This question is only asked if the answer to question H4 is no.) If the answer to this question is yes, **redesign** is compulsory. If the answer is no, the default action is **no scheduled maintenance**, but redesign may be desirable.

| 5 | B | 2 | Y | Y | | | N | N | N | | Y | |
| 2 | A | 5 | Y | Y | | | N | N | N | | N | |

Is a combination of tasks technically feasible and worth doing?
Yes if a combination of any **two or more** proactive tasks will reduce the risk of the failure to an acceptable level (this is very rare). If the answer is no, **redesign** is compulsory.

| 1 | A | 5 | Y | N | N | Y | N | N | N | | | |
| 1 | B | 3 | Y | N | N | N | N | N | N | | | |

-------------- In these two cases, the consequences of the failure are purely economic and no suitable proactive task has been found. As a result, the initial default decision is **no scheduled maintenance**, but redesign may be desirable.

Proposed Task

If a proactive task or a failure-finding task has been selected during the decision-making process, a description of the task should be recorded in the column headed 'proposed task'. Ideally, the task should be described as precisely on the decision worksheet as it will be on the document which reaches the person doing the task. If this is not possible, then the task should at least be described in enough detail to make the intent absolutely clear to whoever writes up the detailed task description.

For example, consider a situation where an on-condition task has been specified for a rolling element bearing. Chapter 7 explained how such bearings can suffer from a variety of potential failure conditions, including noise, vibration, heat, wear and so on. Many machines have more than one and often several such bearings. Consequently, at the very least, the 'proposed task' should specify which bearing is to be checked for what condition. In other words, if a particular bearing is to be checked for noise, the proposed task should read 'check bearing X for audible noise', and not just 'check bearing'.

This issue is discussed in more detail in the next chapter.

If the decision process calls for a design change, then the proposed task should provide a brief description of the design change. The actual form of the new design should be left to the designers.

For example if the RCM process reveals (say) that the fastening mechanism of a guard has to be redesigned for safety reasons, the 'proposed task' should state something like 'more secure fastening mechanism required for guard'. Do not simply write 'redesign required'. On the other hand, it should be left to the designers to decide exactly what sort of fastening mechanism will be used.

This issue is also discussed further in the next chapter.

Finally, if a decision has been taken to allow the failure to occur, in most cases the words 'no scheduled maintenance' should be recorded in the 'proposed task' column. The only exception is hidden failure where 'the risk/reliability profile is such that failure-finding is not required', as explained on page 185.

Initial Interval

Task intervals are recorded on the decision worksheet in the 'Initial Interval' column. We have seen that they are based on the following:

- on-condition task intervals are governed by the *P-F interval*
- scheduled restoration and scheduled discard task intervals depend on the *useful life* of the item under consideration.
- failure-finding task intervals are governed by the *consequences of the multiple failure*, which dictate the availability needed, and the *mean time between occurrences of the hidden failure*.

When completing the decision worksheet, record each task interval on its own merits – in other words, without reference to any other tasks. This is because the reason for doing a task at a particular frequency can change over time – indeed the reason for doing the task at all could disappear. So if the frequency of task X is based on the frequency of task Y and task Y is later eliminated, the frequency of task X becomes meaningless.

As explained in the next chapter, if we are confronted with a number of tasks which need to be done at a wide range of different frequencies, the time to consider consolidating them into a smaller number of work packages is when compiling maintenance schedules. However, the initial task frequencies should always remain on the decision worksheet to remind us how the schedule frequencies were derived (in other words, to preserve the 'audit trail'.).

Note also that task intervals can be based on any appropriate measure of exposure to stress. This includes calendar time, running time, distance travelled, stop-start cycles, output or throughput, or any other readily measurable variable which bears a direct relationship to the failure mechanism. However, calendar time tends to be used where possible because it is the simplest and cheapest to administer.

Can Be Done By

The last column on the decision worksheet is used to list who should do each task. Note that the RCM process considers this issue one failure mode at a time. In other words, it does not approach the subject with any preconceived ideas about who should (or should not) do maintenance work. It simply asks who has the competence and confidence to do *this task* correctly.

The answer could be anyone at all. Tasks might be allocated to maintainers, operators, insurance inspectors, the quality function, specialist technicians, vendors, structural inspectors or laboratory technicians.

A sometimes controversial issue which arises at this stage concerns simple high-frequency on-condition and failure-finding tasks. It sometimes makes sense to allocate these tasks to maintainers, but in many cases, using maintainers to do these tasks has the following drawbacks (especially if they are skilled tradespeople):

- if the task interval is short, the inspection frequency will be very high – sometimes more than once per shift. This can lead to so many high-frequency tasks that maintainers do little more than travel from one task to the next. This travelling time plus the cost of planning and controlling the tasks makes the use of maintainers for this purpose expensive, often to the point where it is simply not worth using them in this capacity.

- many skilled people find high-frequency tasks boring and are often reluctant to do them at all.

- skilled craftspeople are very scarce in many parts of the world, so it is often difficult to spare them for this kind of work in the first place.

A second option is to use operators to do high frequency tasks. This option can be attractive because it is usually more economical and organisationally easier to use people who are near the equipment most of the time to do high frequency tasks. Operators are also often more highly motivated to look after 'their' machines. However, three conditions must be satis-fied before operators can be used with confidence to do these tasks:

- they must be properly trained in how to recognize the appropriate potential failure conditions in the case of on-condition tasks, and must be properly trained to do high-frequency failure-finding tasks safely

- they must have access to simple and reliable procedures for reporting any defects which they do find. (The design of these procedures is discussed in more detail in Chapter 11)

- they must be sure that action will be taken on the basis of their reports, or that they will receive constructive feedback in cases of misdiagnosis.

Using operators for this purpose can also have profound implications in terms of industrial relations and reporting relationships, so it is an issue which needs to be handled with care.

In general, as with most of the other decisions in the RCM process, who exactly is in the best position to do each task is best decided by the people who know the equipment best. This issue is discussed at greater length in Chapter 13.

10.3 Completing the Decision Worksheet

To illustrate how the decision worksheet is completed, we consider three failure modes which have been discussed at length in previous chapters. These are:

- the bearing which seizes on a pump with no standby, as discussed on pages 105 and 106

- the bearing which seizes on an identical pump which does have a standby, as discussed on pages 108 and 109

- the failure of the standby pump set as a whole, as discussed on pages

RCM II DECISION WORKSHEET © 1990 ALADON LTD										

SYSTEM		System Nº		Facilitator:		Date	
SUB-SYSTEM		Sub-system Nº		Auditor:		Date	

Information reference			Consequence evaluation				H1 S1 O1 N1	H2 S2 O2 N2	H3 S3 O3 N3	Default action			Proposed task	Initial interval	Can be done by
F	FF	FM	H	S	E	O				H4	H5	S4			
STAND-ALONE PUMP															
1	A	1	Y	N	N	Y	Y						Check main pump bearing for noise	Weekly	Craftsman
1	A	2		....etc											
DUTY PUMP WITH STANDBY															
1	A	1	Y	N	N	N	N	N	N				No scheduled maintenance		
1	A	2		....etc											
STANDBY PUMP															
2	A	1	N			N	N	N	N		Y		Switch on standby pump instead of duty pump and ensure that standby is capable of filling tank. When the test is complete, switch back to the duty pump	4 weekly	Operator

Figure 10.8: An RCM decision worksheet with sample entries

118 and 179.

The associated decisions are recorded on the decision worksheet shown in Figure 10.8. Please note three important points about this example:

- the first two pumps could suffer from many more failure modes than the failure under consideration. Each of these other failures would also be listed and analyzed on its own merits.

- a number of other preventive tasks could have been chosen to anticipate the failure of the bearing – the decisions in the example are for the purpose of illustration only.

- the standby pump is treated as a 'black box'. In practice, if such a pump were known to suffer from one or more dominant failure modes, these failures would be analyzed individually.

In essence, the RCM worksheets not only show *what* course of action has been selected to deal with each failure mode, but they also show *why* it was selected. This information is invaluable if the need to do any maintenance task is challenged at any time.

The ability to trace each task right back to the functions and desired performance of the asset also make it a simple matter to keep the maintenance program up to date. This is because users can readily identify and reassess tasks which are affected by a change in the operating context of the asset (such as a change in shift arrangements or a change in safety regulations), and avoid wasting time reassessing tasks which are unlikely to be affected by the change.

10.4 Computers and RCM

The information contained in the RCM and Decision Worksheets lends itself readily to being stored in a computerised database. In fact, if a large number of assets are to be analyzed, it is almost essential to used a computer for this purpose. A computer can also be used to sort the proposed tasks by interval and skillset, and to generate a variety of other reports (failure modes by consequence category, tasks by task category, and so on). Finally, storing the analyses in a database makes it infinitely easier to revise and refine the analyses as more is learned and as the operating context changes (as it surely will – see part 7 of Chapter 13).

However, note that a computer should only ever be used to store and sort RCM information, and perhaps to assist with the more complex failure-finding interval calculations. For reasons discussed in Chapter 14, computers should never be used to drive the RCM process.

11 Implementing RCM Recommendations

11.1 Implementation – The Key Steps

We have seen how the formal application of the RCM process ends with completed decision worksheets. These specify a number of *routine tasks* which need to be done at regular intervals to ensure that the asset continues to do whatever its users want it to do, together with the *default actions* which must be taken if an appropriate routine task cannot be found.

The people who participate in this process learn a great deal about how the asset works and about how it fails. This on its own frequently causes the participants to change their behavior in ways which often lead directly to remarkable improvements in asset performance. However, in order to derive the maximum long-term benefit from RCM, steps must be taken to implement the recommendations on a formal basis. These steps should ensure that:

• all the recommendations are approved formally by the managers with overall responsibility for the assets

• all routine tasks are described clearly and concisely

• all actions which call for once-off changes (to designs, to the way the asset is operated or to the capability of operators and maintainers) are identified and implemented correctly

• routine tasks and operating procedure changes are incorporated into appropriate work packages

• the work packages and once-off changes are implemented. Specifically, this in turn entails:

  - incorporating the work packages into systems which ensure that they will to be performed by the right people at the right time and that they will be done correctly

  - ensuring that any faults found are dealt with speedily.

These steps are summarized in Figure 11.1 opposite. The most important of them are discussed in more detail in the rest of this chapter.

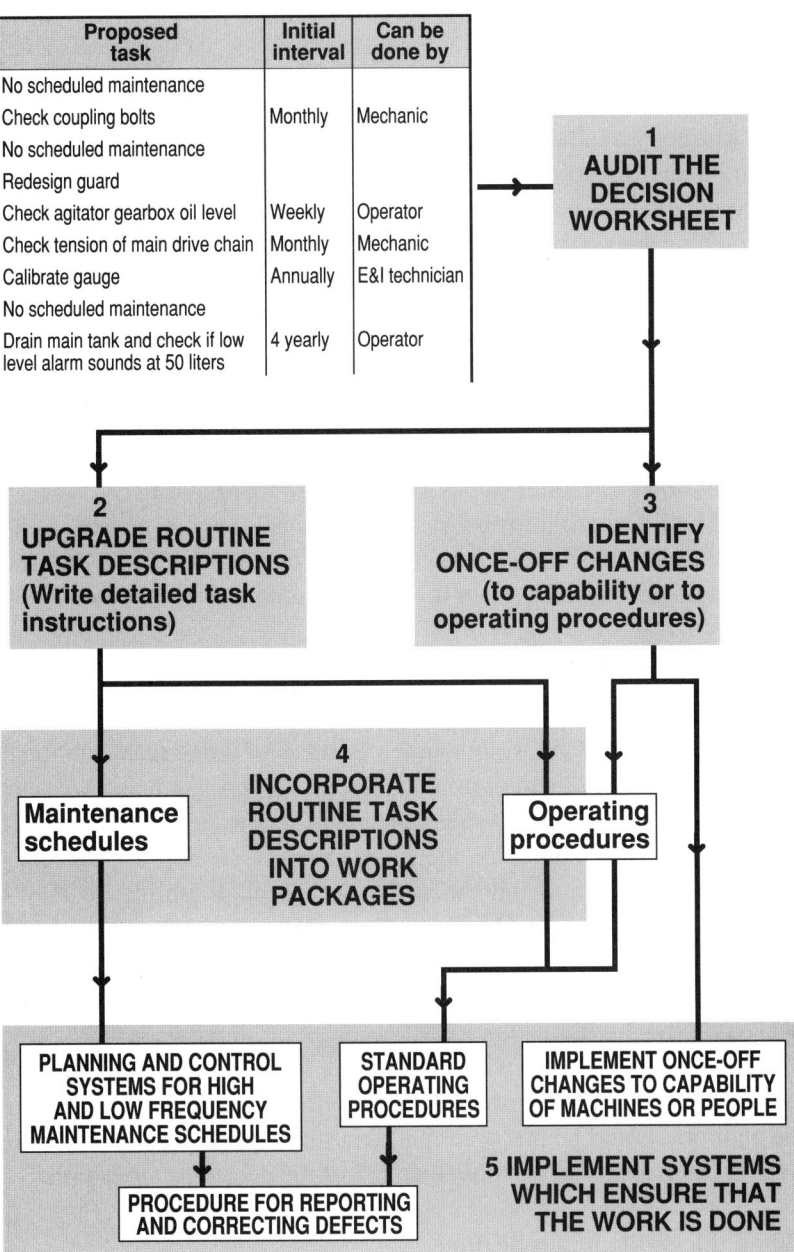

Figure 11.1: After RCM

11.2 The RCM Audit

If it is correctly applied, the RCM process provides the most robust framework currently available for formulating asset management strategies. These strategies profoundly affect the safety, environmental integrity and economic well-being of the organization using the assets. However, if something does go badly wrong in spite of the best efforts of the people applying the process, every decision will be subjected to a thorough and often intensely adversarial review by organizations ranging from regulatory authorities through insurers and shareholders to representatives of victims (or their survivors). As a result, any organization which uses RCM should take great care to ensure that the people who apply it know what they are doing, and also to satisfy itself that their decisions are *sensible* and *defensible*. The latter step is known as the RCM audit.

RCM audits entail a formal review of the contents of the RCM Information and Decision Worksheets. This section of this chapter looks at who should do the audit, when it should be done and what it entails.

Who should do the audit

Senior management bears the overall responsibility for the asset if something goes badly wrong, so it is in the interests of themselves and their employers to satisfy themselves that reasonable steps are being taken to prevent such occurrences. As mentioned in Chapter 1, senior managers do not necessarily have to do the audits themselves, but may delegate them to anyone in whose judgement they have enough confidence. However, if this is done, it should always be understood that the auditors are acting on behalf of senior management, so the latter still bear the ultimate responsibility for the decisions. (Whoever carries out the audits should also be thoroughly trained in RCM.)

If the auditors disagree with any findings or conclusions, they should discuss the matter with the people who performed the analysis. In so doing, the auditors should be prepared to accept that they themselves may be wrong. (In most cases, no more than 5% of the decisions are queried.)

When the audit should be done

Audits should be carried out as soon as possible after each review has been completed (preferably within two weeks), for three reasons:

- the people who did the analysis are keen to see the results of their efforts put into practice. (If this happens too slowly, they start to lose interest, and more seriously, they begin to question whether management was serious about involving them in the first place.)
- people can still recall easily why they made specific decisions
- the sooner the decisions are implemented, the sooner the organization derives the full benefits of the exercise.

When overall agreement is reached about each analysis, the decisions are implemented as described in the rest of this chapter.

What the audit entails

An RCM analysis needs to be audited from the point of view of *method* and *content*. When reviewing the *method*, the auditor seeks to ensure that the RCM process has been correctly applied. When reviewing the *content*, the auditor seeks to ensure that the correct information has been gathered and conclusions drawn both about the asset itself and the process of which it forms part. Issues which most often need attention are as follows:

Levels of analysis
The analysis should be carried out at the right level. The most common fault is to analyze assets at too low a level, and the usual symptom is large numbers of items with only one or two functions defined per item.

Functions
All the functions of the asset should be clearly and correctly described. Key points to look for include the following:

- by and large, each function statement should define only one function, although it may incorporate more than one performance standard. As a rule, each function statement should contain only one verb (unless it is a protective device)
- performance standards should be quantified, and should indicate what the asset *must be able to do in its present operating context* rather than its rated capacity (what it can do)
- all protective devices should be listed and their functions correctly described ('to do X *if* Y occurs')
- the functions of all gauges and indicators should be listed, together with desired levels of accuracy.

Functional failures
All the functional failures associated with each function should be listed (usually complete failure plus the negative of each performance standard in the function statement).

Failure modes
Ensure that failure modes which have happened or which are reasonably likely have not been omitted. Failure mode descriptions should also be specific. In particular,

• they should include a verb, not just specify a component

• the verb should be a word other than *fails* or *malfunctions* unless it is appropriate to treat the failure of a sub-assembly as single failure mode (option 3 on page 87)

• switch and valve failures should indicate whether the item fails in the *open* or *closed* position

Failure modes should relate directly to the functional failure under consideration, and failure modes and effects should not be transposed, as in:

Failure Mode	Failure Effect
Motor trips out	Pump impeller jammed by rock

Another common mistake is to combine two substantially different failure modes in one description, as follows:

Wrong	Right
1 Screens damaged or worn	1 Screens damaged
	2 Screens worn.

Failure effects
Failure effect descriptions should make it possible to decide:

• whether (and how) the failure will be evident to the operating crew

• whether (and how) the failure poses a threat to safety

• what effect (if any) the failure has on production or operations (output, product quality, customer service).

Failure effects should not incorporate actual 'consequence words' like *'This failure affects safety'* or *'This failure is evident'*. However, they should list likely total downtime as opposed to repair time, and should indicate what must be done to rectify the failure (replace, repair, reset, etc)

Finally, auditors should satisfy themselves that anything which is said to be 'analyzed separately' actually is analyzed separately.

Consequence evaluation
Special care should be taken to ensure that the hidden function question (question H on page 200) has been answered correctly. In particular, the correct meanings should have been attached to the terms *on its own* and *under normal circumstances* in this question, as explained on pages 124 and 126. Special attention should also be paid to the evaluation of the safety and environmental consequences of evident failures, and to the effectiveness of any tasks which might have been selected to manage failures in these two categories.

Task selection
Any tasks which have been selected should not only satisfy the criteria for technical feasibility as explained in chapters 6, 7 and 8, but they should also address the consequences of the failure. Key points to look out for:

• if the answer to question H is 'No' and the answer to question H4 is 'No', then question H5 must be answered. If the answer to H5 is 'Yes', the proposed task should *not* be 'no scheduled maintenance'

• if the answer to question S or E is 'Yes', the proposed task should *not* be 'no scheduled maintenance'

• if the failure has operational or non-operational consequences, the task must be cost-effective.

Proposed tasks or default actions should be described in enough detail to leave the auditor in no doubt as to what is intended. In particular, routine task descriptions should not simply list the type of task ('scheduled on-condition tasks' or 'scheduled failure-finding', etc).

The task description should also relate directly and solely to the failure mode in question. It should not incorporate a combination of tasks because this usually signifies two different failure modes (unless the answer to question S4 is yes). For example:

Wrong	Right
Inspect chain for wear and adjust tension	Adjust tension of chain
	or
	Inspect chain for wear

Initial interval
Task intervals should clearly have been set according to the criteria set out in Chapters 6, 7 and 8. In particular, look out for a tendency to confuse P-F intervals with useful life in on-condition task intervals.

11.3 Task Descriptions

Before any task reaches the person who has to do it, it must be described in enough detail to leave no doubt at all as to what is to be done. Clearly, the degree of detail required will be influenced by the overall level of skill and experience of the workers involved. However, bear in mind that the more that is left out of a task description, the greater the chance that someone will miss out a key step or choose to do the wrong task altogether. In this context, special care needs to be taken with the description of any failure-finding task which calls for a hazardous situation to be simulated in order to test the function of a protective device.

Task descriptions should also explain what action must be taken if a defect is encountered. (For instance, should the defect be reported to a supervisor or to the maintenance department – or should it be rectified immediately?) Instructions like 'check component A for condition B and replace if necessary' should be used with caution, because the 'check' part of the task might only take a few seconds, while the 'replace' part could take several hours. This can play havoc with the duration of planned downtime. Instructions of this sort should in fact be written as 'check component A for condition B and report defects to supervisor'. Only use 'if necessary' for quick servicing routines, such as 'check gearbox oil level using dipstick and top up with Wonderoil Type 900 if necessary'.

Examples of the right and the wrong way to specify tasks are shown in Figure 11.2 below:

Wrong	*Right*
Check coupling	Check feedscrew coupling for loose bolts and replace if necessary *or* Visually check agitator coupling flange for cracks and report defects to the maintenance supervisor ... *etc*
Calibrate gauge	Fit 0 - 20 bar test gauge to test point and check if reading on pressure gauge PI1204 is within 0.5 bar of the reading on the test gauge when the test gauge reads 8 bar. Arrange to replace out-of-spec gauges when plant is shut down for cleaning *or* Remove pressure gauge PI1204 to workshop and calibrate following procedure in manual 27A

Figure 11.2: *Task descriptions*

Pages 206 and 207 explained that each task should be defined as clearly as possible on the decision worksheet. This saves the duplication of effort which occurs if detailed procedures have to be written up later by someone else. It also reduces the possibility of transcription errors. However, if time does not permit the procedures to be specified during the RCM analysis, then they must be specified later. As mentioned below, this can often be done as part of an ISO 9000-type initiative.

Note that if detailed task descriptions are to be prepared later, this should ideally be done by someone who participated in the original RCM analysis. If this is not possible, the third party should understand clearly that he or she is being asked to define the tasks on the decision worksheet in more detail, and *not* to re-audit the analysis.

Basic information
In addition to a clear description of the task itself, the document on which the task is listed should also clearly state the following:

- *a description of the asset* to which it applies together with an equipment number where relevant
- *who should do the task* (operator, electrician, fitter, technician, etc)
- *the frequency* with which the task is to be done
- *whether (and if necessary, how) the equipment should be stopped* and/ or isolated while the task is done, together with any other safety precautions which must be taken
- *special tools and prescribed spares.* Listing these items can save much unproductive walking to and fro after the job has started.

ISO 9000 and RCM
A major objective of RCM is to identify what work people should be doing. (In other words, to ensure that 'they do the right job'.) On the other hand, a major thrust of quality systems like ISO 9000 is to define what people should be doing as clearly as possible in order to minimize the chance of errors. (In other words, to ensure that 'they do the job right'.)

This suggests that the process of transferring tasks from RCM decision worksheets to end-user documents can be seen as the point where the output of an RCM analysis becomes the input to an ISO 9000 procedure writing exercise. It also suggests that if both initiatives are to be undertaken, it makes sense to apply RCM first.

11.4 Implementing Once-off Changes

At the end of a typical RCM analysis, it is not unusual to find that between 2% and 10% of the failure modes default to redesign. Part 2 of Chapter 9 mentioned that in the context of RCM, *redesign* means a once-off change in any of the following three areas:

• a change to the *physical configuration* of an asset or system
• a change to a *process* or *operating procedure*
• a change to the capability of a person, usually by *training*.

Once they have been accepted by the auditors, these changes need to be implemented as thoroughly and as quickly as possible. Key issues in each of these three areas are discussed below

Changes to the physical asset
All modifications should be:

• *properly justified.* Chapter 9 explained that modifications should be justified in terms of their consequences. Modifications intended to deal with single or multiple failures which have safety or environmental consequences should reduce the risk (frequency and/or severity) of the consequences to a level which is acceptable. Figure 9.2 showed an algorithm which can be used to justify modifications intended to deal with failures that only have economic consequences

• *correctly designed* by suitably qualified engineers. As a rule, attempts should not be made to redesign assets during the RCM process, but the designer should consult afterwards with the people who did the review in order to develop a correctly focused specification

• *properly implemented.* Steps must be taken to ensure that modifications are carried out as intended in terms of time, cost and quality, and that all drawings, manuals and parts lists are updated correctly

• *properly managed.* Modifications should not interfere with essential routine maintenance activities in other parts of the plant, and the maintenance requirements of every modified item of equipment should be correctly assessed and implemented.

Changes to the way in which the plant is operated
Once-off changes to the way in which plant must be operated are handled in the same way as routine tasks which are incorporated into operating procedures, as explained in the next part of this chapter.

Changes to the capability of people

As explained in Chapter 4, the RCM process frequently reveals failure modes caused by slips or lapses on the part of operators or maintainers (skill-based human errors). These immediately become apparent to any operators or maintainers who participate directly in the process, and they usually modify their behavior appropriately as soon as they learn what they are doing wrong.

However, we also need to ensure that people who have not participated directly in the process acquire the relevant skills. In most cases, the most efficient way to do this is to revise or extend existing training programs, or to develop new programs. In most organizations, this will be done in consultation with the training department.

11.5 Work Packages

Once the maintenance procedures have been fully specified, they need to be packaged in a form which can be planned and organized without too much difficulty, and which can be presented in a neat and compact form to the people who will be doing the tasks. This can be done in two ways:

- high-frequency maintenance procedures to be done by operators can be incorporated into the operating procedures of the equipment
- the balance of the maintenance routines are packaged into separate schedules and checklists.

Standard Operating Procedures

The previous part of this chapter mentioned that any changes which must be made to the way in which an asset is operated should be documented in standard operating procedures, or SOP's. (In situations where they do not exist already, it will almost certainly be necessary to develop them in order to ensure that the changes are implemented.) In many cases, SOP's are also the simplest and cheapest way to manage high frequency tasks which need to be done by operators, as illustrated in Figure 11.3 overleaf.

As a rule, tasks should only be incorporated into operating procedures if they need to be done at intervals of one week or less. Tasks which need to be done by operators at longer intervals should be packaged into separate schedules and planned, organized and controlled in the same way as maintenance schedules, as described in Part 6 of this chapter.

Proposed task	Initial interval	Can be done by	Standard operating procedure WIDGET WASHING MACHINE
No scheduled maintenance			At the start of the shift:
Check coupling bolts	Monthly	Mechanic	• Fill feed hopper
No scheduled maintenance			• Open air valve and wait until
Redesign guard			pressure reaches 50 psi
Check agitator gearbox oil level	Weekly	Operator	• *(Monday mornings only)* Check agitator gearbox oil level
Check tension of main drive chain	Monthly	Mechanic	using dipstick and report if it is
Calibrate gauge	Annually	E&I technician	below level 2
No scheduled maintenance			• Press start button
Drain main tank and check if low level alarm sounds at 50 liters	4 yearly	Operator	• Open detergent valve • Start widget feed • *etc*

Figure 11.3:
Transferring a task from a decision worksheet to an SOP

Maintenance Schedules

A maintenance schedule is a document listing a number of maintenance tasks to be done by a person with a specified level of skill on a specified asset at a specified frequency. Figure 11.4 shows the relationship between these schedules and the RCM decision worksheets:

Proposed task	Initial interval	Can be done by	Maintenance Schedule WIDGET WASHING MACHINE	
			Interval	Done by
No scheduled maintenance			MONTHLY	MECHANIC
Check coupling bolts	Monthly	Mechanic	*Stop machine and follow lock-out*	
No scheduled maintenance			*procedure X, then*	
Redesign guard				
Check agitator gearbox oil level	Weekly	Operator	1 Visually check main drive	
Check tension of main drive chain	Monthly	Mechanic	coupling for loose bolts and tighten if necessary	
Calibrate gauge	Annually	E&I technician	2 Check tension of main drive	
No scheduled maintenance			chain and adjust tensioner if	
Drain main tank and check if low level alarm sounds at 50 liters	4 yearly	Operator	free play halfway between sprockets exceeds 10 mm	

Figure 11.4:
Transferring tasks from a decision worksheet to a maintenance schedule

Compiling schedules from RCM decision worksheets is a fairly straightforward process. However, a few additional factors need to be taken into account as explained in the following paragraphs.

Consolidating frequencies
In Chapter 7 it was mentioned that if a wide range of different task intervals appear on a decision worksheet, they should be consolidated into a smaller number of work packages when compiling the schedules based on the worksheets. Figure 11.5 gives an extreme example of the variety of task intervals which could appear on a decision worksheet, and how they might be consolidated into a smaller number of schedule frequencies.

Intervals of tasks on decision worksheets	Intervals of maintenance schedules
Daily	Daily
Weekly	Weekly
2-weekly	
Monthly	Monthly
6-weekly	
2-monthly	
3-monthly	3-monthly
4-monthly	
6-monthly	6-monthly
9-monthly	
12-monthly	12-monthly

Figure 11.5:
Consolidating task frequencies

The most expensive tasks, in terms of the direct cost of doing them and the amount of downtime needed to do them, tend to dictate basic schedule intervals. However, planning is simplified if schedule intervals are multiples of one another, as shown in the example.

Note also that if a task frequency is changed in this fashion, it should *always* be incorporated into a schedule of a *higher* frequency. Task intervals should never be arbitrarily increased, because doing so could move an on-condition task frequency outside the P-F interval for that failure, or it could move a scheduled discard task past the end of the 'life' of the component.

Contradictions
When a low frequency schedule incorporates a higher frequency schedule, should the latter be incorporated as a global instruction, or should it be rewritten in full? In other words, should (say) an annual schedule include an instruction like 'do the three monthly schedule', or should all the tasks in the three monthly schedule be written out in the annual schedule?

In fact it is wise to rewrite the schedules in order to avoid the problem of contradictions.

For instance, consider what could happen in a situation where a three monthly schedule includes the instruction 'check gearbox oil and top up if necessary', and the annual schedule for the same machine starts with the instruction 'do the three monthly schedule', and later says 'drain, flush and refill gearbox'.

Too many anomalies and contradictions of this nature rapidly erode the credibility of the system in the eyes of the people doing the work, so it is worth taking a little extra time to ensure that they don't occur.

Adding tasks
When compiling schedules on the basis described above, there is often a great temptation to start adding tasks to the completed schedule. This is most often done on the basis that 'when we do A and B, we might as well do X, Y and Z'. This should be avoided for the following reasons:

• extra tasks increase the routine workload. If too many tasks are added, the workload is increased to the point where there is either insufficient labor to do all the tasks, or the equipment cannot be released for the amount of time required to do them, or both

• the people doing the schedules soon realize that X, Y and Z are not strictly necessary, and *they judge the schedule as a whole accordingly.* As a result, they start looking for reasons why they cannot do the schedule as a whole. When they find them, tasks A and B are also not done and the whole maintenance program begins to fall apart.

This problem is common in shutdowns. Many shutdown tasks are done, not because they are really needed, but because the plant is stopped and it is possible to 'get at' the equipment. This adds greatly to the cost and sometimes to the duration of the shutdown. Unnecessary work also leads to an increase in infant mortality when the plant starts up again.

(This does not mean that people who do routine tasks should concentrate only on the specified tasks and ignore any other potential and functional failures which they may encounter. Of course they should keep their eyes and ears open. The point is that the schedule itself should only specify what really needs to be done at that frequency.)

11.6 Maintenance Planning and Control Systems

High- and Low-frequency Maintenance Schedules

Once the tasks have been grouped into sensible work packages, the next step is to set up planning and control systems which ensure that they are done by the right person at the right time. A key factor which influences the design of such systems is the frequency of the schedules.

In particular, high- and low-frequency schedules are handled differently because both the work content and the planning horizons differ. High-frequency schedules are defined as schedules performed at intervals of *up to one week*. These schedules usually consist of simple on-condition and failure-finding tasks. They have a low work content, and hence can be done quickly. Most of them can also be done while the plant is running, so they can be done at more or less any time. These two factors mean that the associated planning systems can be kept very simple.

However, high-frequency schedules also exist in large numbers, so if careful thought is not applied to their administration they can easily get out of hand. For example, daily schedules which have to be done for 350 days of the year on 1000 items of plant could generate 350 000 instructions annually if each schedule is issued separately (either electronically or on paper) every time it has to be done. This is clearly nonsense, and the problems it creates are a common reason why high-frequency schedules are often administered badly or not at all.

But high-frequency tasks are the backbone of successful routine maintenance, so some way must be found to ensure that they are done without creating an excessive administrative burden.

Low-frequency schedules are those done at intervals of *a month or longer*. Their longer planning horizon makes them less amenable to simple planning systems of the type used for high-frequency schedules. They usually have a higher work content so more time is needed to do them, and the plant usually has to be stopped while they are done. As a result, they need more complex planning and control systems.

The next sections of this chapter suggest some of the options which can be used to manage both types of schedules, under the following headings:
- schedules done by operators
- 'schedules' done by the quality function
- high-frequency schedules done by maintenance people
- low-frequency schedules done by maintenance people.

Schedules done by Operators

From the maintenance viewpoint, the most valuable attribute of operators is that they are near the equipment for much of the time. As discussed on page 209, this puts them in an ideal position to do many on-condition and failure-finding tasks. These are often very high-frequency tasks – some will be daily or even once or twice per shift – so special care must be taken to keep the associated administrative systems as simple as possible.

Simple reminder systems which can be used for operator tasks instead of formal checksheets include:

- incorporating the maintenance checks into standard operating procedures, as discussed earlier
- mounting the schedule permanently onto a wall or on a control cabinet where the operators can see it easily
- training the operators in such a way that the inspections become second nature (a high-risk approach which is not usually recommended).

Formal written checklists should only be used for operator checks when the failure consequences are likely to be particularly severe, and there is reason to doubt whether the tasks will be done without a formal reminder. The checklists can be the same as those described later for high-frequency tasks done by maintenance people.

Schedules and Quality Checks

We have seen how more and more performance standards incorporate product quality standards. This means that more and more potential and functional failures can be revealed by product quality checks. These checks are often being done already (for example, using SPC as discussed on pages 151 and 152). Key points to note are as follows:

- quality checks must be recognized as a valid and valuable source of maintenance information
- steps must be taken to ensure that quality-related potential failures are attended to as soon as they are noticed. This issue is discussed in more detail later.

High-frequency Schedules done by Maintenance

Despite all the earlier comments about the merits of using operators to do high-frequency maintenance work, many of these tasks still need to be done by maintenance people. These usually need to be more formally planned than operators' checks, because maintenance people cover more machines spread over a wider area than operators, and they usually do a wider variety of tasks.

One approach is to divide the plant into sections, and prepare a checklist of the type shown in Figure 11.6 for each section.

MAINTENANCE CHECKLIST	PLANT SECTION		TO BE DONE BY	WEEK ENDING DATE
	Boiler House		Mechanic	

ITEM Nº	DESCRIPTION	SCHEDULE	M	T	W	T	F	S	S	REMARKS
03030401	Coal handling system	M-265								
03030402	Boiler Nº 1	M-388								
03030402	Boiler Nº 1	M-389								
03030403	Boiler Nº 2	M-388								
03030403	Boiler Nº 2	M-389								
03030404	Ash handling system	M-539								
03030405	Feed water heater	M-462								
03030406	Flue gas system	M-391								

ALLOCATED TO	TIME	COMPLETED BY	SUPERVISOR

Figure 11.6: A checklist for high-frequency maintenance schedules

Note the following points about this type of checklist:

* the checklist only lists the *schedules* to be done, not individual tasks. The schedules are issued separately, often in book form, and bound in plastic covers for protection. In this way, only one checklist is issued per section per week, rather than dozens of schedules every day.

* roughly the same amount of work should be planned for each day, and it should not exceed between half an hour and an hour per day.

* the checklist shown can be used to plan at intervals between daily and weekly. Jobs can be planned for alternate days and twice per week, so the checklist encompasses a wider range of the shorter P-F intervals.

* the checklist can start and finish on any five or seven day cycle - it is not essential to stick to the Monday/Sunday cycle shown in the example.

* the checklists embody the schedule plans and they are issued automatically every week, so there is no need for any sort of planning system.

* the checklists are not used for any tasks which are to be done at intervals of longer than a week.

* each checklist involves one or two documents per week per section. This amounts to no more than fifty documents per week for a facility containing 1000 items subject to these checks.

Some high-frequency tasks require readings to be taken, either manually (logging a meter reading) or electronically (vibration analysis). Readings of this nature are tasks, while the checklist described above is designed for complete schedules. This can cause problems, especially if we start issuing a separate document for each of these records alongside the checklists. This should be avoided, because the numbers of documents simply start climbing again. Possible alternatives are as follows:

* develop a special document for *all* the readings in each section, and attach this one document to the checklist for that section each week

* use one person to take all such readings in the entire plant

* ask the people taking the readings to record only those readings which are outside acceptable limits in the remarks column of the checklist (unless the readings are recorded automatically, as in the case of certain condition monitoring devices)

* automate the recording process.

Issuing high-frequency schedules

The checklists are issued to the relevant maintainer the week before they are to be done. Preferably, they should be the first activity done by that person each day. Note the following additional features of a well-run checklist system:

- if the maintainer cannot complete the planned tasks on any day, the tasks are done the following day. If the maintainer is continually unable to complete the prescribed checks, something is fundamentally wrong and the situation should be investigated

- the maintainer notes any potential or functional failures in the remarks column of the checklist – not on the schedules themselves

- the maintainer initiates corrective action at the end of each daily round. In some facilities, this may be the responsibility of the maintainer, in others, he or she may have to work through a supervisor. The action will vary from arranging for the plant to stop at once to arranging for the fault to be corrected at the next shutdown. This decision is based on the possible consequences of the failure and the nett P-F interval. (Note that these issues should have been considered as part of the RCM process when the routine task was originally specified)

- as in the case of operators, it is important that action is either taken or the maintainer told why action is unnecessary or being deferred, or the maintainers also lose interest in the system

- at the end of each cycle, the completed checklists can be stored as a record that the tasks were done, so it is not necessary to re-enter them into a history recording system. However, problems which are encountered and the action taken to deal with them should be documented, as discussed in Chapter 14.

Controlling high-frequency schedules

A problem associated with most checklist systems is the 'tearoom tick syndrome'. This means that people indicate that the checklist has been done when in fact it has not. To avoid this problem, supervisors should conduct random over-inspections. These entail doing the schedules on the checklist in the company of the maintainer who normally does it. If the checklist is not being done correctly, unreported failures soon become apparent, and the supervisor takes appropriate action.

Low-frequency Schedules done by Maintenance

We have seen that high-frequency schedules can be planned, organized and controlled using one carefully structured checklist. In contrast, the long planning horizon associated with low-frequency schedules means that the steps needed to plan, organize and control them are carried out separately. What is more, the procedures used to *plan* schedules based on elapsed time differ markedly from those used for schedules based on running time, but similar procedures can be used to *organize* and *control* the two types of schedule. As a result, we consider the planning process separately in the following paragraphs, but consider the subsequent steps together.

Elapsed time planning
The principles of elapsed time planning are well known, and are used for many purposes in addition to maintenance planning. For low-frequency schedules, elapsed time planning is usually based on a planning board similar to that shown in Figure 11.7 (or its computerised equivalent).

ITEM Nº	DESCRIPTION	1	2	3	4	5	6	7	8	9	10	11	12	----	47	48	49	50	51	52

Figure 11.7: A typical low-frequency planning board

Most of these systems use an overall planning horizon of one year, divided into 52 weeks. However, bear in mind when setting up such systems that some failure-finding tasks in particular can have cycle times of up to ten years, and the planning horizon of any associated planning system must accommodate such tasks. When setting up these systems, note also low-frequency schedules nearly always involve equipment stoppages, and these can have operational consequences in exactly the same way as the stoppages which they are supposed to prevent, so special care must be taken to minimize these consequences. Points to watch for include:

• peaks and troughs in the production cycle. The most time consuming schedules should be planned for periods of lowest activity, in order to minimize their effect on operations

• two machines which require the same special resource at the same time (such as a crane)

- cases where it is only possible to do a schedule if other machines are stopped at the same time. This applies especially to services like steam raising plant and air compressors.

On the other hand, wherever such constraints permit, try to spread the routine maintenance workload as evenly as possible over the year in order to stabilize labor requirements.

A final point about elapsed-time-based low-frequency schedule planning is that it looks deceptively simple to use computers for this purpose. However, bear in mind that the issues discussed above introduce a wide range of completely unrelated constraints into the calendar time planning process. For this reason, take great care when designing or acquiring calendar-time-based systems which plan schedules on the basis of predetermined parameters, or which automatically re-plan schedules that have not been done. The author has encountered a number of such systems which simply move schedules from week to week, regardless of policy constraints. This becomes chaotic, especially when schedules that should only be done in the low season are gradually moved into the middle of the high season and so on.

Running time planning
Running time planning involves the following steps:

- the number of cycles each machine has completed in each period are recorded (they can be measured in terms of time, distance travelled, units of output, etc)
- this record is fed into the planning system
- the cumulative total of hours run is updated to reflect the time run since the last schedule was done.

Manual running time systems can range from sophisticated boards costing hundreds – even thousands – of pounds to counters which move along pieces of string. If possible, these systems should count *down* to zero, so that planners can see at a glance how much time is left before schedules are due. This also provides visual early warning of peaks that could overload the workshop.

Running time planning systems lend themselves readily to the use of computers because they entail processing and storing large quantities of data. Also the dynamism of running time systems means that they have fewer policy constraints than elapsed time systems.

However, if the collection of the run time data is not automated it can be expensive and prone to errors, so if computers are to be used for running time planning, data capture should also be automated if possible. The system should also be designed to provide a continuously updated forecast of the scheduled workload on each workshop as far as possible into the future. This gives managers time to smooth any peaks and troughs which appear in the forecast.

Organizing low-frequency schedules
Most planning systems start organizing low-frequency schedules the week before the schedules are due (except for shutdown schedules). The organizing process usually contains the following elements:

• a list is prepared which shows the schedules due the following week. They are usually separated by craft and plant section

• meetings are held with the operations department to agree on which day and at what time the schedules will be done (especially those which require equipment downtime)

• the schedules themselves are issued to the relevant supervisors, who plan who will do them and arrange any other resources which may be needed as they would for any other incoming maintenance job.

Controlling low-frequency schedules
Low-frequency schedules are subject to the same performance controls as any other type of maintenance work. This applies to the time taken to do the schedules, standards of workmanship, and so on.

Two additional factors need to be considered. Firstly, the planning system should indicate when any schedule is overdue. As mentioned earlier, such schedules should not be reprogrammed automatically, but should be managed on an exception basis.

Finally, maintenance schedules should be reviewed continuously in the light of changing circumstances (especially circumstances which affect the consequences of failure) and new information. In this context, bear in mind that the more everyone associated with the equipment is involved in determining its maintenance requirements to begin with, the more they are likely to offer thoughtful and constructive feedback about these requirements in future. This issue is discussed in more detail in the next chapter.

11.7 Reporting Defects

In addition to ensuring that the tasks are done, we also need to ensure that any potential failures which are found are rectified before they become functional failures, and that hidden functional failures are rectified before the multiple failure has a chance to occur. This means that anyone who might discover a potential or functional failure must have unrestricted access to a simple, reliable and direct procedure for reporting it immediately to whoever is going to repair it.

This communication takes place instantaneously if the person who operates the machine is also the person who maintains it. The speed and accuracy of the response to defects which can be achieved under these circumstances are a major reason why people who operate machines should also be trained to maintain them (or vice versa). A second benefit of this approach is that formal defect reporting systems are only needed for failures which the operator/maintainer is unable to deal with on his own.

If this organization structure is not possible or not practical, the next best way to ensure that defects are attended to quickly is to allocate maintenance people permanently to a specific asset or group of assets. Not only do such people get to know the machines better, which improves their diagnostic skills, but the speed of response also tends to be quicker than it would be if they work in a central workshop. It is also still possible to keep the defect reporting systems simple and informal.

If it is not possible to organize close maintenance support of either sort, then it becomes necessary to implement more formal defect reporting systems. In general, the further away the maintenance function is from the assets it is to maintain – in other words, the more heavily centralized it is – the more formal the defect reporting process becomes. This is also true of defects which can only be dealt with during major shutdowns.

Basically, formal defect reporting systems enable anyone to inform the maintenance department in writing (electronically or manually) about the existence of a potential or functional failure. The chief criteria of such systems should always be simplicity, accessibility and speed.

Manual defect reporting systems are usually based on simple job cards of the type shown in Figure 11.8. (These job cards can also be used by the maintenance department to plan and record work, but this aspect of their use is beyond the scope of this book.) If a computerised defect reporting system is used, the screen is formatted in much the same way as the card.

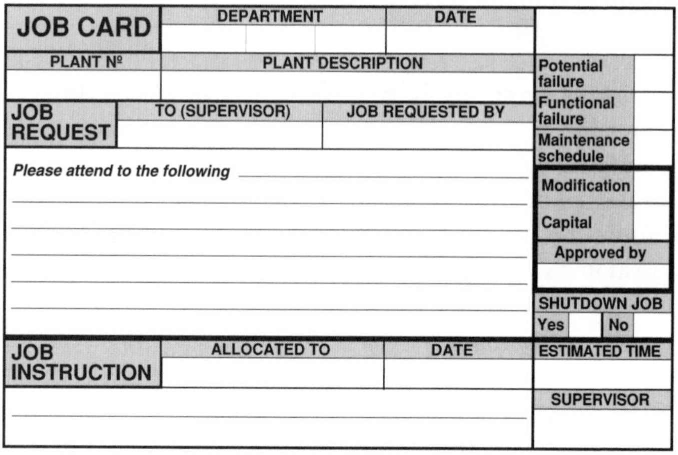

JOB CARD	DEPARTMENT	DATE	
PLANT Nº	PLANT DESCRIPTION		Potential failure
			Functional failure
JOB REQUEST	TO (SUPERVISOR)	JOB REQUESTED BY	Maintenance schedule
Please attend to the following			Modification
			Capital
			Approved by
			SHUTDOWN JOB
			Yes / No
JOB INSTRUCTION	ALLOCATED TO	DATE	ESTIMATED TIME
			SUPERVISOR

Figure 11.8: A typical job request

The final point about systems of this sort is that people must be properly motivated to use them. This means that defects which are reported must be acted upon, or the user must be told why if no action is taken. Nothing will kill such a system more quickly than if defects are reported and nothing apparently happens.

12 Actuarial Analysis and Failure Data

12.1 The Six Failure Patterns

Throughout this book, numerous references have been made to the six patterns of failure shown again in Figure 12.1 below. Frequent use has also been made of terms like *age*, *life* and *MTBF*. This chapter explores these concepts and the relationship between them in more detail. It also considers what role (if any) technical history records and other failure data play in formulating maintenance policies.

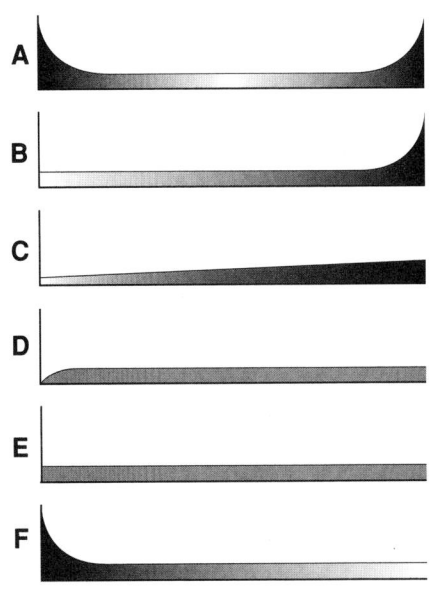

We start with a detailed look at failure patterns B and E, because they represent the most widely held views of age-related and random failure. Next we review patterns C and F, then take a look at patterns D and A. Part 2 of this chapter summarizes the uses and limitations of failure data.

Figure 12.1: Six patterns of failure

Failure Pattern B

Chapters 1 and 6 mentioned that failure pattern B depicts age-related failures. Chapter 6 explained that although these failures are the result of a more-or-less linear process of deterioration, there will still be considerable differences in the behavior of any two components that are subject to the same nominal stresses. Figure 12.2 shows how this behavior translates into failure pattern B.

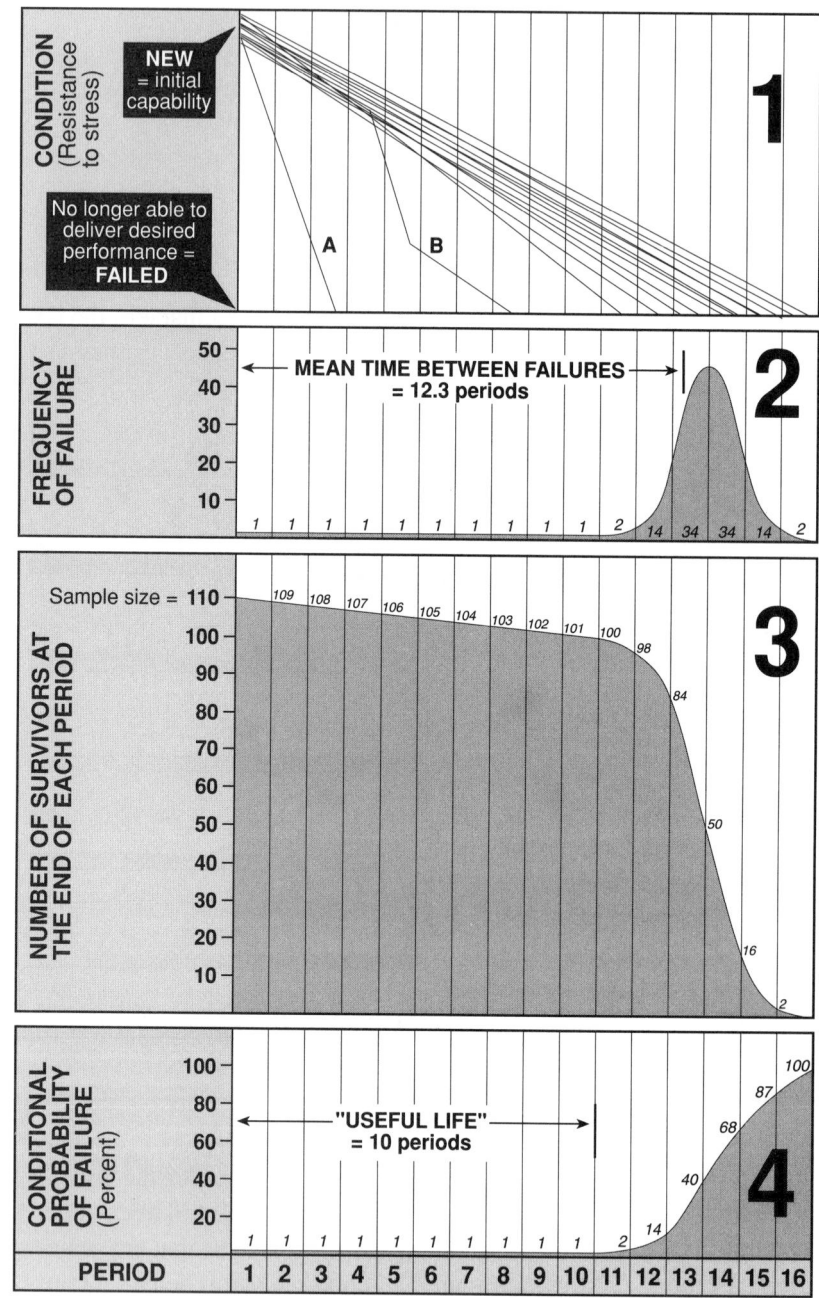

Figure 12.2: Failure pattern B

An example of a component which might behave as shown in Figure 12.2 is the impeller of a pump which is used to pump a moderately abrasive liquid. Part 1 of Figure 12.2 shows the wear-out characteristics of a dozen such impellers. Ten of them deteriorate at roughly the same rate, and last between 11 and 16 periods before failing. However, two of the impellers fail much sooner than expected, 'A' perhaps because it was not properly case hardened and 'B' because the properties of the liquid changed for a while, causing it to wear more quickly than usual. *Note that this failure distribution only applies to impellers which fail due to wear. It does not apply to impellers which fail for other reasons.*

In Part 2 of Figure 12.2, the distribution of failure frequencies is plotted against operating age for a large sample of components. It shows that apart from a few 'premature' failures, the majority of the components are likely to conform to a normal distribution about one point.

For example, assume that we have accumulated actual failure data for a sample of 110 impellers, all of which have failed due to wear. Ten of these impellers failed prematurely, one in each of the first ten periods. The other 100 impellers all failed between periods 11 and 16, and the frequency of these failures conforms to a normal distribution. (For a normal distribution, the failure frequencies in the last six periods would be roughly as shown if they are rounded to the nearest whole number.) On the basis of these figures, the mean time between failures of the impellers due to wear is 12.3 periods.

Part 3 of Figure 12.2 shows the survival distribution of the impellers based on this frequency distribution.

For example, 98 impellers lasted for more than 11 periods, and 16 impellers lasted for more than 14 periods.

Part 4 of Figure 12.2 is failure pattern B. It shows the probability that any impeller which has survived to the beginning of a period will fail during that period. This is known as the *conditional probability of failure*.

Allowing for a small degree of rounding error, this shows for instance that there is a 14% chance that an impeller which has survived to the beginning of period 12 will fail in that period. Similarly, 14 out of the sixteen impellers which make it to the beginning of period 15 will fail in that period – a conditional probability of failure of 87%.

The frequency curve in Part 2 and the probability curve in Part 4 are depicting the same phenomenon, but they differ markedly in the way they show it. In fact, the conditional probability of failure curve provides a better illustration of what is really happening than the frequency curve, because the latter could deceive us into thinking that things are getting better after the peak of the frequency curve.

These curves illustrate a number of additional points, as follows:

- the frequency and conditional probability curves show that the word 'life' can actually have two quite distinct meanings. The first is the mean time between failures (which is the same as the *average life* if the whole sample has run to failure). The second is the point at which there is a rapid increase in the conditional probability of failure. For want of any other term, this has been named the *useful life*.

- if we were to plan to overhaul or replace components at the mean time between failures, half would fail before they reached it. In other words, we would only be preventing half of the failures, which is likely to have unacceptable operating consequences. Clearly, if we wish to prevent most of the failures, we would need to intervene at the end of the 'useful life'. Figure 12.2 shows that the useful life is shorter than the mean time between failures – if the bell curve is wide, it can be very much shorter.

  As a result, it can only be concluded that *the mean time between failures is of little or no use in establishing the frequency of scheduled restoration and scheduled discard tasks* for items which conform to failure pattern B. The key variable is the point at which there is a rapid increase in the probability of failure.

- if we do replace the component at the end of its useful life as defined above, the average service life of each component would be shorter than if we let it run to failure. As discussed on page 137, this would increase the cost of maintenance (provided that there is no secondary damage associated with the failures).

  For instance, if we were to replace all the surviving impellers in Figure 12.2 at the end of period 10, the average service life of the impellers would be about 9.5 periods, instead of 12.3 periods if they were allowed to run to failure.

- The fact that there are two 'lives' associated with pattern B-type failures means that we must take care to specify which one we mean whenever we use the term 'life'.

  For example, we might phone the manufacturer of a certain component to ask what its 'life' is. We may have in mind the useful life, but if we don't spell out exactly what we mean, he might in all good faith give us the mean time between failures. If this is then used to establish a replacement frequency, all kinds of problems arise, often resulting in wholly unnecessary unpleasantness.

These issues apart, perhaps the biggest problem associated with pattern B is that very few failure modes actually behave in this fashion. As mentioned in Chapters 1 and 6, it is much more common to find failure modes which show little or no long-term relationship between age and failure.

Failure Pattern E

Figure 7.9 on page 156 illustrated three components which failed on a random basis. A number of reasons why failures can occur on this basis were discussed in Chapter 7. This part of this chapter explores some of the quantitative aspects of random failure in more detail, and goes on to review some of the implications of failure pattern E. To start with, Figure 12.3 overleaf shows the relationship between the frequency and conditional probability of random failures.

In part 1 of Figure 12.3, the dotted lines represent a number of components – in this case, ball bearings – which fail at random. As in Figure 7.9, each failure is preceded by a (somewhat elongated) P-F curve.

Random failure means that the probability that an item will fail in any one period is the same as it is in any other. In other words, the conditional probability of failure is constant, as shown in Part 2 of Figure 12.3.

For example, if we accept the empirical evidence that rolling element bearings usually conform to a random failure pattern – a phenomenon first observed by Davis[1952] – the conditional probability of failure is constant as shown in Figure 12.3, Part 2. Specifically, this shows that there is 10% probability that a bearing which has made it to the beginning of any period will fail during that period.

Part 3 of Figure 12.3 shows how a conditional probability of failure which is constant translates into a survival distribution which is exponential.

For example, if we started with a sample of 100 bearings and the probability of failure in the first period is 10%, then 10 bearings would fail in period 1 and 90 bearings would survive for more than one period. Similarly, if there is a 10% probability that the bearings which survive beyond the end of period 1 will fail in period 2, then 9 bearings would fail in period 2, and 81 bearings would make it to the beginning of period 3. Part 3 of Figure 12.3 shows how many bearings would survive to the beginning of each subsequent period for the first sixteen periods.

Theoretically, this process of decay would continue until infinity. In practice, however, we usually stop at unity – in other words, when the survival curve drops below one.

In the example shown in Figure 12.3, a rate of decay of 10% per period means that unity is reached after about 43 periods. This suggests that one lone bearing might last for 43 periods, but the vast majority will have failed long before then.

Finally, Part 4 of Figure 12.3 shows the frequency curve derived from the survival curve in Part 3. This curve is also exponential. (The shape of this frequency curve often causes it to be confused with failure pattern F, which is a conditional probability curve based on a different frequency distribution.)

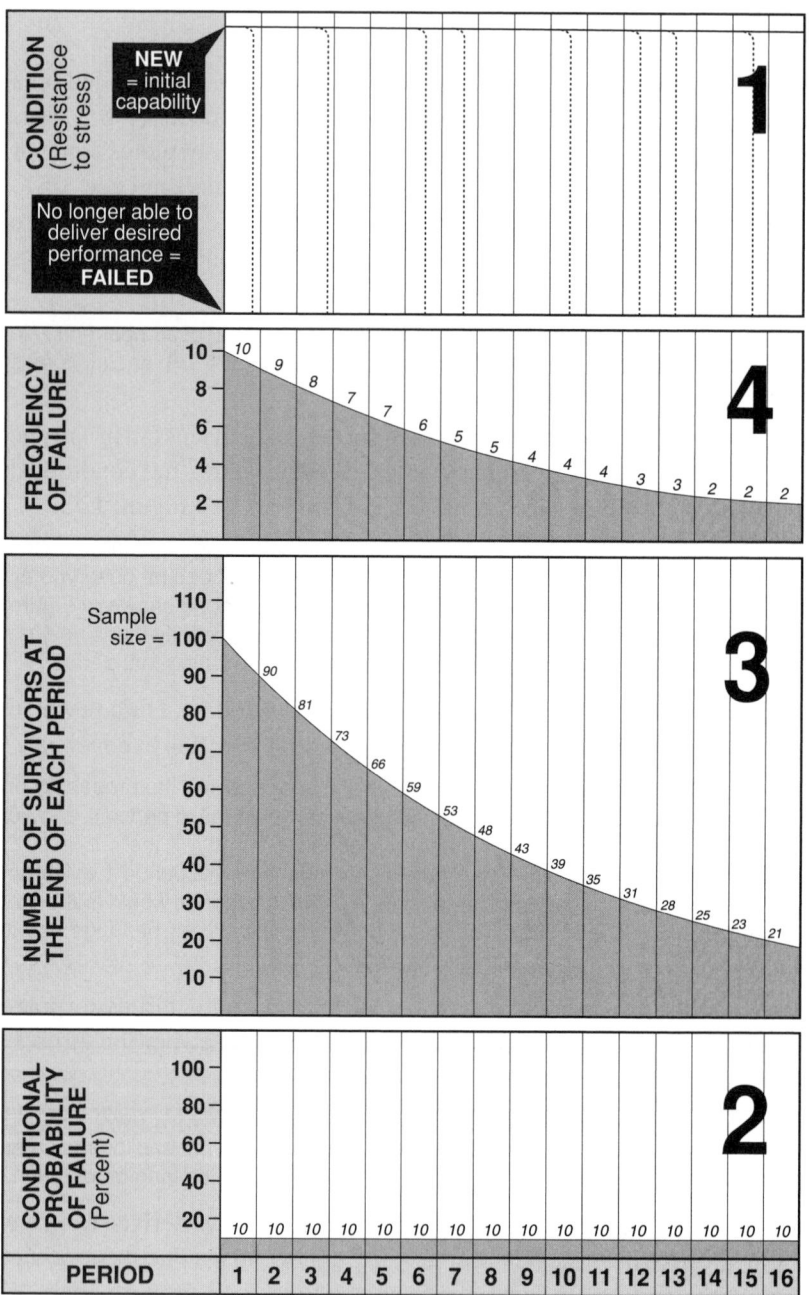

Figure 12.3: Failure pattern E

The fact that frequency and survival curves both carry on declining indefinitely means that the conditional probability curve also remains flat indefinitely. In other words, at no stage does Pattern E show a significant increase in the conditional probability of failure, so at no stage can an age be found at which we should contemplate scheduled rework or scheduled discard. Further points about Pattern E are as follows:

- *MTBF and random failures:* despite the fact that it is impossible to predict how long any one item which conforms to failure pattern E will last (hence the use of the term 'random' failure), it is still possible to compute a mean time between failures for such items. It is given by the point at which 63% of the items have failed.

  For example, Part 3 of Figure 12.3 indicates that 63% of the items have failed about half way through period 10. In other words, the MTBF of the bearings in this example is 9.5 periods.

  The fact that these items have a mean time between failures but do not have a 'useful life' as defined earlier means that we must be doubly careful when talking about the 'life' of an item.

- *comparing reliability:* the MTBF provides a basis for comparing the reliability of two different components which both conform to failure pattern E, even though the failure is 'random' in both cases. This is because the item with the higher MTBF will have a lower probability of failure in any given period.

  For example, assume that Brand X bearings conform to the failure distribution shown in Figure 12.3. If the conditional probability of failure of Brand Y is only 5% in each period, they would only be half as likely to fail and so would be considered much more reliable.

In the case of items which conform to failure pattern B, a more reliable component has a longer 'useful life' than one which is less reliable. So in simple language it could be said of the pattern B components that one type *lasts longer* than the other, while in the case of the Pattern E components, one type *fails less often* than the other.

(In practice, the reliability of bearings is measured by the 'B10' life. This is the life below which a bearing supplier guarantees that no more than 10% of his bearings will fail under given conditions of load and speed. This corresponds to one period on Part 2 of Figure 12.3. It also suggests that if a bearing conforms to a truly exponential survival distribution, then the MTBF of bearings due to 'normal wear and tear' should be about 9.5 times the B10 life.

So if bearing Brand Y is twice as reliable as Brand X, the B10 life – which is also known as the L10 life or the N10 life – of Brand Y will be twice as long as that of Brand X. This is useful when making procurement decisions about bearings, but it still does not tell us how long any one bearing will last in service.)

- *P-F curves and random failures:* Figure 7.9 on page 156 and Part 1 of Figure 12.3 both show random failures preceded by P-F curves. This is not meant to suggest that *all* failures which happen on a random basis are preceded by such a curve. In fact a great many failure modes which conform to pattern E are not preceded by any sort of warning, or if they are, the warning period is often much too short to be of any use. This is especially true of most of the failures which affect light current electrical and electronic items.

This does not detract in any way from the validity of the analysis. It simply means that no form of preventive maintenance – on-condition, scheduled restoration or scheduled discard – is technically feasible for these components, and they have to be managed on an appropriate default basis as discussed in Chapters 8 and 9.

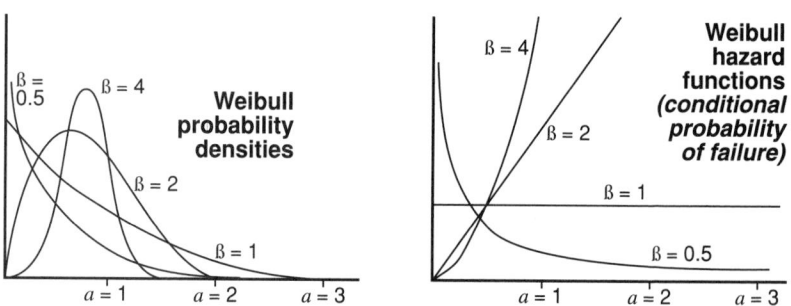

Figure 12.4: *Weibull distributions*

A Note on Weibull Distributions

At this stage, it is worth commenting on the Weibull distribution. This distribution is widely used because it has a great variety of shapes which enable it to fit many kinds of data, especially data relating to product life. The Weibull frequency distribution (or more correctly, *probability density function*) is:

$$f(t) = (\beta/a^\beta)t^{\beta-1}\exp[-(t/a)^\beta]$$

ß is called the *shape parameter* because it defines the shape of the distribution. *a* is the *scale parameter*. It defines the spread of the distribution and corresponds to the 63rd percentile [100 (1 - e^{-1})] of the cumulative distribution. The Weibull probability density function and corresponding conditional probability curves are shown in Figure 12.4. (This shows that the conditional probability of failure is also known as the 'hazard rate'.) When ß = 1, the Weibull distribution is the exponential distribution. When ß is between 3 and 4, it closely approximates the normal distribution. Later in this chapter we see how it describes other failure patterns.

Failure Pattern C

Failure pattern C shows a steadily increasing probability of failure, but no one point at which we can say "that's where it wears out". This part of this chapter looks at a possible reason why pattern C occurs, and then shows how it is derived.

The possible cause of pattern C which we consider is fatigue. Classical engineering theory suggests that fatigue failure is caused by cyclic stress, and that the relationship between cyclic stress and failure is governed by the *S-N curve*, as shown in Figure 12.5.

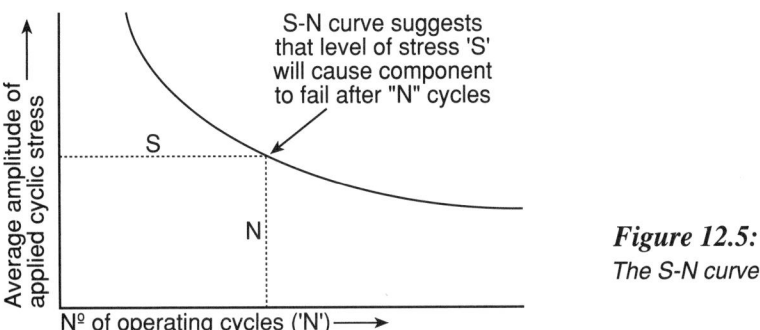

Figure 12.5:
The S-N curve

Figure 12.5 suggests that if the S-N curve is known, then we should be able to predict the life of the component with great accuracy for a given amplitude of cyclic stress. However, this is not so in practice because the average amplitude of the cyclic stress is not constant, and the ability of the component to withstand the stress – in other words, the location of the S-N curve – will not be exactly the same for every component.

Part 1 of Figure 12.6 overleaf suggests that the average amplitude of the applied stress might conform to a normal distribution about some

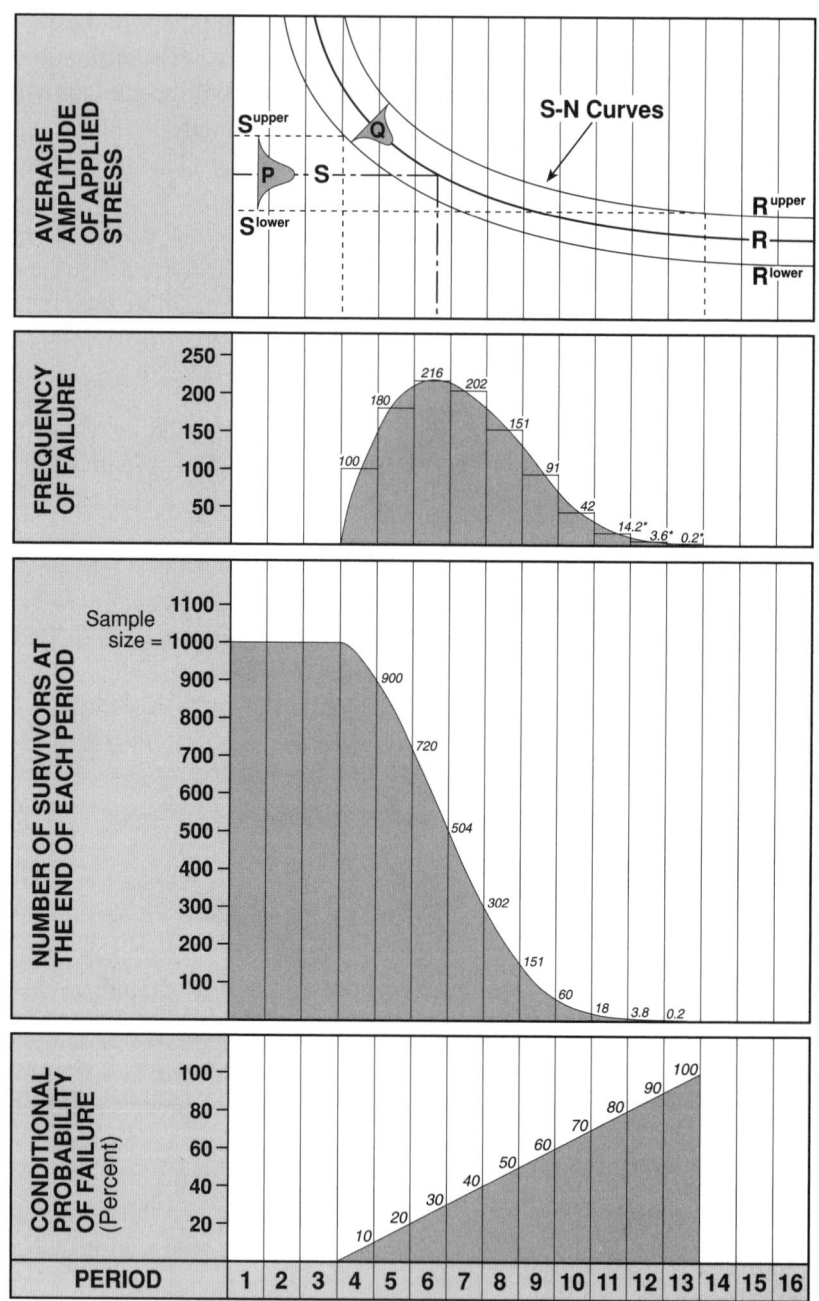

Figure 12.6: Failure pattern C (shifted)

mean, which is designated by 'S' in Figure 12.6. This distribution is shown by curve P. Similarly, the distribution of the S-N curves might be designated by normal curve Q. The combination of these two curves will be such that the ages at which failure occurs will conform to a distribution skewed to the left. How much it is skewed depends on the shape of the S-N curve itself. For the sake of argument, Part 2 of Figure 12.6 suggests that it will conform to a Weibull distribution with shape parameter ß = 2. (Strictly speaking, this should be called a 'shifted' Weibull distribution because it does not start at time zero.)

On the basis of this distribution, Part 2 of Figure 12.6 goes on to suggest how many failures might occur in each period if we were to test a sample of 1000 components to failure. (The fact that the numbers marked with an asterisk are not integers explains why this curve should be called a probability density rather than a frequency distribution.)

Part 3 of Figure 12.6 translates Part 2 into a survival curve, while Part 4 shows the conditional probability of failure based on the preceding two curves. Both of the latter curves are derived in the same way as the corresponding curves in Figure 12.2.

Further points about failure pattern C include the following:

- the shifted Weibull distribution means that the conditional probability curve starts at a point to the right of time t = 0. Figure 12.6 shows that this is the point where there is 'a rapid increase in the conditional probability of failure', which is of course the useful life as defined earlier. In Figure 12.6 this is three periods. However, earlier depictions of pattern C show a conditional probability of failure starting above zero. This might occur in practice if a failure mode led to a *truncated* Weibull distribution (one that hypothetically starts to the left of time t = 0) with a shape parameter of ß = 2, as shown in Figure 12.7.

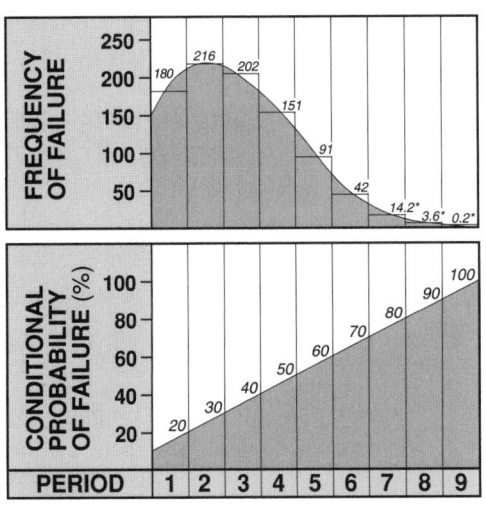

Figure 12.7:
*Truncated Weibull
distribution and failure
pattern C*

- the slope of pattern C appears to be quite steep in these examples. However, bear in mind that the actual slope is governed by the Weibull scale parameter *a*, which can be measured in anything ranging from weeks to decades (or even centuries), so the slope of pattern C can vary from quite steep to almost flat.

- pattern C is not only associated with fatigue. For instance, it has been found to fit the failure of the insulation in the windings of certain types of generators.

- conversely, not all fatigue-related failures necessarily conform to failure pattern C.

   For instance, if curve P in Figure 12.6 were skewed towards the S^{lower} limit and curve Q were skewed towards the R^{upper} limit, the failure frequency curve would be biased further towards the right. This would give a Weibull shape parameter greater than 2, which tends towards a normal distribution and so gives a conditional probability of failure curve which resembles pattern B.

   On the other hand, if the S^{lower} limit is below the point at which R^{lower} becomes asymptotic, then the frequency distribution will develop a long 'tail' on the right. This corresponds to a Weibull distribution where ß is between 1 and 2, which in turn generates failure pattern D.

   Finally, the discussion on page 159 mentioned that a large number of factors influence the rate at which fatigue failures develop in ball bearings. This would make the spread of any distribution very wide, which would in turn lead to an almost flat conditional probability curve. Add to this the variety of additional bearing failure modes listed on page 159 which have the same symptoms as fatigue, and the overall probability density effectively becomes fully exponential, which leads to failure pattern E as we have seen.

   So fatigue could manifest itself as failure pattern B, C, D or even E.

Failure Pattern D

As mentioned above, failure pattern D is the conditional probability curve associated with a Weibull distribution whose shape parameter ß is greater than 1 and less than 2.

Failure Pattern F

Pattern F is perhaps the most interesting, for two reasons:
- it is the only pattern where the probability of failure actually *declines* with age (apart from A, which is a special case)
- it is the most common of the six patterns, as mentioned on page 13.

For these reasons, it is worth exploring in more detail the factors which give rise to this pattern.

The shape of failure pattern F indicates that the highest probability of failure occurs when the equipment is new or just after it has been overhauled. This phenomenon is known as *infant mortality*, and it has a wide variety of causes. These are summarized in Figure 12.8 and discussed in the subsequent paragraphs.

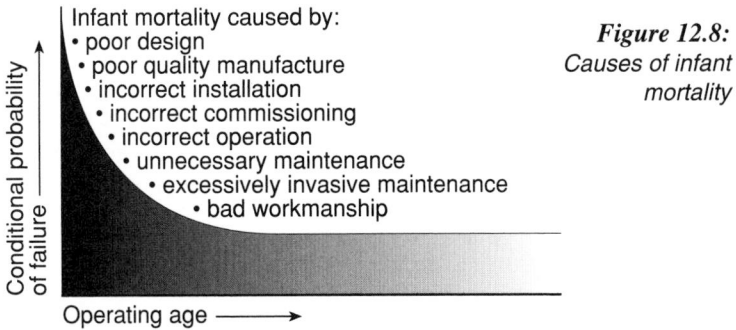

Infant mortality caused by:
- poor design
- poor quality manufacture
- incorrect installation
- incorrect commissioning
- incorrect operation
- unnecessary maintenance
- excessively invasive maintenance
- bad workmanship

Figure 12.8: Causes of infant mortality

Conditional probability of failure

Operating age ⟶

Design

Infant mortality problems attributable to design occur when part of an item is simply incapable of delivering the desired performance, and hence tends to fail soon after being put into service. When they affect an existing asset, these problems can only really be solved by redesign, as discussed in Chapter 9. They can be forestalled to some extent by

- using proven technology. *The author encountered one company which professed to be "in a headlong rush to be second" in adopting new technology, because it found that being first usually means a huge investment in 'de-bugging' new equipment - an involuntary investment made in the form of equipment downtime. On the other hand, being second can be competitively disadvantageous in the long term.*

- using the simplest possible equipment to fulfil the required function *on the premise that bits which aren't there can't fail.*

Manufacture and installation

Infant mortality attributable to equipment manufacture occurs either because the manufacturer's quality standards are too loose, or because the parts concerned have been badly installed. These problems can only be solved by rebuilding the affected assemblies or replacing the affected parts. Two ways to forestall these problems are:

- to implement suitable SQA (Supplier Quality Assurance) and PQA (Project Quality Assurance) schemes. Such schemes usually work best when they are run by someone other than the prime contractor.
- to request extended warranties, perhaps with the full-time on-site support of the vendor's technicians until the equipment has been working as intended for a specified period.

Commissioning
Commissioning problems occur either when equipment is set up incorrectly, or when it is started up incorrectly. These problems are minimized if care is taken to ensure that everyone involved in commissioning knows exactly how the plant is supposed to work, and is given enough time to ensure that it does so.

Routine maintenance
A great deal of infant mortality is caused by routine maintenance tasks which are either unnecessary, or unnecessarily invasive. The latter are tasks which disrupt or disturb the equipment, and so needlessly upset basically stable systems. The way to avoid these problems is to stop doing unnecessary tasks, and in cases where scheduled maintenance is necessary, to select tasks which disturb the equipment as little as possible

Maintenance workmanship
Clearly, if something is badly put together it will fall apart quickly. This problem can only be avoided by ensuring that anyone who is called upon to do a preventive or corrective maintenance task is trained and motivated to do it correctly the first time.

Infant mortality and RCM
The above discussion suggests that infant mortality problems are usually solved by once-off actions rather than by scheduled maintenance (with the exception of a few cases where it may be feasible to use on-condition tasks to anticipate failures). However, despite the minimal role played by *routine* maintenance, using RCM to analyze a new asset before putting it into service still leads to substantial reductions in infant mortality for the following reasons:

- a detailed study of the functions of the asset usually reveals a surprising number of design flaws which, if not corrected, would make it impossible for the asset to function at all

- craftsmen and operators learn exactly how the asset is supposed to function, and so are less inclined to make mistakes which cause failures

- many weaknesses which would otherwise lead to premature failures are identified and dealt with *before* the asset enters service

- routine maintenance is reduced to the essential minimum, which means fewer de-stabilizing interventions, but this essential minimum ensures that the early life of the asset is not plagued by failures which could have been anticipated or prevented.

Failure Pattern A

It is now generally accepted that failure pattern A – the bathtub curve – is really a combination of two or more different failure patterns, one of which embodies infant mortality and the other of which shows increasing probability of failure with age. Some commentators even suggest that the central (flat) portion of the bathtub constitutes a third period of (random) failure between the other two, as shown in Figure 14.9.

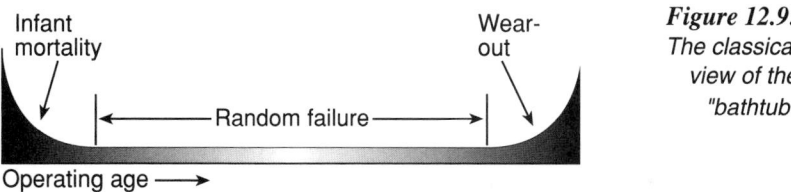

Figure 12.9:
The classical view of the "bathtub"

This means that failure pattern A actually depicts the conditional probability of two or more different failure modes. From the failure management viewpoint, each of these must be identified and dealt with in the light of its own consequences and its own technical characteristics.

Similar conclusions can be drawn about failure pattern B as it is shown in Figure 12.2. This is because the failures which occur between periods 11 and 16 are caused by 'normal' wear, while those occurring between periods 1 and 10 are caused by other 'random' factors which still cause the impeller to wear out, but cause it to do so faster than normal.

This starts to raise a number of questions about the meaning of these patterns, which are considered at length in the next part of this chapter.

12.2 Technical History Data

The Role of Actuarial Analysis in Establishing Maintenance Policies

A surprising number of people believe that effective maintenance policies can only be formulated on the basis of extensive *historical* information *about failure*. Thousands of manual and computerised technical history recording systems have been installed around the world on the basis of this belief. It has also led to great emphasis being placed on the failure patterns described in Part 1 of this chapter. (The fact that the bathtub curve still appears in nearly every significant text on maintenance management is testimony to the almost mystical faith which we place in the relationship between age and failure.)

Yet from the maintenance viewpoint, these patterns are fraught with practical difficulties, conundrums and contradictions. Some of these are summarized below under the following headings:
- complexity
- sample size and evolution
- reporting failure
- the ultimate contradiction.

Complexity
Most industrial undertakings consist of hundreds, if not thousands of different assets. These are made up of dozens of different components, which between them exhibit every extreme and intermediate aspect of reliability behavior. This combination of complexity and diversity means that it is simply not possible to develop a complete analytical description of the reliability characteristics of an entire undertaking – or even any major asset within the undertaking.

Even at the level of individual functional failures, a comprehensive analysis is not easy. This is because many functional failures are caused not by two or three but by two or three *dozen* failure modes. As a result, while it may be fairly easy to chart the incidence of the functional failures, it is a major statistical undertaking to isolate and describe the failure pattern which applies to each of the failure modes which falls within the envelope of each functional failure. What is more, many failure modes have virtually identical physical symptoms, which makes them easy to confuse with each other. This in turn makes sensible actuarial analysis almost impossible.

Sample size and evolution

Large industrial processes usually possess only one or two assets of any one type. They also tend to be brought into operation in series rather than simultaneously. This means that sample sizes tend to be too small for statistical procedures to carry much conviction. For new assets which embody high levels of leading-edge technology, they are always too small.

These assets are also usually in a continuous state of evolution and modification, partly in response to new operational requirements and partly in an attempt to eliminate failures which either have serious consequences or which cost too much to prevent. This means that the amount of time which any asset spends in any one configuration is relatively short.

So actuarial procedures are not much use in these situations because the database is both very small and constantly changing. (As discussed later, the main exception is undertakings which use large numbers of identical components in a more-or-less identical manner.)

Reporting failure

The problem of analyzing failure data is further complicated by differences in reporting policy from one organization to another. One area of confusion is the distinction between potential and functional failures.

For instance, in the tire example discussed on pages 160 and 161, one organization might classify and record the tires as 'failed' when they are removed for retreading after the tread depth drops below 3 mm. However, as long as the tread depth is not allowed to drop below 2 mm, this 'failure' is actually a potential failure as defined in Chapter 6. So other organizations might choose to classify such removals as 'precautionary', because the tires have not actually failed in service, or even as 'scheduled', because the tires are 'scheduled' for replacement at the earliest opportunity after the potential failure has been discovered. In both of the latter cases, it is likely that the removals will not even be reported as failures.

On the other hand, if for some reason the tread depth does drop below 2 mm, then there is no doubt that the tire has failed.

Similar differences might be caused by different performance expectations. Chapter 3 defined a functional failure as the inability of an item to meet a desired standard of performance, and these standards can of course differ for the same asset if the operating context is different.

For instance, page 50 gives the example of a pump which has failed if it is unable to deliver 800 liters per minute in one context and 900 liters per minute in another.

This shows that what is a failure in one organization – or even one part of an organization – might not be a failure in another. This can result in two quite different sets of failure data for two apparently identical items.

Further differences in the presentation and interpretation of failure data can be caused by the different perspectives of the manufacturers and users of an asset. The manufacturer usually considers it his responsibility to provide an asset capable of delivering a warranted level of performance (if there is one) under specific conditions of stress. In other words, he warrants a certain basic design capability, and often makes this conditional upon the performance of certain specified maintenance routines.

On the other hand, we have seen that many failures occur because users operate the equipment beyond its design capabilities (in other words, the 'want' exceeds the 'can', as discussed on pages 61 – 64.) While users are naturally inclined to incorporate data about these failures in their own history records, manufacturers are naturally reluctant to accept responsibility for them. This leads many manufacturers to 'censor' failures caused by operator error from failure data. As Nowlan and Heap[1978] put it, the result is that users talk about what they actually saw, while the manufacturer talks about what they should have seen.

The ultimate contradiction (The Resnikoff Conundrum)

An issue which bedevils the whole question of technical history is the fact that if we are collecting data about failures, it must be because we are not preventing them. The implications of this are summed up most succinctly by Resnikoff[1978] in the following statement:

*"The acquisition of the information thought to be most needed by maintenance policy designers – information about critical failures – is in principle unacceptable and is evidence of the failure of the maintenance program. This is because critical failures entail potential (in some cases, certain) loss of life, **but there is no rate of loss of life which is acceptable to (any) organization as the price of failure information to be used for designing a maintenance policy.** Thus the maintenance policy designer is faced with the problem of creating a maintenance system for which the expected loss of life will be less than one over the planned operational lifetime of the asset. This means that, both in practice and in principle, the policy must be designed without using experiential data which will arise from the failures which the policy is meant to avoid."*

Despite the best efforts of the maintenance policy designer, if a critical failure should happen to occur, Nowlan and Heap[1978] go on to make the following comments about the role of actuarial analysis:

"The development of an age-reliability relationship, as expressed by a curve representing the conditional probability of failure, requires a considerable amount of data. When the failure is one which has serious consequences, this body of data will not exist, since preventive measures must of necessity be taken after the first failure. Thus actuarial analysis cannot be used to establish the age limits of greatest concern – those necessary to protect operating safety."

In this context, note also the comments made on page 139 about safe-life limits and test data. These data are usually so scanty that the safe-life limit (if there is one) is established by dividing the test results by some conservatively large arbitrary factor rather than by the tools of actuarial analysis.

The same limitation applies to failures which have really serious operational consequences. The first time such a failure occurs, immediate decisions are usually made about preventive or corrective action without waiting for the data needed to carry out an actuarial analysis.

All of which brings us to the ultimate contradiction concerning the prevention of failures with serious consequences and historical information about such failures: *that successful preventive maintenance entails preventing the collection of the historical data which we think we need in order to decide what preventive maintenance we ought to be doing.*

This contradiction applies in reverse at the other end of the scale of consequences. Failures with minor consequences tend to be allowed to occur precisely because they don't matter very much. As a result, large quantities of historical data will be available concerning these failures, which means that there will be ample material for accurate actuarial analyses. These may even reveal some age limits. However, because the failures don't matter much, it is highly unlikely that the resulting scheduled restoration or scheduled discard tasks will be cost-effective. So while the actuarial analysis of this information may be precise, it is also likely to be a waste of time.

The chief use of actuarial analysis in maintenance is to study reliability problems on the middle ground, where there is an uncertain relationship between age and failures which have significant economic consequences but no safety consequences. These failures fall into two categories:

- those associated with large numbers of identical items whose functions are to all intents and purposes identical, and whose failure might only have a minor impact when taken singly but whose cumulative effect can be an important cost consideration.

Examples of items which fall into this category are street lights, vehicle components (especially from large fleets) and many of the components used by the armed forces and in the electricity, water and gas *distribution* industries. Items of this type are used in sufficient numbers for precise actuarial analyses to be carried out, and detailed cost-benefit studies are justified (in many cases, if only to minimize the amount of travelling involved in maintaining the items).

- the second category of failures which merit actuarial investigation are those which are less common but are still thought to be age-related, and where both the cost of any preventive task and the cost of the failure are very high. As mentioned on page 134, this applies especially to gradually increasing failure probabilities typified by failure pattern C.

The way forward
The above paragraphs indicate that except for a limited number of fairly specialized situations, the actuarial analysis of the relationship between operating age and failure is of very little use from the maintenance management viewpoint. Perhaps the most serious shortcoming of historical information is that it is rooted in the past, whereas the concepts of anticipation and prevention are necessarily focused on the future.

So a fresh approach to this issue is needed – one which switches the focus from the past to the future. In fact, RCM is just such an approach. Firstly, it deals with the specific problems identified above as follows:

- *defining failure:* by starting with the definition of the functions and the associated performance standards of each asset, RCM enables us to define with great precision what we mean by 'failed'. By distinguishing clearly between built-in capability and desired performance, and between potential failures (the *failing* state) and functional failures (the *failed* state), it eliminates further confusion.

- *complexity:* RCM breaks each asset down into its functions and each function into functional failures, and only then identifies the failure modes which cause each functional failure. This provides an orderly framework within which to consider each failure mode. This in turn makes them much easier to manage than if we were to start out at the failure mode level (which is the starting point of most classical FMEA's and FMECA's).

- *evolution:* by providing a comprehensive record of all the performance standards, functional failures and failure modes associated with each asset, RCM makes it possible to work out very quickly how any change to the design or to the operating context is likely to affect the asset, and to revise maintenance policies and procedures only in those areas where changes need to be made

• *the ultimate contradiction:* RCM deals with the ultimate contradiction in several ways. Firstly, by obliging us to complete the Information Worksheet described in Chapter 4, it focuses attention on *what could happen.* Contrast this with the actuarial emphasis on *what has happened.* Secondly, by asking how, and how much, each failure matters as set out in Chapter 5, it ensures that we focus on failures which have serious consequences and that we do not waste time on those that don't. Finally, by adopting the structured approach to the selection of proactive tasks and default actions described in Chapters 6 – 9, RCM ensures that we do what is necessary to prevent serious failures from happening, and as far as humanly possible, avoid having to analyze them historically at all.

Secondly, the RCM process focuses attention on the information needed to support specific decisions. It does not ask us to collect a whole lot of data in the hope that they will eventually tell us something. This point is discussed in more detail in the next section of this chapter.

Specific Uses of Data in Formulating Maintenance Policies

In spite of all the above comments, the successful application of RCM needs a great deal of information. As explained at length in Chapters 2 to 9, much of this information is descriptive or qualitative, particularly on the RCM Information Worksheet. However, in view of the emphasis that has been placed on quantitative issues in this chapter, Table 12.1 summarizes the principal types of quantitative data which are used to support different stages of the maintenance decision process. It does so under the following headings:

• *datum:* the piece of information of interest.

• *application:* a very brief summary of the use to which each datum is put. Note that some are used in conjunction with others to reach a final decision, and that many are only used when qualitative data are not strong enough to make an intuitive decision possible.

• *comments:* where each datum is most likely to be found. Note that in some cases, they are established by the user of the asset.

• *pages:* refers to the pages in this book where the use of each datum is discussed at greater length.

DATUM	APPLICATION	COMMENTS	PAGES
Desired standards of performance	These standards define the objectives of maintenance for each asset. They cover output, product quality/customer service, energy efficiency, safety and environmental integrity	Set by the users of the assets (and by regulators for environmental and some safety standards)	22 - 27 47 - 52
ASSESSING OPERATIONAL AND NON-OPERATIONAL FAILURE CONSEQUENCES			
Downtime	Assessing whether each failure will affect production/operations, and if so how much	Not the same thing as MTTR ('mean time to repair')	76
Cost of lost production	Used together with downtime to evaluate total cost of each failure which affects operations	Only needed when the cost-benefit of scheduled maintenance is not intuitively obvious	105 - 106
Cost of repair	Used together with MTBF to evaluate cost effectiveness of scheduled maintenance	Only needed when the cost-benefit of scheduled maintenance is not intuitively obvious (Operational and non-operational consequences only)	108 - 109
Mean time between failures	Used with downtime, cost of lost production (if any) & repair cost to compare cost of scheduled maintenance with the cost of a failure over a period of time	Only needed when the cost-benefit of scheduled maintenance is not intuitively obvious (Operational and non-operational consequences only)	105 - 106 108 - 109
ASSESSING SAFETY AND ENVIRONMENTAL FAILURE CONSEQUENCES			
Tolerable risk of a single failure	Used to assess whether scheduled maintenance is worth doing for failures which could have a direct adverse effect on safety or the environment	Almost always assessed by the users of the assets/likely victims on an intuitive basis	98 - 101
ESTABLISHING ON-CONDITION TASK FREQUENCIES			
Potential failure	Point at which imminent failure becomes detectable	Based on the nature of the P-F curve and the monitoring technique: usually quantified for performance monitoring, condition monitoring and SPC	144 -145
P-F interval	Used to establish the frequency of on-condition tasks	'How quickly it fails': very seldom formally recorded	145 - 149 162 - 165

Table 12.1: Summary of key maintenance decision-support data

DATUM	APPLICATION	COMMENTS	PAGES
SCHEDULED RESTORATION AND SCHEDULED DISCARD TASK FREQUENCIES			
Age at which there is a rapid increase in the conditional probability of failure	Used to establish the frequency of most scheduled restoration and scheduled discard tasks	'Useful life': Based on formal records if these are available: more often based on consensus of people who have the most knowledge of the asset	236 - 238
Actuarial analysis of relationship between age and failure	Optimizing restoration/discard intervals for large numbers of identical parts whose failure is known to be age-related, or for expensive pattern C-type failures	Worth doing for no more than 1 - 2% of failure modes in most industries: needs extensive and reliable historical data: used for failure modes which have operational and non-operational consequences only	253 - 254
HIDDEN FAILURE CONSEQUENCES AND FAILURE-FINDING TASK FREQUENCIES			
Tolerable probability of a multiple failure	Used to establish maintenance policies for protected systems	Set by the users of the asset: only used when a rigorous analysis is to be done	118 - 120 179 - 182
Mean time between failures of a protected function (M_{TED})	Used together with 'tolerable probability of a multiple failure' to determine the desired availability of a protective device	Based on past and anticipated future performance of the protected function: Only used to support a rigorous analysis – not needed for the intuitive approach *(see below)*	179 - 182
Desired availability of a protective device	Used together with the MTBF of the protective device to establish a failure-finding task interval	Derived from the two above variables if the task frequency is to be derived on a rigorous basis: otherwise set directly by the users of the asset on the basis of an intuitive assessment of the risks of the multiple failure	118 175 - 179
Mean time between failures of a protective device (M_{TIVE})	Used with desired availability to establish a failure-finding task interval	Based on records of *failures found* if these are available: if not, any suitable data source should be used to begin with (including educated guesses) but a suitable database should be started immediately	175 - 182

A number of final points concerning quantitative data are reviewed under the following headings:
* management information
* a note on the MTBF
* technical history.

Management information

Table 12.1 only describes data which are used directly to formulate policies designed to deal with specific failure modes. It does not include data used to track the overall performance of the maintenance function and usually classified as 'management information'. Examples of such information are plant availability statistics, safety statistics and information about expenditure on maintenance against budgets.

Monitoring the overall performance of the maintenance function is of course an essential aspect of maintenance management. This topic is discussed in more detail in Chapter 14.

A note on the MTBF

In recent times, the concept of the 'mean time between failures' seems to have acquired a stature which is quite disproportionate to its real value in maintenance decision-making. For instance, it has nothing to do with the *frequency* of on-condition tasks, and nothing to do with the *frequency* of scheduled restoration and scheduled discard tasks. However, it does have certain very specific uses. Table 12.1 mentions three of these:

* to establish the *frequency* of failure-finding tasks.

* to help decide whether scheduled maintenance is *worth doing* in the case of failure modes which have operational or non-operational consequences only. (In other words it helps us to decide *whether* such tasks need to be done, but not *how often* they need to be done.)

* to help establish the *desired availability* of a protective device.

In the first case, the MTBF is always needed to make the appropriate decision, but in the second two it is only used if the nature and consequences of the failures are such that a rigorous analysis must be carried out.

The MTBF also has a number of uses outside the field of maintenance policy formulation, as follows:

* *in the field of design:* to carry out a detailed cost-justification of a proposed modification, as mentioned briefly on page 195

- *in the field of procurement:* to evaluate the reliability of two different components which are candidates for the same application, as mentioned on page 241
- *in the field of management information:* as discussed in Chapter 14, one way to assess the overall effectiveness of a maintenance program is to track the mean time between unanticipated failures of any asset.

A detailed exploration of the first two of these issues is beyond the scope of this book. The third is dealt with in chapter 14.

Technical history
Together with the above comments about the MTBF, Table 12.1 can be used to help decide what sorts of data really need to be recorded in a technical history recording system.

Perhaps the most important information which needs to be recorded on a formal basis is **what is found** *each time a failure-finding task is done.* Specifically, we need to record whether the item was found to be fully functional or whether it was in a failed state. Such records enable us to determine the mean time between failures of the protective device (M_{TIVE} on page 177), and hence to check the validity of the associated failure-finding task interval. *This information should be recorded for **all** hidden functions – in other words, for all protective devices which are not fail-safe.*

In addition to hidden failures, Table 12.1 identifies two further areas where historical failure data can be used to make (or to validate) decisions about maintenance policies, as follows:

- *the occurrence of failure modes which have significant operational consequences.* This information can be used to compute the mean time between the failures in order to assess the cost effectiveness of scheduled maintenance. However, as mentioned in Table 12.1, this only needs to be done if the cost-benefit of proactive action is not intuitively obvious. If it is, such action – be it scheduled maintenance or redesign – would be taken and so there should be no more failures to record (except perhaps as potential failures if the proactive action is an on-condition task).

Table 12.1 mentions that in rare cases, it may also be worth capturing these data in order to carry out full actuarial analyses with a view to optimizing scheduled restoration and scheduled discard frequencies.

- *the mean time between failures of a protected function (M$_{TED}$ on page 179)*. This is needed if a failure-finding interval is to be determined on a rigorous basis. It can be determined by recording the number of times a protective device is called upon to function by the failure of the protected function.

For instance, a record can be made every time the overpressurization of a boiler causes a relief valve to start passing.

If any of these data are to be captured, the failure reporting systems should be designed to identify the datum which is required – usually the *failure mode* – as precisely as possible. This can be done by asking the person who does the task (or who discovers the failure in the case of failure-finding) either to:

- complete a suitably designed form which is then used to enter the data into a manual or computerised history recording system, or

- enter the data directly if an on-line computer system is used to store it.

In most organizations, the records themselves can be stored in:

- a simple proprietary PC-based database, or

- a specialized computerised or manual maintenance history recording system.

The design of such systems is also beyond the scope of this book. However, Table 12.1 suggests that if technical history recording systems are used to capture specific data for specific reasons, rather than to record everything in the hope that it will eventually tell us something, they become useful and powerful contributors to the practice of maintenance management rather than the expensive white elephants that so many of them tend to be.

13 Applying the RCM Process

13.1 Who Knows?

The seven basic questions which make up the RCM process have been considered at length in Chapters 2 to 10. After looking more deeply at the information needed to answer the questions, Chapter 12 concluded that in most industries, historical records are seldom (if ever) comprehensive enough to be used for this purpose on their own. Yet the questions must still be answered, so the required information still has to be obtained from somewhere.

More often than not, 'somewhere' actually turns out to be 'someone' – someone who has intimate knowledge and experience of the asset under consideration. There are also occasions when the information-gathering process reveals widely differing viewpoints which have to be reconciled before decisions can be made.

Later sections of this chapter describe how small groups can be used to gather the information, reconcile differing views and make the decisions. However, before considering these groups, this part of this chapter reviews the information needed to answer each question, and considers who is most likely to possess it. It does so with reference to earlier sections of the book where the questions have been discussed in detail.

- *What are the functions and associated performance standards of the asset in its present operating context?*
RCM is based on the premise that every asset is acquired to fulfil a specific function or functions, and that maintenance means doing whatever is necessary to ensure that it continues to perform each function to the satisfaction of its users. In most cases, the most important representatives of the users are *operations and production managers*. In order to ensure that RCM generates a maintenance program which delivers what these managers want, they need to participate actively in the entire process. (In areas such as safety, hygiene or the environment, the advice of appropriate *specialists* may also be needed.)

However, we have also seen that the built-in capability of the asset – what it *can* do – is the most that maintenance can actually deliver. *Maintenance and design people, often at supervisory levels,* tend to be the custodians of this information, so they too are a key part of this process.

If this information is shared at a single forum, maintainers begin to appreciate much more clearly what operators are trying to achieve, while users gain a clearer understanding of what maintenance can – and cannot – deliver.

• *In what ways does it fail to fulfil its functions?*

The example on page 51 showed why it is essential that the performance standards used to judge functional failures should be set by *maintenance and operations people* working together.

• *What causes each functional failure?*

Chapter 4 explained how maintenance is really managed at the failure mode level. It went on to stress the importance of identifying the causes of each functional failure. The example on page 72 showed how these causes are often most clearly understood by the shop-floor and supervisory people who work most closely with each machine (especially the *craftsmen and technicians* who have to diagnose and repair each failure). In the case of new equipment, a valuable source of information about what can fail is a *field technician who is employed by the vendor* and who has worked on the same or similar equipment.

• *What happens when each failure occurs?*

Part 5 of Chapter 4 lists a wide variety of information which needs to be recorded as failure effects. These include:

- the evidence that the failure has occurred, which is most often obtained from the *operators* of the equipment
- the amount of time the machine is usually out of action each time the failure occurs, again obtained from *operators or first-line supervisors*
- the hazards associated with each failure, which may need *specialist advice* (especially concerning such issues as the toxicity and flammability of chemicals, or the hazards associated with mechanical items such as pressure vessels, lifting equipment and large rotating components)
- what must be done to repair the failure, which is usually obtained from the *craftsmen or technicians* who carry out the repairs.

- *In what way does each failure matter?*
Failure consequences are discussed at length in Chapter 5 and summarized in the four questions at the head of Figure 10.1 on pages 200/201. The assessment of failure consequences can only be done in close consultation with production/operations people, for the following reasons:

- *safety and environmental consequences:* if the effects of a failure mode are explained reasonably thoroughly, it is usually quite easy to assess whether it is likely to affect safety or the environment. The main difficulty in this area lies in deciding what level of risk is acceptable. The discussion about who should evaluate risk on page 101 suggests that this decision should be made by a group consisting of the *likely victims of the failure, the people who would bear the responsibility if it were to occur, and if necessary, an expert on the specific characteristics of the failure.*

- *hidden failures:* The analysis of hidden functions requires at least four items of information, especially if a rigorous approach is used to determine failure-finding task intervals (see Chapter 8). This information is summarized below:
  - *evidence of failure:* the first question on the RCM Decision Diagram asks if the loss of function caused by this failure mode on its own *will become evident* to the operating crew under normal circumstances. This question can only be answered with assurance by consulting *the operating crew* concerned.
  - *normal circumstances:* as explained on page 126, different people can attach quite different meanings to the term 'normal' in the same situation, so it is wise to ask this question in the presence of the *operators and their supervisors.*
  - *acceptable probability of a multiple failure:* this should also be established by the group discussed on page 101
  - *the mean time between failures of a protected function:* this is needed if the desired availability of a protected device is to be determined on a rigorous basis. If this information has not been recorded in the past, it can sometimes be obtained by asking the *operators* of the equipment how often the protective device is called upon to operate by the failure of the protected function.

- *operational consequences:* a failure has operational consequences if it affects output, product quality or customer service, or if it leads to an increase in costs other than the direct costs of repair. Clearly, the people who are in the best position to assess these consequences are *operations managers and supervisors*, perhaps with help from *cost accountants*.

- *non-operational consequences:* the people who are usually in the best position to assess direct repair costs are *first- and second-line maintenance supervisors*

• What can be done to predict or prevent each failure?

The information needed to assess the technical feasibility of different types of proactive tasks was discussed in Chapters 6 to 9, and the key questions are summarized on page 205. If clear actuarial data are not available to provide answers, then the questions must again be answered on the basis of judgement and experience, as follows:

- *on-condition tasks:* pages 154 and 155 stressed how important it is to consider as many different potential failures as possible when seeking on-condition tasks. The monitoring possibilities range from sophisticated condition monitoring techniques through product quality and primary effects monitoring to the human senses, so we should consult *operators, craftsmen, supervisors* and, if necessary, *specialists in the different techniques*.

   A similar group would need to consider the duration and consistency of the associated P-F intervals, as explained on pages 164 and 165.

   The amount of time needed to avoid the consequences of the failure (in other words, the nett P-F interval) is established jointly by *maintenance and operations supervisors*

- *scheduled restoration and scheduled discard:* in the absence of suitable historical data, the people who are usually most likely to know whether any failure mode is age-related, and if so whether and when there is a point at which there is a rapid increase in the conditional probability of failure, are again the *operators, craftsmen and supervisors* who are closest to the asset.

   Whether it is possible to restore the original resistance to failure of the asset is usually decided by *maintenance supervisors* or in doubtful cases, by *technical specialists*.

• *What if a suitable proactive task cannot be found?*
The two default actions which need active consideration are failure-finding tasks and redesign:

• *failure-finding:* If the frequency of a failure-finding task is to be established without performing a rigorous analysis of the protected system, the desired availability of the protective device should be determined by a *group of the sort described on page 101.*

In the absence of formal records, the MTBF of the protective device can be derived initially either by asking the manufacturer of the device for this information, or by asking anyone who might have done any functional checks in the past what they found when they did the checks. As mentioned on page 183, this is usually an *operator or maintainer.*

Maintenance craftsmen and supervisors are usually the people who are best qualified to assess whether it is possible to do a failure-finding task in accordance with the criteria set out on page 185.

• *redesign:* the question of redesign is discussed at length in Chapter 9. Note that the formal RCM process is only meant to identify situations where redesign is either compulsory or desirable. RCM review groups should not attempt to develop new designs during RCM meetings for two reasons:

- the design process requires skills which are usually not present at an RCM forum.

- done properly, developing even one new design takes a great deal of time. If this time is spent during RCM review meetings, it slows down and can even paralyze the rest of the program. (This is not to suggest that designers should not consult the users and maintainers of the assets – just that it should not be done as part of the RCM review process.)

The above paragraphs demonstrate that it is impossible for one person, or even for a group of people from one department, to apply the RCM process on their own. The diversity of the information which is needed and the diversity of the people from whom it must be sought mean that it can only be done on the basis of extensive consultation and cooperation, especially between production/operations and maintenance people. The most efficient way to organize this is to arrange for the key people to apply the process in small groups.

13.2 RCM Review Groups

In the light of the issues raised in Part 1 of this chapter, we now consider who should participate in a typical RCM review group, what each group actually does, and what the participants get out of this process.

Who should participate
The people mentioned most frequently in Part 1 of this chapter were first-line supervisors, operators and craftsmen. This suggests that a typical RCM review group should include the people shown in Figure 13.1.

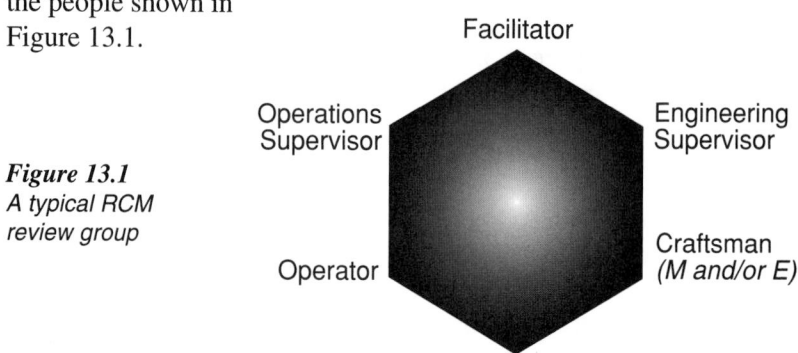

Facilitator

Operations Supervisor

Engineering Supervisor

Figure 13.1
A typical RCM review group

Operator

Craftsman *(M and/or E)*

External Specialist (if needed) *(Technical or Process)*

In practice, the places on every group do not have to be filled by exactly the same people as shown in Figure 13.1. The objective is to assemble a group which can provide most if not all of the information described in Part 1 of this chapter. These are the people who have the most extensive knowledge and experience of the asset and of the process of which it forms part. To ensure that all the different viewpoints are taken into account, this group should include a cross-section of users and main-tainers, and a cross-section of the people who do the tasks and the people who manage them. In general, it should consist of not less than four and not more than seven people, the ideal being five or six.

The group should consist of the same individuals throughout the analysis of any one asset. If the faces present at each meeting change, too much time is lost going over ground which has already been covered for the benefit of the newcomers.

As suggested in Part 1 of this chapter, 'specialists' can be specialists in any of the following:

- some aspect of the process. These usually tend to be dangerous or environmentally sensitive issues.
- a particular failure mechanism, such as fatigue or corrosion.
- a specific type of equipment, such as hydraulic systems.
- some aspect of maintenance technology, such as vibration analysis or thermography.

Unlike other group members, specialists only need to attend meetings at which their speciality is under discussion.

What each group does
The objective of each group is to use the RCM process to determine the maintenance requirements of a specific asset or a discrete part of a process. Under the guidance of a facilitator, the group analyzes the context in which the asset is operating, and then completes the RCM Information Worksheet as explained in Chapters 2 to 4. (The actual writing is done by the facilitator, so the group members do not have to handle any paper if they don't wish to.) They then use the RCM Decision Diagram shown on pages 200 and 201 to decide how to deal with each of the failure modes listed on the Information Worksheet. Their decisions are recorded on RCM Decision Worksheets as explained in Chapter 10.

The watchword throughout this process is *consensus*. Each group member is encouraged to contribute whatever he or she can at each stage in the process, as shown in Figure 13.2. Nothing should be recorded until it has been accepted by the whole group. (As discussed in Part 3 of this chapter, the facilitator has a crucial role to play in this aspect of the process.)

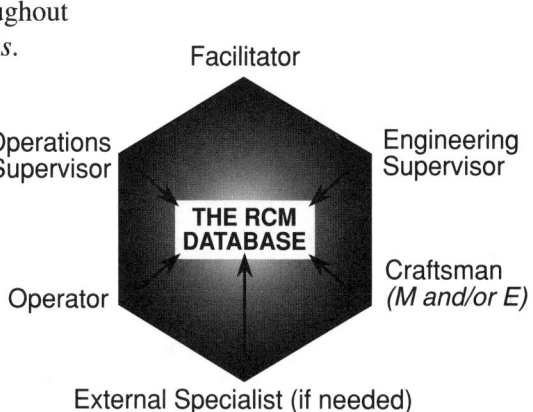

Facilitator

Operations Supervisor

Engineering Supervisor

THE RCM DATABASE

Operator

Craftsman
(M and/or E)

External Specialist (if needed)
(Technical or Process)

Figure 13.2
The flow of information into the RCM database

This work is done at a series of meetings which last for about three hours each, and each group meets at an average rate of anywhere from one to five times per week. If the group includes shift workers, the meetings need to be planned with special care.

The asset should be subdivided and allocated to groups in such a way that any one group can complete the entire process in not less that five and not more than fifteen meetings – certainly no more than twenty.

What participants get out of the process
The flow of information which takes place at these meetings is not only *into* the database. When any one member of the group makes a contribution, the others immediately learn three things:

• more about the asset, more about the process of which it forms part and more about what must be done to keep it working. As a result, instead of having five or six people who each know a bit – often a surprisingly little bit – about the asset under review, the organization gains five or six experts on the subject.

• more about the objectives and goals of their colleagues. In particular, maintenance people learn more about what their production colleagues are trying to achieve, while operations people learn much more about how maintenance can – and cannot – help them to achieve it.

• more about the individual strengths and weaknesses of each team member. On balance, much more tends to be learned about strengths than weaknesses, which has a salutary effect on mutual respect as well as mutual understanding.

In short, participants in this process gain a much better understanding of
• what each group member (themselves included) should be doing
• what the group is trying to achieve by doing it and
• how well each group member is equipped to make the attempt.
This changes the group from a collection of highly disparate individuals from two notoriously adversarial disciplines (operations and maintenance) into a team.

The fact that they have each played a part in defining the problems and identifying solutions also leads to a much greater sense of ownership on the part of the participants. For instance, operators start talking about 'their' machines, while maintenance people are much more inclined to offer constructive criticism of 'their' schedules.

This process has been described as 'simultaneous learning', because the participants identify what they need to learn at the same time as they learn it. (This is much quicker than the traditional approach to training, which starts with a training needs analysis, proceeds through the development of a training program and ends with the presentation of training courses – a process which can take months.)

One limitation of group learning in this fashion is that unless specific steps are taken to disseminate the information further, the only people who benefit directly are the members of each group. Two ways to overcome this problem are as follows:

• to ensure that anyone in the organization can gain access to the RCM database at any time

• to use the output of the RCM process to develop formal training courses.

The RCM meetings also provide a very efficient forum for key people to learn how to operate and maintain *new* equipment, especially if one of the vendor's field technicians attends meetings held during the final stages of commissioning. The RCM process provides a framework for such technicians to transfer everything they know about the asset to the other group members in an orderly and systematic fashion. The RCM worksheets enable the organization to capture the information in writing for dissemination to anyone else who needs to know.

13.3 Facilitators

Part 2 of this chapter mentioned that the facilitator has a crucial role to play in the implementation of RCM. The primary function of an RCM facilitator is to facilitate the application of the RCM philosophy by asking questions of a group of people chosen for their knowledge of a specific asset or process, ensuring that the group reaches consensus about the answers, and recording the answers.

Of all the factors which affect the ultimate quality of the analysis, the skill of facilitator is the most important. This applies both to the *technical* quality of the analysis, and to:
• the pace at which the analysis is completed
• the attitude of the participants towards the RCM process.
To achieve a reasonable standard, an RCM facilitator has to be competent in 45 key areas. These can be divided into 5 main skillsets, as follows:

- applying the RCM logic
- managing the analysis
- conducting the meetings
- time management
- administration, logistics and managing upwards.

Key points about each skillset are discussed in the following paragraphs.

Applying the RCM Logic

The facilitator must ensure that the RCM process is applied correctly by the review group. This entails ensuring that all the questions embodied in the RCM process are asked correctly in the correct sequence, that they have been correctly understood by all the group members and that the group reaches consensus about the answers.

Managing the Analysis

By and large, the following decisions are made by the facilitator and/or the facilitator alone does the work.

- *Prepare for meetings:* Prior to the first meeting, the facilitator should collect basic information about the asset/process. This includes flow diagrams, operating manuals, history records – if any – and electrical, hydraulic and pneumatic circuit drawings.

- *Select levels of analysis/define boundaries:* The equipment to be analyzed by each review group will be identified during the planning phase. However, it may become necessary to group the equipment differently in order to carry out a sensible analysis. This means that the final decision about equipment grouping/levels of analysis is made by the facilitator, who then has to define the boundaries of the analysis accordingly.

- *Handle complex failure modes appropriately:* Decide when to choose which of the four options listed in Chapter 4 part 7 (pages 86 - 88) when listing failure modes

- *Know when to stop listing failure modes:* Knowing when to stop listing the failure modes that might cause each functional failure is one of the key elements of successful facilitating, and requires careful judgment. Moving on too soon to the next functional failure means that critical failure modes may be overlooked or that failure effects are inadequately described. Listing too many failure modes leads to analysis paralysis.

- *Interpret and record decisions with a minimum of jargon:* As a rule, the facilitator physically records the decisions of the group. In so doing, care must be taken to ensure that all technical terms used will be understood by everyone on the site (including auditors, design engineers and senior managers).

- *Recognize when the group doesn't know:* The facilitator has to distinguish between uncertainty (the group is not 100% sure, but sure enough to make a viable decision) and ignorance (the group simply doesn't know enough to make a viable decision).

- *Curtail attempts to redesign the asset in RCM meetings:* Attempts to redesign the asset is the biggest single time waster in RCM review meetings. The facilitator should simply note that redesign is compulsory/may be desirable, and may jot down a suggestion if the answer seems obvious. The redesign process itself should be carried out elsewhere (This is not to say that the RCM group cannot get involved in the redesign process - they should – it simply means that they shouldn't do so in the RCM meeting.)

- *Complete the RCM Worksheets:* Whether they are stored manually or electronically, the RCM Information and Decision Worksheets should be completed in a way which is clear and readable. Abbreviations should be avoided, and they should contain a reasonable minimum of spelling mistakes and grammatical errors.

- *Prepare an audit file:* As discussed in Chapter 11, managers with overall responsibility for each asset need to audit the analyses carried out by the review groups. Before this can be done, the facilitator needs to prepare the RCM Worksheets in a clear, coherent fashion. This usually entails binding them into a formal document called the audit file. This file should also contain enough background information – schematic drawings, known failure data, even photographs of the equipment – to enable the auditors to do their job properly.

- *Enter RCM data into computerised database:* This is done either by a typist or by the facilitator. Exactly who depends on the keyboard skills and computer literacy of and the amount of time available to the facilitator. (The data should only be entered directly into a computer during the meetings if the facilitator can type at least as fast as he or she can write, and if what is being typed can be displayed in a way which can be read easily and immediately by every group member. As discussed in Part 8 of this chapter, the computer should never be used to 'ask the questions'.)

Conducting the Meetings

The following points deal with the way in which the facilitator interacts with participants at meetings on a purely human level.

- *Set the scene*: At the first meeting of each group, the facilitator must agree basic meeting norms with the group (issues such as use of names, dress, punctuality, etc.) and ensure that every group member understands the scope and objectives of the exercise and why he or she has been asked to take part. At the start of all subsequent meetings, the facilitator should briefly recap what has been done to date and provide a brief agenda for this meeting. The facilitator should also ensure that the group has enough materials (drafts of completed worksheets, etc) to enable them to keep track of the process.

- *The conduct of the facilitator:* How the facilitator conducts him or herself in meetings has a profound effect on the way the other group members behave. In particular, the facilitator should set a good example by displaying a positive attitude to the process, take care to preserve the dignity of group members and provide positive feedback in response to positive contributions.

- *Ask the RCM questions in order:* Once the meetings are under way, the key role of the facilitator is to ask the questions required by the RCM process. It is essential to avoid any tendency to skip questions or to take answers for granted. (In particular, take care not to ignore or overlook questions designed to establish whether any task is worth doing.)

- *Ensure that each question has been correctly understood:* In spite of the fact that they should all have attended a basic RCM training course, group members are not as familiar with the RCM process as the facilitator. As a result, they often misunderstand the questions, especially in the early stages, and the facilitator must be alert to such misunderstandings. Common mistakes were discussed in Chapter 11 part 2.

- *Encourage everyone to participate:* Everyone who has something to contribute should do so. This entails encouraging reticent people to participate, while ensuring that dominant personalities do not take over the meetings to the exclusion of everyone else. Interest can be sustained and participation encouraged by asking group members to do small tasks between meetings such as clarifying technical points (perhaps by calling a vendor, measuring a dimension, checking out a quality standard etc).

- *Answering the questions:* Facilitators should avoid what often becomes a strong temptation to answer the RCM questions directly. However, it is legitimate to clarify doubtful answers by further questioning.

- *Secure consensus:* One of the most important functions of the facilitator is to ensure that the group reaches consensus. Consensus does not mean that decisions are made by casting a vote. It also does not mean that everyone must agree completely with every decision. It does mean that everyone is prepared to accept the majority view. (If a group simply cannot reach consensus, the facilitator should ask someone whose expertise is respected by all the group to counsel them further, and if necessary to make the final judgment.)

- *Motivate the group:* As discussed above, one of the most important factors which affects the attitude of the group is the attitude of the facilitator. Other motivational issues which the facilitator may need to deal with are waning enthusiasm, especially if a large number of meetings is needed to review a big asset, and scepticism, where group members don't believe that their recommendations will be taken seriously by management.

- *Manage disruptions appropriately:* All meetings occasionally suffer from disruptions. However in the case of RCM, the group is trying to do a great deal of work which requires intense concentration, so interruptions can be especially unwelcome. Three areas that usually need special care are digressions, personality clashes and grievances which are not related to the RCM process.

- *Coach the group or individual members:* It is sometimes necessary for the facilitator to provide formal coaching to individuals or to the group as a whole in some element of the RCM philosophy. However, coaching is inefficient and time-consuming, so it should not be seen as a substitute for formal training in RCM.

Time Management

RCM is a resource intensive process – sufficiently so for management at all levels to be concerned about the amount of time and effort it takes to complete each analysis. Both the resources required to apply RCM and the duration of each project are profoundly affected by the pace at which facilitators conduct meetings and the way they manage their time outside meetings. As a result, facilitators need to develop their time management skills every bit as much as their skills in any other aspect of RCM.

Five overall key measures of time management effectiveness include:

• *Pace of working:* A number of people are present at each RCM meeting, so the amount of time spent in these meetings has the greatest impact on the total number of man-hours spent on the RCM process. Slow progress at meetings also means that more meetings need to be held, which could delay the project completion date. As a result, this is the most important of the five measures of time effectiveness.

• *Total number of meetings held:* The total number of meetings needed to perform a complete analysis should be estimated as part of the RCM project planning phase. A second measure of time effectiveness is to compare the actual number of meetings held with this estimate. However, estimates can themselves be inaccurate, so it is usually acceptable for a facilitator to complete any one analysis within 20% of the estimated number of meetings (with due allowance for the learning process in the case of new facilitators)

• *Actual completion date versus target completion date:* The completion date of each set of meetings should also be determined during the RCM project planning phase. The facilitator should go to great lengths to achieve this date. Completion of the meetings is usually delayed either because the number of meetings required exceeds the estimate or because meetings are not held as planned. If either of these problems occurs, every effort should be made to recover lost ground, if necessary by scheduling extra meetings

• *Time spent preparing for audit:* As explained earlier, the facilitator needs to prepare an RCM audit file after the meetings have been completed. Since recommendations cannot be implemented until they have also been audited, this step should also be carried out as quickly as possible. An experienced facilitator should be able to have an analysis ready for final audit no more than two weeks after the last meeting of the review group.

• *Time outside meetings:* Facilitators are also scarce and expensive resources, so they owe it to themselves and to their employers to use their own time as effectively as possible. In the RCM context, this means that the amount of time facilitators spend on administrative work outside meetings should be about the same as the time spent in the meetings themselves.

Administration, Logistics and Managing Upwards

This part of this chapter deals with activities where the facilitator interacts with people (usually managers) who are not members of review groups. These interactions involve making decisions, providing information or getting work done. Who actually does each task may vary from place to place, but regardless of who is supposed to do it, the facilitator still plays a major part in ensuring that it actually gets done. As a result, facilitators tend to be judged on progress in these areas as much as in any other:

- *Set up the RCM project as a whole:* This consists of the following steps:
  - decide which assets (or which parts of which assets) are to be analyzed using the RCM process
  - establish the objectives of each analysis, and agree when and how their achievement is to be measured
  - estimate how many RCM meetings will be needed to review each asset
  - decide how the assets are to be divided among different review groups
  - decide who will audit each analysis.

  These steps are usually carried out in close consultation with the RCM project manager and the asset manager. If RCM is new to the business unit, this phase also tends to be done with assistance from experienced consultants (especially in estimating the numbers of meetings)

- *Plan the project:* Before starting each analysis, each of the following must be planned in detail:
  - decide who is going to participate in each review group
  - arrange training in RCM for group members and auditors who have not yet been trained
  - decide when, where and at what time every meeting is to be held
  - decide when the analysis will be audited
  - decide when to hold the top management presentation.

  These steps are also usually carried out in consultation with the RCM project manager and the asset manager.

- *Communicate the plans:* Participants and their bosses should receive written notice of initial plans for training courses and meetings. Any subsequent revisions to these plans should also be communicated in good time. Auditors need to be reminded about forthcoming audits. Once the meetings are under way, the RCM project manager should ensure that people actually attend planned meetings. Attendance norms should be clearly defined, well publicized and strictly adhered to.

- *The meeting venue:* An RCM meeting room should be big enough for people to sit around a table without touching each other, and it should be reasonably close to the group members' normal workplace. It should also be quiet, reasonably secluded, well lit and adequately ventilated. It should not be interrupted by phone calls or pagers. A flip chart or white board are usually essential. Whether or not refreshments are provided at meetings depends on organizational norms.

- *Communicate urgent findings:* Appropriate managers should be told before the audit about findings or recommendations that may be of special interest to them, or which may need urgent attention (such as serious safety or environment hazards.) This ensures that potentially dangerous problems are dealt with quickly, and also helps to sustain the interest of the people who are providing the resources for the project.

- *Communicate progress:* Keep management informed about progress against plan. Bring to their attention problems which you cannot solve yourself and which are impeding or threaten to impede progress, such as sustained absenteeism from meetings, seriously counterproductive behavior, excessive interruptions, etc.

- *Ensure that RCM worksheets are audited:* The facilitator should usually attend audit meetings in person, to answer queries, note corrections and (if required) to provide guidance to the auditors on the RCM process (although the auditors must undergo formal training in RCM before attempting to audit an RCM analysis). The facilitator must also ensure that consensus is achieved between the auditors and the review group during the audit process. This entails reporting audit findings back to the group, and ensuring that differences are resolved. Finally, the facilitator must update the worksheets to incorporate the results of the audit.

- *Top management presentation:* A short, high-quality summary of at least one major RCM analysis should be presented to the senior managers of each business unit in which the process is applied. It should show how the initial objectives of the analysis have been or will be achieved, and what had to be done to achieve them.

- *Implementation:* Ensuring that RCM decisions are implemented is usually the overall responsibility of the asset manager, although the facilitator will need to remain involved. The key elements of the implementation process were discussed in Chapter 11.

• *A living program:* After completing each analysis, the facilitator should work with the RCM project manager and the asset managers to set up meetings to reappraise and where necessary update the analysis. These meetings should be held at intervals of nine to twelve months, and ideally should be facilitated by the original facilitator. This issue is discussed in more detail in Part 5 of this chapter

Who should facilitate

Facilitators should have a strong technological background, should be highly methodical and be natural consensus builders. They can work as facilitators on a full-time or part-time basis, They should also have a reasonable understanding of the process and of the technology embodied in the assets under review, but should *not* be experts on either subject. This whole approach is based on the notion that the other group members are the experts in these areas. (It may also explain why process experts and line maintenance managers and supervisors should participate in the process as group members, but should not do so as facilitators.)

The field in which a facilitator *should* of course be an expert is RCM, which means that appropriate training will usually be required. In order to secure the highest possible level of 'ownership' of and long-term commitment to the conclusions drawn during the process, the facilitator should also be a full-time employee of the organization which will be operating and/or maintaining the asset in the long term. (This is one of many reasons why it is strongly recommended that outsiders should not be used as RCM facilitators.)

13.4 Implementation Strategies

Broadly speaking, the group approach to RCM described above can be applied in one of three ways, as follows:

• the task force approach
• the selective approach
• the comprehensive approach.

Key aspects of each of these approaches are discussed in the following paragraphs.

The Task Force Approach

Organizations which have assets or processes which are suffering from intractable problems with serious consequences often adopt a 'task force' approach to RCM. This approach entails training a small group (the 'task force') to carry out a comprehensive RCM analysis of the affected system. Each task force consists of members drawn from the same disciplines as the groups described in Part 2 of this chapter. They often work full-time on the review project until it is complete, and the group is then disbanded.

* The main advantages of this approach are that it is *quick*, because only one or two groups have to make their way up the RCM learning curve, it is *easy to manage*, because only a small number of people are involved, and if it is successful – which is usually the case – it can yield *substantial returns* (in terms of improved plant performance) *for a relatively small investment.*

* The main disadvantages of this approach are that *it does nothing to secure the long-term involvement and commitment* of all the people in the organization to the results, so the results are much *less likely to endure*, and because it is narrowly focused, it *does little to foster best practice* across the entire organization

The Selective Approach

In addition to acute problems which might lend themselves to the task force approach, most organizations also have some assets which are more susceptible than others to chronic problems which are difficult to identify. These problems usually manifest themselves as downtime, poor product quality, poor customer service or excessive maintenance costs. Other areas might be confronted with unacceptable safety or environmental hazards which need to be tackled on a systematic basis.

Given hundreds if not thousands of items to choose from in a large undertaking, it makes sense to start applying a technique with the power of RCM in areas where the worst of these problems are encountered. Once these have been dealt with, a decision is taken as to whether RCM will be used to analyze assets with less serious problems, and so on.

The author has found that in most cases, the simplest, quickest and most effective way to identify where physical assets are causing the most serious problems (especially in terms of failure consequences) is to ask their users. This usually means production or operations managers at all levels.

If the worst problems are not immediately obvious, or if it is not possible to achieve consensus about where to start on an informal basis, then it is sometimes necessary to decide on a more formal basis where RCM should be applied. This can be done in three stages:

- identify 'significant' assets. These are assets which are most likely to benefit from the RCM process.
- rank the assets which are significant in descending order of importance.
- decide whether to use a 'template' approach for very similar assets.

Significant assets

An asset is judged to be significant if it could suffer from any failure mode which on its own:

- could threaten safety or breach any known environmental standard
- would have significant economic consequences.

Items are also judged to be significant if they contain hidden functions whose failures would expose the organization to a multiple failure with significant safety, environmental or operational consequences. Conversely, for any item to be classified as non-significant, we must be sure that:

- none of its failure modes will affect safety or the environment
- none of its failure modes will have significant operational consequences
- it does not contain a hidden function whose failure exposes the organization to the risk of a significant multiple failure.

The process of identifying significant items is quick, approximate and conservative. In other words, if it is not certain that any asset is not significant in the sense defined above, then it should be subjected to a full RCM review. Note that the assessment of significance can be done at any level, on the understanding that this may not be the level at which the RCM analysis is eventually conducted.

When making decisions about significance, note also that the RCM process is applied to any asset in its operating context. This context is a function of the process or system of which the asset forms part, so any asset should only be analyzed in the context of a specific process or system (such as a packing line, a rolling mill or a crane). The selection of significant items should never be based on generic items or components (all pumps, all bearings, all relief valves), because these would necessarily have to be taken out of context.

In the civil aviation industry, a surprisingly high percentage of items can be classified as non-significant in the sense described above. However, for thirty years this industry has been designing aircraft specifically to avoid or minimize the consequences of failure, so there is a very high (but still not infallible) level of redundancy built into their assets.

Assets in other industries, however, tend to enjoy a much lower level of redundancy, so a rather higher proportion of items end up being classified as significant, especially if due consideration is given to failures that could affect safety or the environment. This means that most organizations will still be confronted with a large number of items which should be analyzed. If the answer is not self-evident, the next question which needs to be answered systematically becomes: "Where do we start?"

Ranking significant items in order of importance
A large number of techniques have been developed which attempt to provide a systematic, usually quantitative basis for deciding what assets are likely to benefit most from the application of analytical processes such as RCM. Sometimes called 'criticality assessments', most of these techniques use some variation of a concept known as the 'probability/risk number', or PRN

A PRN is derived by attaching a numerical value to the probability of failure – or failure rate – of an asset (the higher the probability, the higher the value), and another value to severity of the consequences of the failure (again, the more serious the failure, the higher the value). The two numbers are multiplied to give a third, which is the PRN. Assets with the highest PRN's are analyzed first, then those with lower scores and so on until assets are encountered where the likely return does not justify detailed analysis.

More sophisticated variations of this process build up composite PRN's by attaching different numerical weightings to different categories of failure consequences (typically, high for safety or environmental consequences, intermediate for operational, and lower for direct repair costs). If hard data about historical failure rates and costs are available, these rankings can be further refined using Pareto analysis.

Systematic rankings of this sort can be useful in helping to clarify and build consensus about what assets really matter and about where large, complex systems are particularly vulnerable. However, the criteria and the relative weightings used to assess severity and probability vary widely from company to company, so most criticality assessment processes use scales and values which are unique to specific organizations.

Templating
Another way to reduce the investment in RCM is to use the analysis of one asset as a 'template' for another. For reasons which were stressed repeatedly throughout Chapters 2, 3, 4 and 5, this approach can only be applied to assets or processes which are very similar, if not identical, and which are operating in virtually the same context.

When this approach is adopted, an RCM group carries out a comprehensive, zero-based analysis of the first item or process in a series of very similar items or processes, and then uses this analysis as the basis for a review of the other items in the series. To do this, the group ask if the functions and performance standards of each subsequent item differ in any way from those listed on the worksheets for the zero-based item. The differences (if any) are recorded on the worksheets for the second item, and the analysts move on to compare the functional failures in the same way, and so on until they have completed the entire analysis.

If the items are technically virtually identical and the operating context is very similar, this approach can save considerable amounts of time and effort because in most cases, a substantial proportion of the analysis remains unchanged for the subsequent items.

However, while it is technically appealing, templating can also have quite serious motivational drawbacks. This is because the operators and maintainers of the subsequent assets are asked to accept decisions made by others, which naturally reduces their sense of ownership. In extreme cases, the latter people may even reject the initial analysis out of hand because "it was not invented here". This phenomenon has led some organizations not to use templating at all, but to start all analyses from a zero-base.

(Interestingly, this can lead to some quite different maintenance programs as different groups select different methods of dealing with the same failure. One way in which this can occur quite legitimately was explained in Figure 7.8 on page 154.)

Advantages and disadvantages of the selective approach
Typically, organizations which adopt the selective approach apply RCM to between 20% and 40% of their assets.

- The main advantage of this approach is that the investment is only made where it will yield *quick and* (usually) *measurable returns.* Because RCM is only applied to part of the facility, the overall *project is less costly* and hence *easier to manage* than if an entire facility is analyzed.

- The main disadvantage of this approach is that it places much *greater emphasis on the technical and operational performance of the equipment than on the people* on whom the equipment ultimately depends in the long-term (the operators and maintainers).

The Comprehensive Approach

The third approach to the application of RCM places at least as much emphasis on improving the knowledge and motivation of individuals and on improving teamwork between the users and the maintainers of the assets as it does on the performance of the assets themselves. Two ways in which this is often done are:

- to analyze all the assets on the site in one short, intensive campaign. Campaigns of this nature usually last from six to eighteen months on most sites. Up to twenty or even more groups can be active at once, working under the direction of anywhere from three or four to thirty or forty facilitators. As soon as a group completes the analysis of their asset or process, a new group is activated. In this way, the entire campaign is finished quickly and the organization enjoys the benefits equally quickly. In fact this is an excellent way to achieve massive and lasting step changes in maintenance performance for companies that need to do so in a hurry.

  However, this approach is highly resource intensive, so it needs a great deal of careful planning and management attention. It should not really be considered if a number of other initiatives are to be undertaken in parallel with RCM.

- a second possibility is still to review all the equipment on the site, but to do so in stages. Perhaps four or five groups are activated at a time, working under the direction of one or two facilitators. On this basis, it could take five to ten years to analyze all the equipment on a large site (three to four on a smaller one). The organization still derives all the benefits of RCM, but it takes much longer to do so. This approach is less disruptive in the short term, but if expectations are not very carefully managed, it could be seen to be 'dragging on forever', and hence could become demotivating. On the other hand, it means that RCM can be applied in parallel with other initiatives and vice versa.

Since the people who could benefit from this approach often substantially outnumber the assets, it is usually necessary to analyze most if not all of the assets so that everyone can take part in the process.

- The main disadvantages of this approach are that it is *slower*, because more people have to become familiar with the RCM methodology, and it is *more difficult to manage*, because many more people are involved.
- The main advantage is that it secures much more *broadly-based long-term ownership of maintenance problems and their solutions*. This not only improves individual motivation and teamwork, but it also ensures that the results of the exercise are *far more likely to endure*. (Best practice becomes 'part of the way we do things around here'.)

Deciding which approach to use

If it is to be applied correctly, RCM requires a substantial commitment of resources. If the comprehensive approach described above is applied, it needs the whole-hearted involvement and cooperation of large numbers of people. As a result, it is wise to decide in stages which approach should be used.

Since managers have to commit the resources to RCM, it makes sense to start by giving them the opportunity to learn what RCM is all about, to assess for themselves what resources are required to apply it and to judge for themselves what potential benefits it offers in their areas of responsibility. The best way to do this is usually to arrange for them to attend an introductory training course.

If the response is favorable, the next step is to run one or two pilot projects. These enable the organization to gain first-hand experience of the dynamics of the whole RCM process, what it achieves, and what resource commitments are needed to achieve it.

However, before undertaking any pilot project, it is essential to assess the resources required to do it relative to the likely benefits, and to plan the project as thoroughly as possible, This should be done in close consultation with the managers of the area where a pilot project is likely to be undertaken, and entails the following steps:

- confirm the scope of the project and define the objectives (now state and desired end state)
- estimate time needed to review equipment in each area
- identify project manager and facilitator(s)
- identify participants (by title and by name)
- plan training for participants and facilitators
- plan date, time and location of each meeting.

When the pilot project(s) are complete, the participants are in a position to evaluate the results for themselves and to decide whether, where, and how quickly RCM should be applied to the remaining assets in the organization. Chapter 14 explains that RCM yields substantial returns, but that the nature of these returns varies widely from one organization to another. As a result, the best time to decide which approach to adopt is after a small number of pilot projects have been completed and the organization is able to judge for itself what returns RCM offers in relation to what inputs.

13.5 RCM in Perpetuity

The application of RCM leads to a much more precise understanding of the functions of the assets which have been reviewed, and a much more scientific view of what must be done to cause them to continue to fulfil their intended functions. However, the analysis will not be perfect – and never will be perfect – for two reasons:

- the evolution of a maintenance policy is inherently imprecise. Numerous decisions have to be made on the basis of incomplete or non-existent hard data, especially about the relationships between age and failure. Other decisions have to be made about the likelihood and the consequences of failure modes which haven't happened yet, and which may never happen. In an environment like this, it is inevitable that some failure modes and effects will be overlooked completely, while some failure consequences and task frequencies will be assessed incorrectly.

- the assets and the processes of which they form part will be changing continuously. This means that even parts of the analysis which are completely valid today may become invalid tomorrow.

The people involved in the process will also change. This is partly because the perspectives and priorities of those who take part in the original analysis inevitably change with time, and partly because people simply forget things. In other cases, people leave and their places are taken by others who need to learn why things are as they are. All these factors mean that the validity of the RCM database and people's attitudes towards it will inevitably deteriorate if no attempt is made to prevent this from happening.

One way to do this is to use the RCM process to analyze all significant unanticipated failure modes which occur after the initial analysis has been completed. This can be done by convening an ad-hoc group which uses RCM to determine the most effective way of dealing with the failure.

The results of their deliberations should be woven into the RCM database for the affected asset. The ad-hoc group itself should include as many as possible of the people who carried out the original analysis.

A second – and much surer – way to ensure that RCM databases remain current *in perpetuity* is to ask the original groups to review the database for 'their' asset on a formal basis once every nine to twelve months. Such a review meeting need not last for more than one afternoon. Specific questions which should be considered include the following:

- has the *operating context* of the equipment changed enough to change any of the decisions made during the initial analysis? (Examples include a change from single shift operation to double-shifting, or vice-versa.)

- have any *performance expectations* changed enough to necessitate revisions to the performance standards recorded on the RCM worksheets?

- since the previous meeting, have any *failure modes* occurred which should be recorded on the Information Worksheets?

- should anything be added to or changed in the descriptions of *failure effects?* (This applies especially to the evidence of failure and estimates of downtime.)

- has anything happened to cause anyone to believe that *failure* consequences should be assessed differently? (Possibilities here include changes to environmental regulations, and changed perceptions about tolerable levels of risk.)

- is there any reason to believe that any of the *tasks* selected initially is not in fact technically feasible or worth doing?

- has any evidence emerged which suggests that the *frequency* of any task should be changed?

- has anyone become aware of a *proactive technique* which could be superior to one of those selected previously? (In most cases, 'superior' means 'more cost effective', but it could also mean technically superior.)

- is there any reason to suggest that a task or tasks should be *done by* someone other than the person selected originally?

- has the asset been *modified* in a way which adds or subtracts any functions or failure modes, or which changes the technical feasibility of any tasks? (Special attention should be paid to control systems and protective systems.)

If such reviews are carried out regularly, they only take a small fraction of the time needed to set up the database to begin with, but they ensure that the organization continues to enjoy the benefits of the original exercise in perpetuity. These benefits are discussed in more detail in Chapter 14.

13.6 How RCM Should Not be Applied

If it is applied correctly, RCM yields results very quickly. However, not every application of RCM yields its full potential. Some even achieve little or nothing. In the author's experience, some of the main reasons why this happens are technical in nature, but the majority are organizational. The most common are discussed in the following paragraphs.

The analysis is performed at too low a level.
The problems which arise if an RCM analysis is performed at too low a level were listed in detail Part 7 of Chapter 4. Most important among these are that the analysis takes far longer than it should, it results in a massive increase in paperwork and the quality of the decisions deteriorates. As a result, people start finding the process tedious and lose interest, it costs much more than it should and it does not achieve as much as it could.

Too hurried or too superficial an application.
This is usually the result of insufficient training, or too heavy an emotional investment in the status quo on the part of key participants. It often results in a set of tasks which are almost the same as they were to begin with.

Too much emphasis on failure data
There is often a tendency to over-emphasize the importance of data such as MTBF's and MTTR's. This issue is discussed at length in Chapter 12. Such data are nearly always over-emphasized at the expense of properly defined and quantified performance standards, the thorough evaluation of failure consequences and the correct use of data such as P-F intervals.

Asking a single individual to apply the process
One of the least effective ways to apply RCM is to ask a single individual to apply the process on his or her own. In fact, no matter how much effort a single individual applies to the development of a maintenance program (whether using RCM or any other technique) the resulting schedules nearly always die when they reach the shop floor, for two main reasons:

- *technical validity:* no one individual can possibly have an adequate understanding of the functions, the failure modes and effects and the failure consequences of the assets for which his or her program is being developed. This leads to programs which are usually generic in nature, so people who are supposed to do them often see them as being incorrect if not totally irrelevant

- *ownership:* people on the shop floor (supervisors and craftsmen) tend to view the schedules as unwelcome paperwork which appears from some ivory tower and disappears after it is signed off. Many of them learn that it is more comfortable just to sign off the schedules and send them back than it is to attempt to do them. (This leads to inflated schedule completion rates which at least keeps the planners happy.) The main reason for the lack of interest is undoubtedly sheer lack of ownership.

The only way around the problems of technical invalidity and lack of ownership is to involve shop floor people directly in the maintenance strategy formulation process as discussed earlier in this chapter.

Done correctly, this not only produces schedules with a much higher degree of technical validity than anything that has gone before, but it also produces an exceptionally high level of ownership of the final results.

Using the maintenance department on its own to apply RCM
In many organizations, an almost impenetrable divide still exists between the maintenance and production functions. This often leads the maintenance people in such organizations to try to apply RCM on their own. In fact, as Chapter 2 made clear, maintenance is all about ensuring that assets continue to function to standards of performance required by their users. We have seen that the 'users' are nearly always production or operations people. If these people are not closely involved in helping to define functions and performance standards, two problems usually arise:

- the maintenance people do it for them. In the author's experience, this nearly always leads to large numbers of inaccurate function statements and performance standards, and consequently to distorted or inappropriate programs designed to preserve those functions

- there is little or no 'buy-in' to the maintenance program on the part of the users, who after all are the 'customers' of the maintenance service. This in turn means that users understand less clearly why it is in their own interests to release machines for essential maintenance, and also why operators need to be asked to carry out certain maintenance tasks.

In addition to defining what they want the asset to do, users also have a vital contribution to make to the rest of the strategy formulation process. As explained in part 1 of this chapter, by participating in the FMEA, they learn a great deal about failure modes caused by human error, and hence what they must do to stop breaking their machines. They also play a key role in evaluating failure consequences, and they have invaluable personal experience of many of the most common warnings of failure. All this is lost if they do not participate in the process.

In short, from a purely technical point of view, it is rapidly becoming apparent that it is virtually impossible to set up a viable, lasting maintenance program in most industrial undertakings without involving the users of the assets. (This focus on the user – or customer – is of course the essence of TQM.) If their involvement can be secured at all stages in the process, that notorious barrier rapidly starts to disappear and the two func-tions start to work, often for the first time ever, as a genuine team.

Asking manufacturers or equipment vendors to apply RCM on their own
A universal feature of traditional asset procurement is the insistence that the equipment manufacturer should provide a maintenance program as part of the supply contract for new equipment. Apart from anything else, this implies that manufacturers know everything that needs to be known to draw up suitable maintenance programs.

In fact, as explained on Page 78, equipment manufacturers usually possess surprisingly little of the information needed to draw up truly context-specific maintenance programs. They also have other agendas when specifying such programs (not least of which is to sell spares). What is more, they are either committing the users' resources to doing the maintenance (in which case they don't have to pay for it, so they have little interest in minimizing it) or they may even be bidding to do the maintenance themselves (in which case they may be keen to do as much as possible).

This combination of extraneous commercial agendas and ignorance about the operating context means that maintenance programs specified by manufacturers often embody a high level of over-maintenance (sometimes ludicrously so) coupled with massive over-provisioning of spares.

Most maintenance professionals are aware of this problem. However, despite this awareness, most of us still persist in demanding that manufacturers provide these programs, and then accept that they must be followed in order for warranties to remain valid (and so bind ourselves contractually to doing the work, at least for the duration of the warranty period).

None of this is meant to suggest that manufacturers mislead us deliberately when they put together their recommendations. In fact, they usually do their best in the context of their own business objectives and with the information at their disposal. If anyone is at fault, it is really us – the users – for making unreasonable requests of organizations which are not in the best position to fulfil them.

A small but growing number of users solve this problem by adopting a completely different approach to the development of maintenance programs for new assets, by involving the manufacturers' field technicians in a user-driven RCM analysis, as discussed on page 78.

In this way, the *user* gains access to the most useful information that the manufacturer can provide, while still developing a maintenance program directly suited to the context in which the equipment will actually be used. The *manufacturer* may lose a little in up-front sales of spares and maintenance, but will definitely gain all the long-term benefits associated with improved equipment performance, lower through-life costs and a much better understanding of the real needs of his customer. A classic win-win situation.

Using outsiders to apply RCM

It is wise to steer clear of the temptation to use third parties to formulate maintenance strategies. In this context, they suffer from most of the shortcomings which apply to single individuals, maintenance departments on their own and manufacturers/equipment vendors as discussed above. In addition, most outsiders know little about the dynamics of the organization for which the schedules are being written, such as the operating context of each asset, the risks which the organization is prepared to tolerate and the skills of the operators and maintainers of the assets. This often results in generic analyses which contain many more assumptions than if the analysis is facilitated by informed insiders. What is more, after the initial analyses have been completed, outsiders more often than not move on to other organizations. After they have gone, there is often no-one left with a sufficiently strong sense of ownership of the analyses and their outcomes to ensure that they stay alive in the sense discussed in part 5 of this chapter.

Finally, the fact that most outsiders are usually working under contract introduces commercial constraints which can distort the RCM process if they are not managed very carefully indeed. In particular, the need to finish contracts on time and on budget creates additional time pressures that can cause too many decisions to be taken too quickly. These could have devastating consequences years, even decades, after the contracts are complete.

On the other hand, if RCM is applied by properly trained insiders, their own jobs – indeed, their own lives – often quite literally depend on the *long-term* validity of each analysis. As a result, they will naturally be more inclined (and less constrained) to take whatever extra time is needed to ensure that *all* reasonably foreseeable risks are dealt with appropriately.

Using computers to drive the process.
Chapter 10 mentioned that computerised databases should be used to store and sort the information generated by the RCM process. However, as with so much in the world of information technology, it is easy to succumb to the temptation to go beyond what computers *should* be used for, and to focus on their apparently 'nice to have' uses.

For instance, it is tempting to computerise RCM algorithms such as the main decision diagram on pages 200 and 201. This is often done by creating a screen which asks (say) question H, and setting up the system so that a 'no' answer brings up another screen (or window) that asks question H1 while a 'yes' answer leads to one that asks question S, and so on. This is done in the utterly mistaken belief that a succession of screens will somehow speed up or 'streamline' the process. In fact, there is simply no way that referring to a succession of twelve to twenty screens is quicker than reading a single sheet of paper, so using a computer in this fashion actually slows the process down.

Using a computer inappropriately to drive the process can also have a strong negative influence on perceptions of RCM. Too much emphasis on a computer means that RCM starts being seen as a mechanistic exercise in populating a database, rather than exploring the real needs of the asset under review. For this reason the author agrees with Smith[1993] when he says that there is no "software code to do the engineering thinking for us", and that the computer "doesn't replace the need for solid engineering know-how and judgement". In short, RCM is thoughtware, not software.

Conclusion
These comments suggest that the surest way to achieve most if not all of the positive benefits of RCM is to apply the process at the right level, and to do so on a formal basis using groups of properly trained people who represent the operations and maintenance functions, and who have an intimate first-hand knowledge of the equipment under review.

13.7 Building Skills in RCM

RCM provides a common framework which enables people from diverse backgrounds to achieve consensus about a wide range of highly technical issues. However, this process itself embodies many concepts which are new to most people. They need to learn what these are and how they fit together before they can use the process successfully. (Some people who have been steeped in traditional approaches to maintenance also need to unlearn a great deal.)

The best way to ensure that large numbers of people acquire the relevant skills quickly is to provide suitable training. The most appropriate mix of courses for people at different levels is as follows:

• *maintainers and operators:* a course in the basic principles of RCM. Such a course should incorporate a variety of case studies and practical exercises which enable delegates to gain an appreciation of how the theory works in practice.

• *maintenance managers, engineers, operations managers, supervisors and senior technicians:* a course which covers the same ground as the course for craftsmen and operators, but which also explains what must be done to manage the implementation of RCM.

• *facilitators:* facilitators should be introduced to RCM on an introductory course such as the one described above, and then undergo at least ten more days of intensive formal training before starting to work with groups. Thereafter, most facilitators require further mentoring from a skilled RCM practitioner for a period of a few months after their formal training program, before they become fully competent in all 45 of the key skill areas listed in part 3 of this chapter.

(For a description of a comprehensive array of training and other support services which meet all the above requirements, see the worldwide website **http://www.aladon.co.uk**).

14 What RCM Achieves

14.1 Measuring Maintenance Performance

As discussed at length in Chapter 11, the application of RCM results in three tangible outcomes, as follows:

- maintenance schedules to be done by the maintenance department
- revised operating procedures for the operators of the asset
- a list of areas where once-off changes must be made to the design of the asset or the way in which it is operated, in order to deal with situations where the asset cannot deliver the desired performance in its current configuration.

Two other less tangible outcomes which were mentioned in Chapter 13 are that participants in the process learn a great deal about how the asset works, and also tend to function better as teams.

Achieving all these outcomes requires a great deal of time and effort, especially if RCM is applied as described in Chapter 13. However, if RCM is applied correctly, it yields returns which far outweigh the costs involved. Most applications pay for themselves in a matter of months, although some have paid for themselves in two weeks or less. The wide variety of ways in which RCM pays for itself are discussed at length in part 4 of this chapter. In order to place this discussion in perspective, we first need to consider different ways in which it is possible to measure the performance of the maintenance function.

Maintenance performance can be considered from two quite distinct viewpoints. The first focuses on how well maintenance ensures that assets continue to do what their users want them to do. This is usually referred to as maintenance *effectiveness*, and it is likely to be of most interest to the users or 'customers' of the maintenance service. The second viewpoint concentrates on how well maintenance resources are being used. This is referred to as maintenance *efficiency*. It is usually of more interest to managers who are directly responsible for maintenance. These two issues are considered separately in the next two sections of this chapter.

14.2 Maintenance Effectiveness

Chapters 1 and 2 emphasized that the objective of maintenance is to ensure that any physical asset *continues* to fulfil its intended functions to the standards of performance desired by the user. As a result, any assessment of how well maintenance is achieving its objectives must entail an assessment of how well the assets are *continuing* to fulfil their functions to the desired standard. This is influenced in turn by three issues:

• 'continuity' can be measured in several different ways
• users have different expectations of different functions
• individual assets can have more than one and often several functions, as explained in Chapter 2.

These issues are considered in more detail in the following paragraphs.

Different Ways of Measuring Maintenance Effectiveness

The primary function of any highly mechanized and fully loaded manufacturing facility is to produce at least as many units of saleable product as it was expected to produce when it was built. ('Fully loaded' means that it is operating seven days per week/24 hours per day and that there is a ready market for every unit which the facility can produce.) In this context, any failure which reduces output results in lost sales.

In cases like these, the simplest overall measure of the operational performance of the facility as a whole is *total output per period.*

If the facility is not producing what the users – usually the owners – feel it should be producing on a regular basis, they will not be satisfied until the situation is put right. At least until then, the *users* will be inclined to judge effectiveness in terms of total output against targets. This should be recognized when setting up any system for tracking maintenance effectiveness

All is not necessarily well if overall output is on target. A facility which is producing the right number of units could still be experiencing problems which affect safety, product quality, operating costs, environmental integrity, customer service, and so on, so these also need to be measured and dealt with appropriately.

There are a great many ways in which we can measure how effectively an asset is fulfilling its functions. Five of the most common are as follows:

• *how often it fails.* This is the most widely understood meaning of the term 'reliability'. It is usually measured by 'mean time between failures' or 'failure rate'.

- *how long it lasts.* This is usually thought of as its 'life' or its 'lifespan', at the end of which the item under consideration fails and is either rebuilt or discarded and replaced with a new one. Strictly speaking, this phenomenon should be described as 'durability'.

- *how long it is out of service when it does fail.* This is usually referred to as 'downtime' or 'unavailability', and measures how much of the time the item is incapable of fulfilling a stated function to the satisfaction of the user, in relation to the amount of time the user would like it to be capable of doing so. Unavailability (or the converse, avail-ability) is usually expressed as a percentage.

- *how likely it is to fail in the next period* assuming that it has survived to the beginning of that period. We have seen that this is the conditional probability of failure. This could perhaps be described as a measure of 'dependability', if only to distinguish it from the other three variables. One common variation of this measure is the 'B10 life'. Chapter 12 explained that this is usually measured from the moment the item is put into service, and is the period before which not more than 10% of the items can be expected to fail. (In other words, the conditional probability of failure in the stated period is 10%.)

- *efficiency.* In common business usage, the term efficiency actually has two quite distinct meanings. The first measures output relative to input, while the second measures how well something is performing against how well it should be performing.

For instance, in a power station, energy efficiency measures the amount of energy exported in relation to the amount of energy released by the fuel. Depending on the technology used (coal, gas, combined cycle, etc), this usually ranges from about 35% to about 58%. However, if a station which should average 40% energy efficiency is only averaging 38%, it will be exporting 95% of the energy which it should be exporting. In the context of this book, the first (40%) measure is a functional *performance standard*. As explained in Chapter 3, this is used to judge whether the item has failed. The second (95%) measure is used to judge the *effectiveness* with which the organization is achieving the desired performance on an on-going basis.

'Efficiency' also refers to pace of working, and also does so in two ways – how fast an asset should work relative to the pace at which it could work (desired performance versus initial capability), and how fast it actually works relative to the pace at which it should work (actual performance versus desired performance). We have seen that desired performance must be less than initial capability because allowance must be made for deterioration. So in the context of this chapter, efficiency compares the pace at which an asset actually works with the pace at which it *should* work, not with the pace at which it *could* work.

'Efficiency'-type measurements can also apply in a slightly different fashion to the consumption of maintenance consumables (such as lubricating oil and hydraulic oil) and process consumables (such as solvents and reagents used in chemical plants and in the extraction of minerals).

All five of these measures are valid. It is simply a matter of deciding which is the most appropriate in the context under consideration.

For example, if a turbo-generator set has the lowest energy costs per unit of output among all those used by an electric utility, it is likely that its users would want it to generate (base load) power for as much of the time as possible. In terms of this function, the most appropriate measure of maintenance effectiveness is *availability*. (The operators may occasionally *choose* to run the set at less than full load. They may even choose to shut it down completely from time to time for purely operational reasons. Slowdowns or shutdowns of this nature affect the *utilization* of the asset as opposed to its availability. In essence, availability measures what percentage of time the machine is *available* to fulfil its primary performance requirement, while utilization measures how much it actually fulfils it.)

On the other hand, the generator set might only be used periodically to satisfy peak demands for power (peak loads). In this case, the primary concern of the users will be that the generator comes on stream as soon as it is required, so a primary measure of effectiveness will be how often it does so (or conversely, how often it fails to do so, expressed by a *failure rate).*

When measuring safety , performance is usually measured in terms of number of days or number of manhours worked between lost time incidents (or fatalities). This is a form of 'mean time between failures'. Similar measures are used for environmental incidents.

On the product quality front, a scrap rate of (say) 4% can be seen as a measure of *unavailability*, in the sense that while a machine is producing scrap, it is not 'available' to produce first grade product. (A scrap rate of 4% corresponds to a *yield* of 96%). Scrap rates can also be expressed as (say) 20 parts per million, which is another way of expressing a *failure rate.* Both are valid measures of maintenance effectiveness, especially in highly mechanized or automated processes.

Different Expectations

Every function has associated with it a unique set of continuity (reliability and/or durability and/or availability and/or dependability) expectations.

For instance, two of the functions associated with the bodywork of a car are 'to isolate the occupants of the car from the elements' and 'to look acceptable'. Most car owners expect the bodywork to be able to fulfil the first function throughout the expected life of the car (unless the car is a convertible or unless they open a door or a window). On the other hand, everyone knows that cars get dirty – and hence start 'to look unacceptable' – in the space of a few days or weeks. So in the first case we have a continuity expectation which might be measured in hundreds of thousands of miles or decades, while in the second case, the continuity expectation is measured in hundreds of miles or days.

This issue is complicated by the fact that the loss of nearly every function can be caused by more than one – sometimes dozens – of failure modes. Each failure mode has associated with it a specific failure rate (or MTBF), and each will take the function out of service for an amount of time which is specific to that failure mode. As a result, the continuity characteristics of any function will actually be a composite of the continuity characteristics of all the failure modes which could cause the loss of that function.

For instance, take the function 'to look acceptable' which was mentioned above. In addition to the accumulation of dirt, this function could be lost due to rust or corrosion, fading of the paintwork, external damage (sideswiped in a parking lot) and vandalism, among others. It should also be apparent that some of these failure modes have little or nothing to do with maintenance. For instance, external damage is mainly a function of how this car – or the other vehicle involved – is operated, although design may play a small part by adding rubbing strips to reduce damage and/or by making it easier and cheaper to replace damaged panels. The probability of vandalism is also a function of where the car is used (the operating context), so it is almost completely beyond the control of the designer and the maintainer. The rate of accumulation of dirt is a function of where and when a car is used (road conditions and climatic conditions), and it is managed by a suitable maintenance program (washing the car). Corrosion and fading of paintwork can be influenced substantially at the design stage (although yet again the operating context – climatic conditions and the provision of shelter – and to some extent maintenance activities – polishing and chassis washing – can play a part in moderating the severity and frequency of these failures).

This example leads to two important conclusions, as follows:

• we need a thorough understanding of all the failure modes which are likely to cause each loss of function in order to be able to design, operate and maintain an asset in such a way that the effectiveness expectations which we have of each function will be achieved

• it is unreasonable to hold the maintainer of an asset alone accountable for the achievement of any continuity (reliability/availability/durability/dependability) targets for any asset or any function of any asset. The achievement of these targets is also a function of how it is designed, built and operated. Accountability for achieving the associated targets should be divided jointly between the people responsible for all of these functions. (In other words, 'maintenance' effectiveness as it is being defined in this chapter is not only a measure of the effectiveness of the maintenance department. It measures how effectively *everyone* associated with the asset is playing their part in doing whatever is necessary to ensure that it continues to do what its users want it to do.)

Different Functions

Perhaps the most important point about measuring the effectiveness of maintenance activities is the fact that every asset has more than one and sometimes dozens of functions. As explained above, a unique set of continuity expectations is associated with each function. This means that if an asset has ten functions, then the effectiveness with which the asset is being maintained can be measured in (at least) ten different ways.

For instance, let us consider how maintenance effectiveness might be measured by the owner of a typical suburban gas station. For the purpose of this example, the 'asset' is a storage and pumping system used for gasoline. In this system, unleaded gasoline is stored in an underground tank with a capacity of 50 000 liters. It is periodically filled by a road tanker to a level of 48 000 liters. An upper level switch in the tank switches on a local warning light if the tank has been filled to a level of 48 500 liters, and another switches on another warning light in the main office if the level drops to 5 000 liters. A low level alarm sounds in the office if the tank level drops to 2 000 liters, and a local ultimate high level alarm sounds if the tank level reaches 49 000 liters. The tank is double skinned to ensure that gasoline is contained in the event of a leak in the inner skin. A level indicator indicates the fuel level in the tank.

The tank supplies gasoline to five pumps. Each pump is switched on and off by pressing and releasing a handle in the nozzle. The nozzle also incorporates a pressure switch which trips the pump when the vehicle fuel tank is filled to the tip of the nozzle. A flow meter measures the amount of fuel delivered each time the pump is activated and displays the volume and value of the fuel delivered to the customer. This meter is zeroed each time the nozzle is returned to its cradle.

(This system embodies additional secondary functions which deal with access onto and into the tank, drains, venting, valving, ease of use by a customer, other protection, appearance and so on. These would also be listed in a real-life situation. However, for the purpose of this example, we only consider the functions described above.) On this basis, a list of functions might read as follows:
• to pump between 25 and 40 liters/minute of gasoline to the vehicle
• to indicate volume and value of fuel delivered to customer to within 0.03% of actual volume/value
• to shut off pump when required by customer or when customer's fuel tank is full.
• to contain the gasoline
• to store between 2 000 and 48 000 liters of gasoline
• to switch on a warning light in main office if the tank level drops to 5 000 liters
• to switch on a local warning light if the tank level reaches 48 500 liters
• to sound an alarm in main office if the tank level drops below 2 000 liters
• to sound an alarm if the tank level reaches 49 000 liters
• to contain the contents of the tank in the event of a leak
• to indicate the level of fuel in the tank to within 0.05% of the actual level
When assessing the maintenance effectiveness of this system, the owner of the gas station will have different criteria for each of the above functions. For instance:

- *Function 1: to pump between 25 and 40 liters/minute of gasoline to the vehicle.* This function can fail in three ways with three quite different sets of consequences, so each functional failure needs to be considered on its own merits, as follows:

  - *Functional failure A: fails to pump at all:* Obviously, if a pump isn't working, it cannot be used to pump gasoline. However, there are five pumps in the station so the level of availability required depends on the pattern of demand. For example, the station owner may tell us that he "hardly ever" has all five pumps in use at once – so seldom that we can ignore the possibility. He might also tell us that four pumps tend to be in use simultaneously for a total of not more than one hour a day, and then never for more than about ten minutes at a time. If *each* pump has an average availability of 95%, two pumps will be out of service simultaneously for no more than 2% of the time. In other words, four pumps would be available 98% of the time, while there is a demand for four pumps 4% of the time. Under these circumstances, only a tiny fraction of customers would need to wait for gasoline, and then not for very long. This might tempt the owner to accept an *availability* of 95%. (If he regularly had five or more customers wanting to buy gas at the same time, he would expect a much higher availability. But it may cost him somewhat more to achieve it, especially if he has to pay a premium for rapid response when calling out technicians to deal with failures.)

  - *Functional failure B: pumps less than 25 liters/minute:* Some regular customers might find slow pumps sufficiently irritating to take their business elsewhere, especially if there are faster alternatives nearby. Consequently, the owner is likely to want any of his pumps which wasn't failed completely to pump at the required rate "all the time – or at least, as close to all the time as you can make it". This might turn out to mean (say) 99.8% of the time that the pump is not otherwise out of action – another form of *'availability'*.

  - *Functional failure C: pumps more than 40 liters/minute:* If the pump pumps too fast, it is likely to generate sufficient back pressure to keep tripping the 'tank full' pressure sensing mechanism in the nozzle. Customers would have to learn to throttle back the filling rate by not depressing the handle so much, which many regulars might also find irritating enough to cause them to take their business elsewhere. As a result, the owner is likely to say that he wouldn't want this failed state to occur "too often". He might then quantify this expectation as a *failure rate* - say not more than once in fifty years on any one pump.

- *Function 2: to indicate volume and value of fuel delivered to customer to within 0.03% of actual volume/value.* This function can fail in two ways, as follows:

  - *Functional failure A: indicates that more than 0.03% less fuel has been delivered than actual:* If this happens, the station owner appears to be selling less fuel than he is actually selling, so he loses money. The failure becomes apparent after a while, because the ratio of fuel sold to fuel received will start to decline. Nevertheless, the owner would probably still seek a low *failure rate* – say not more than one in 1 000 years on any one pump. (If the indicator fails completely, it shows that nothing has been delivered. If this happens, one lucky customer might get a free tank of fuel, then the station manager would shut down the affected pump until the problem is rectified.)

- *Functional failure B: indicates that more than 0.03% more fuel has been delivered than actual:* If this happens and it comes to the attention of either the customers or the trading standards authorities (probably both), the station owner would be in serious trouble. Many of his customers would regard him as a crook and take their business elsewhere. The authorities would probably fine him and, depending on the severity of the discrepancy, might even revoke his license to trade (thus putting him out of business). Whatever else happens, his standing in the community would take a beating. The severity of these consequences would lead him to seek a very low *failure rate* – say once in 50 000 years on any one pump. (Whether this is achievable or not is another issue entirely.)

• *Function 3: to shut off pump when required by customer or when customer's fuel tank is full.* This function can also fail in three ways, as follows:

- *Functional failure A: fails to shut off when required by the customer:* If the pump carries on pumping after the customer releases the handle, the back pressure sensor should shut it off when the tank is full. As a result, the customer will end up with much more gas in the tank than he or she wanted. This would almost certainly lead to a row about how much should be paid for and possible loss of a customer. As a result, the station owner would probably require a fairly low *failure rate* – say once in 1 000 years on any one pump.

- *Functional failure B: fails to shut off when tank is full:* Many customers rely on the sensor to tell them when the tank is full. If it fails to do so, the pump should shut off when the customer releases the handle. However, it is likely that the tank will overflow onto the shoes of the customer before he or she is able to react, leading to a lot of unpleasantness and perhaps a demand for compensation. This too would lead the owner to expect a low *failure rate* – say again once in a 1 000 years for any one pump

- *Functional failure C: both local switches unable to switch off pump:* If the sensor and the handle both fail to shut off the pump, it will carry on pumping gasoline all over the forecourt until the electrical supply is shut off at the main circuit breaker. This would create a nasty fire hazard, so the owner would expect a very low *failure rate* – say once in 1 000 000 years. (This is attainable if each switch independently achieves 1 in 1 000.)

• *Function 4: containment:* When asked about this function, the station owner might say something like "we have had one leak in the gasoline system in the last ten years – and that was one too many." Here the user is measuring effectiveness in terms of a *failure rate.* When pressed, he might accept a rate of (say) one in 500 years for a 'small' leak, which he might choose to define as less than 5 liters per hour. (It is highly unlikely that anyone would measure containment in terms of availability, because (say) 99% availability means that the system would be leaking 1% of the time – about 800 hours out of ten years. Even 99.9% still means that it would leak for 80 hours. Clearly this is nonsense.)

• *Function 5: to store between 2 000 and 48 000 liters of gasoline.* This function can also fail in three ways, each of which must again be considered separately, as follows:

- *Functional failure A: level drops below 2 000 liters:* Based on normal patterns of demand, fresh supplies of gasoline are ordered when the tank level approaches 5 000 liters, and we are told that they are nearly always delivered before the level reaches 2 000 liters. If the level in the tank drops much below 2 000 liters, there is a greatly increased chance that the tank will empty, causing the station to lose business. As a result, the station manager expedites the delivery if the level drops to 2 000 liters (as indicated by the low level alarm). He says he needs to expedite deliveries about once a year, which he says is "just about acceptable". Here he is again judging effectiveness in terms of a *failure rate*. (Note that this failed state is caused by increased demand and/or slow delivery. It has nothing to do with the maintenance department in the classical sense. Nonetheless, dealing with this failure can be seen as 'maintenance' because we are seeking to 'cause the business to continue'.)

- *Functional failure B: level rises above 48 000 liters:* The level in the tank is only likely to rise above 48 000 liters if the delivery driver is not paying attention to the tank level indicator when filling the tank or if the level indicator itself has failed. In both cases the warning light comes on at 48 500 liters. We are told that this happens "about once every six months" – another *failure rate* which the people involved might say they accept.

- *Functional failure C: tank contains something other than gasoline:* The tank can only contain something other than gasoline if it is filled with something else – (say) diesel. If this happens, customers could fill their tanks with the wrong fuel and cause serious damage to their engines. The station owner figures that the resulting bad publicity and claims for damages could put him out of business, so he would rather this didn't happen at all. When reminded that 'never' is an unattainable ideal, he might decide to accept a *failure rate* of (say) once in 100 000 years.

• *Function 6: to switch on a local warning light if level drops to 5 000 liters.* The station manager usually logs the level in all the fuel tanks every day in order to track consumption, and orders more fuel when levels approach 5000 liters. The low level warning light serves as a reminder if the level indicator fails or if there is a sudden surge in demand between readings. This light is needed about once every two years (M_{TED} = 2 years). If it does not work when needed, the low level alarm sounds when the level drops to 2 000 liters. If an initial order is placed at this late stage, the tank will almost certainly run dry and the station will be out of gasoline for several hours. The owner says he will accept a mean time between occurrences of this multiple failure (M_{MF}) of 400 years. In the light of this expectation, the formula on page 116 tells us that the maximum unavailability the station can tolerate for the low level warning light is $M_{TED}/M_{MF} = 2/400 = 0.5\%$. This means that the low level alarm is being maintained effectively if its *availability* remains above 99.5%.

• *Function 7: to switch on a local warning light if level rises to 48 500 liters.* The high level warning light is backed up by an audible alarm, so following similar logic to the above example, the owner might come to the conclusion that he will accept an *availability* of 97.5% for this warning light.

• *Function 8: to sound an alarm if the level in the tank drops below 2 000 liters.* If the level in the tank drops to 2 000 liters and the low level warning does not sound, the delivery is not expedited. We are told that under these circumstances, there is a 50% chance that the tank will run dry before the tanker arrives, and the station would be out of gasoline for about one hour on average under such circumstances. This leads the station owner to conclude that he will not accept this multiple failure (level drops below 2 000 liters while low level alarm is failed) more than "once in a hundred years" (M_{MF} = 100 years). As discussed above, M_{TED} is one year, so the station can tolerate a maximum unavailability for the low level alarm of M_{TED}/M_{MF} = 1/100 = 1%. In the light of this objective, the low level alarm is being maintained effectively if its *availability* remains above 99%.

Similar logic would be followed to determine availabilities for functions 9. 10 and 11 in the above example. It would also be used to establish effectiveness measures for the functions of this system that were not included in the above list. However, for the functions discussed, the effectiveness expectations of the gas station owner can be summarized as follows:

Function	Functional Failure	Measure of Effectiveness Availability	Measure of Effectiveness MTBF	Comments
1	A	≥ 95%		Each pump
	B	≥ 99.8%		Each pump
	C		≥ 50 years	Each pump
2	A		≥ 1 000 years	Each pump
	B		≥ 50 000 years	Each pump
3	A		≥ 1 000 years	Each pump
	B		≥ 1 000 years	Each pump
	C		≥ 1 000 000 years	Each pump
4	A		≥ 500 years	Whole system
5	A		≥ 1 year	Tank
	B		≥ 6 months	Tank
	C		≥ 100 000 years	Tank
6	A	≥ 99.5%		L/L warning light
7	A	≥ 97.5%		H/L warning light
8	A	≥ 99%		L/L alarm

The example illustrates several important points about the measurement of maintenance effectiveness, as follows:

• when measuring maintenance performance, we are not measuring *equipment* effectiveness – we are measuring *functional* effectiveness. The distinction is important, because shifting emphasis from the equipment to its functions helps people – maintainers in particular – to focus on what the equipment *does* rather than what it *is*.

- even quite simple assets have a surprisingly large number of functions. Each of these functions has a unique set of performance expectations. Before it is possible to develop a comprehensive maintenance effectiveness reporting system, we need to know what all these functions are, and we must be prepared to establish what *the user* thinks is acceptable or otherwise in each case.

  This means that it is not possible to list a single continuity statement for an entire asset, such as "to fail not more than once every two years" or "to last at least eleven years". We need to be specific about which function must not be lost more than once every two years or must not fail for at least eleven years (or more precisely, which functional failure must not occur more than once every two years, or which functional failure must not occur before eleven years).

- there is often a tendency to focus too heavily on primary functions when assessing maintenance effectiveness. This is a mistake, because in practice apparently trivial secondary functions often embody bigger threats to the organization if they fail than primary functions. As a result, *every* function must be considered when setting up maintenance effectiveness measures and targets.

  For instance, the primary functions listed for the gasoline system are to pump and to store fuel (Functions 1 and 5 respectively). However, two of the highest expectations of the owner centered around two secondary functional failures – 2-B (a failure which could put him out of business) and 3-C (a failure with serious safety implications).

Multiple Performance Standards and the OEE

If a function embodies multiple performance standards, it is tempting to try to develop a single composite measure of effectiveness for the entire function. For instance, the primary function of a machine performing a conversion operation in a manufacturing facility usually incorporates three performance standards, as follows:
- it must work at all
- it must work at the right pace
- it must produce the required quality.

The effectiveness with which it continues to meet each of these expectations is measured by *availability, efficiency* and *yield*. This suggests that a composite measure of the effectiveness with which this machine is fulfilling its primary function on an on-going basis could be determined by multiplying these three variables, as follows:

overall effectiveness = availability x efficiency x yield.

For instance, the primary function of a milling machine might be:
• To mill 101 ± 1 workpieces per hour to a depth of 11 ± .1 mm.
If this machine is out of action completely for (say) 5% of the time, its availability is 95%. If it is only able to produce 96 pieces per hour when it is running, its efficiency is 96%. If 2% of its output are rejects, its yield is 98%. Applying the above formula gives an overall effectiveness of 0.95 x 0.98 x 0.96 = 0.894, or 89.4%.

This particular composite measure is sometimes referred to as 'overall equipment effectiveness', or OEE. Composite measures of this sort are popular because they allow users to assess maintenance effectiveness at a glance. They also seem to offer a basis for comparing the performance of similar assets (so-called 'benchmarking'). However, these measures actually suffer from numerous drawbacks, as follows:

• the use of three variables in the same equation implies that all three have equal weighting. This may not be the case in practice.

For instance, in the milling machine example above, the workpiece may have a work-in-process value of $200 at that point in the process. The organization might be making a gross profit of $100 on a finished product sale price of (say) $500. This means that 1% downtime or 1% loss of efficiency costs the company one sale per hour – a lost profit of $100 per hour. On the other hand, 1% scrap means that the organization has to write off 1 workpiece per hour, representing $200-worth of work-in-process in addition to $100 lost profit – a total loss of $300 per hour. Consequently, the machine in the above example is losing:
(5 x 100) + (4 x 100) + (2 x 300) = $1 500 per hour
due to downtime, slow running and rejects. However, an identical machine producing the same product might suffer from 4% downtime, run at 98% of its rated speed and produce 4% scrap. In this case the 'overall effectiveness' would be 0.96 x 0.98 x 0.96 = 0.903, or 90.3%. This is apparently a better performance than the first machine. However, this machine is losing:
(4 x 100) + (2 x 100) + (4 x 300) = $1 800 per hour
which is actually a significantly worse performance than the first machine!

• It is possible for many assets to operate too fast as well as too slowly. Overspeeding an asset would increase the OEE as defined above, which means that it possible to obtain an apparent improvement in 'overall' performance by forcing the asset to operate in a failed state.

For instance, a primary performance standard of the milling machine was that it should produce 101 ± 1 workpieces per hour. The '+ 1' means that if the machine produces more than 102 units per hour, it is in a failed state (perhaps because it starts going faster than a bottleneck assembly process, leading to a build-up of work-in-process, or because going too fast causes the milling cutter to overheat and damage the workpieces or because it leads to excessive tool wear.) However, if it operates at 103 workpieces per hour, the apparent 'efficiency' is 102%. This increases the 'overall equipment effectiveness' as defined above, at a time when the machine is actually in a failed state. This is clearly nonsense.

- the OEE as defined above only relates to the primary function of any asset. This is misleading, because as in the case of the gasoline storage system, every asset – machine tools included – have many more functions than the primary function, and each of these will have their own unique performance expectations. Consequently, the OEE is not a measure of 'overall' effectiveness at all, but only a measure of the effectiveness with which the primary function of the asset is being fulfilled.

- finally, for the reasons discussed earlier, truly user-oriented maintenance enterprises need to turn their attention away from 'equipment' effectiveness towards functional effectiveness. So if measures of this sort must be used, it is much more accurate to refer to them as measures of 'primary functional effectiveness' (PFE) rather than 'overall equipment effectiveness'.

Conclusion

The two most important conclusions to emerge from part 2 of this chapter are that:

- when evaluating the contribution which maintenance is making to the performance of any asset, the effectiveness with which *each function* is being fulfilled must be measured on an on-going basis. This in turn requires a crystal clear understanding of *all* the functions of the asset, together with an equally clear understanding of what is meant when it is said to be 'failed'.

- the ultimate arbiter of effectiveness is the *user* (whose expectations must in turn be realistic). What users expect will vary – quite legitimately – from function to function and from asset to asset, depending on the operating context.

14.3 Maintenance Efficiency

As mentioned at the beginning of this chapter, maintenance efficiency measures how well the maintenance function is using the resources at its disposal. The large number of ways in which this can be done are generally well understood, so they are only discussed briefly in this part of this chapter for the sake of completeness.

Efficiency measures can be grouped into four categories, These are *maintenance costs, labor, spares and materials* and *planning and control.*

Maintenance costs

The costs referred to in this part of this chapter are the direct costs of maintenance labor, materials and contractors, as opposed to the indirect costs associated with poor asset performance. The latter issues were discussed in Part 3 of this chapter.

In many industries, the direct cost of maintenance is now the third highest element of operating costs, behind raw materials and either direct production labor or energy. In some cases, it has risen to second or even first place. As result, controlling these costs has become a top priority.

Some industries offer scope for substantial reductions in direct maintenance costs, especially those whose processes embody mature or stable technologies and/or which have a large legacy of second generation thinking embodied in their maintenance practices. However, in other industries, especially those which are newly mechanizing or automating their processes at a significant rate, the sheer volume of maintenance work to be done is often growing at such a pace that maintenance costs are likely to *rise* in absolute terms over the next ten years or so. As a result, take care to evaluate the pace and direction of technological change before committing to substantial long-term reductions in total maintenance costs.

The most common ways in which maintenance costs are measured and analyzed are as follows:
• Total cost of maintenance (actual and budgeted)
  - for the entire facility
  - for each business unit
  - for each asset or system
• Maintenance cost per unit of output
• Ratio of parts to labor expenditure.

Labor

The cost of maintenance labor typically amounts to between one third and two thirds of total maintenance costs, depending on the industry and overall wage levels in the country concerned. In this context, maintenance labor costs should include expenditure on contract labor (which is often – incorrectly – grouped under 'spares and materials' because it is bought out). When considering maintenance labor, it is also wise not to make the common mistake of treating maintenance work done by operators as a zero cost "because the operators are there anyway". In using operators for this work, the organization is still committing resources to maintenance, and the cost should be acknowledged accordingly.

Common ways of measuring and analyzing maintenance labor efficiency include the following:
- Maintenance labor cost (total and per unit of output)
- Time recovery (time performing specific tasks as a percentage of total time paid for)
- Overtime (absolute hours and as a percentage of normal hours)
- Relative and absolute amounts of time spent on different categories of work (proactive tasks, default actions and modifications, and subsets of these categories)
- Backlog (by number of work orders and by estimated hours)
- Ratio of expenditure on maintenance contractors to expenditure on full-time maintenance employees.

Spares and materials
Spares and materials usually account for the portion of maintenance expenditure which does not come under the heading of 'labor'. How well they are managed is usually measured and analyzed in the following ways:
- Total expenditure on spares and materials (total and per unit of output)
- Total value of spares in stock
- Stock turns (total value of spares and materials in stock divided by the total annual expenditure on these items)
- Service levels (percentage of requested stock items which are in stock at the time the request is made)
- Relative and absolute values of different types of stocks (consumables, active spares, 'insurance' spares, dead stock).

Planning and control
How well maintenance activities are planned and controlled affects all other aspects of maintenance effectiveness and efficiency, from the overall utilization of maintenance labor to the duration of individual stoppages. Typical measures include:
- Total hours of predictive/preventive/failure-finding maintenance tasks issued per period
- The above hours as a percentage of total hours
- Percentage of the above tasks completed as planned
- Planned hours worked vs unplanned hours
- Percentage of jobs for which the time was estimated
- Accuracy of estimates (estimated hours vs actual hours for jobs which were estimated).

Some of these efficiency measures are useful for making immediate decisions or initiating short-term management action (expenditure against budgets, time recovery, schedule completion rates, backlogs). Others are more useful for tracking trends and comparing performance with similar facilities in order to plan longer term remedial action (maintenance costs per unit of output, service levels and ratios in general). Together, they are a great help in focusing attention on what must be done to ensure that maintenance resources are used as efficiently as possible.

Maintenance efficiency is also quite easy to measure. The issues which it addresses are usually under the direct control of maintenance managers. For these two reasons, there is often a tendency for these managers to focus too much attention on efficiency and not enough on maintenance effectiveness. This is unfortunate, because the issues discussed under the heading of maintenance effectiveness usually have a much greater impact on the overall physical and financial well-being of the organization than those discussed under maintenance efficiency. Truly 'customer'-oriented maintenance managers direct their attention accordingly. As Part 4 of this chapter explains, the greatest strength of RCM is the extent to which it helps them to do so.

14.4 What RCM Achieves

Figure 1.1 in Chapter 1, reproduced as Figure 14.1, shows how expectations of the maintenance function have evolved over the past fifty years.

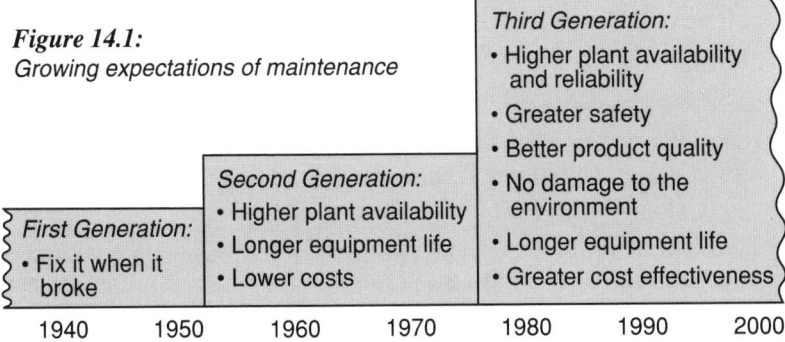

Figure 14.1:
Growing expectations of maintenance

Third Generation:
• Higher plant availability and reliability
• Greater safety
• Better product quality
• No damage to the environment
• Longer equipment life
• Greater cost effectiveness

Second Generation:
• Higher plant availability
• Longer equipment life
• Lower costs

First Generation:
• Fix it when it broke

1940   1950   1960   1970   1980   1990   2000

The use of RCM helps to fulfil all of the Third Generation expectations. The extent to which it does so is summarized in the following paragraphs, starting with safety and environmental integrity.

Greater Safety and Environmental Integrity

RCM contributes to improved safety and environmental protection in the following ways:

* the *systematic review of the safety and environmental implications of every evident failure before considering operational issues* means that safety and environmental integrity become – and are seen to become – top maintenance priorities.

* from the technical viewpoint, *the decision process dictates that failures which could affect safety or the environment **must** be dealt with* in some fashion – it simply does not tolerate inaction, As a result, tasks are selected which are designed to reduce *all* equipment-related safety or environmental hazards to an acceptable level, if not eliminate them completely. The fact that these two issues are dealt with by groups which include both technical experts and representatives of the 'likely victims' means that they are also dealt with realistically.

* the structured approach to protected systems, especially the concept of the hidden function and the orderly approach to failure-finding, leads to substantial improvements in the maintenance of protective devices. This *greatly reduces the probability of multiple failures* which have serious consequences. (This is perhaps the most powerful single feature of RCM. Using it correctly significantly lowers the risk of doing business.)

* involving groups of operators and maintainers directly in the analysis makes them much more sensitive to the real hazards associated with their assets. This makes them *less likely to make dangerous mistakes*, and *more likely to make the right decisions when things do go wrong*.

* the *overall reduction in the number and frequency of routine tasks* (especially invasive tasks which upset basically stable systems) reduce the risk of critical failures occurring either while maintenance is under way or shortly after start-up.

This issue is particularly important if we consider that preventive maintenance played a part in two of the three worst accidents in industrial history (Bhopal, Chernobyl and Piper Alpha). One was caused directly by a proactive maintenance intervention which was currently under way (cleaning a tank full of methyl isocyanate at Bhopal). On Piper Alpha, an unfortunate series of incidents and oversights might not have turned into a catastrophe if a crucial relief valve had not been removed for preventive maintenance at the time.

As mentioned in Part 2 of this chapter, the most common way to track performance in the areas of safety and environmental integrity is to record the number of incidents which occur, typically by recording the number of lost-time accidents per million man-hours in the case of safety, and the number of excursions (incidents where a standard or regulation is breached) per year in the case of the environment. While the ultimate target in both cases is usually zero, the short-term target is always to better the previous record.

To provide an indication of what RCM has achieved in the field of safety, Figure 14.2 shows the number of accidents per million take-offs recorded each year in the commercial civil aviation industry over the period of development of the RCM philosophy (excluding accidents caused by sabotage, military action or turbulence). The percentage of these crashes which were caused by equipment failure also declined. Much of the improved reliability is of course due to the use of superior materials and greater redundancy, but most of these improvements were driven in turn by the realization that maintenance on its own could not extract the required level of performance from the assets as they were then configured. As explained in Chapter 12, this shifted attention from a heavy reliance on fixed time overhauls in the 1960's to doing what ever is necessary to avoid or eliminate the *consequences* of failures, be it maintenance or redesign (the cornerstone of the RCM philosophy). It also reduced the number of crashes which might otherwise have been caused by inappropriate maintenance interventions.

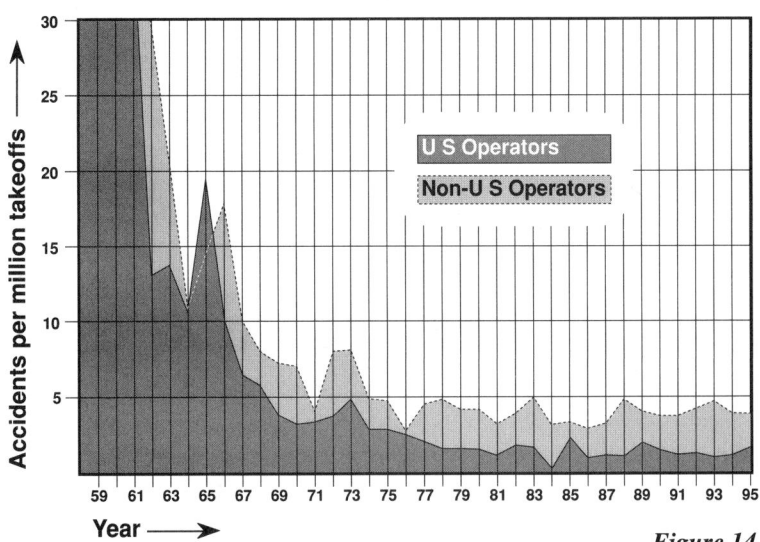

Figure 14.2:
Safety in the civil aviation industry
Source: C A Shifrin: "Aviation Safety Takes Center Stage Worldwide"
Aviation Week & Space Technology: Vol 145 No 19: Pp 46 - 48

Higher Plant Availability and Reliability

The scope for performance improvement clearly depends on the performance at the outset. For example, an undertaking which is achieving 95% availability has less improvement potential than one which is currently only achieving 85%. Nonetheless, if it is correctly applied, RCM achieves significant improvements regardless of the starting point.

For instance, the application of RCM has contributed to the following:
• a 16% increase in the total output of the existing assets of a 24-hour 7-day milk-processing plant. This improvement was achieved in 6 months, and most of it was attributed to an exhaustive RCM review done during this period.
• a 300-ton walking dragline in an open-cast coal mine whose availability rose from 86% to 92% in six months.
• a large holding furnace in a steel mill which achieved 98% availability in its first eighteen months of operation against an expectation of 95%.

Plant performance is of course improved by reducing the number and the severity of unanticipated failures which have operational consequences. The RCM process helps to achieve this in the following ways:

• the *systematic review of the operational consequences of every failure* which has not already been dealt with as a safety hazard, together with the stringent criteria used to assess task effectiveness, ensure that only the most effective tasks are selected to deal with each failure mode.

• the emphasis placed on on-condition tasks helps to ensure that *potential failures are detected before they become functional failures*. This helps reduce operational consequences in three ways:
  - problems can be rectified at a time when stopping the machine will have the least effect on operations
  - it is possible to ensure that all the resources needed to repair the failure are available before it occurs, which shortens the repair time
  - rectification is only carried out when the assets really need it, which extends the intervals between corrective interventions. This in turn means that the asset has to be taken out of service less often.

For instance, the example concerning tires on page 130 shows that the tires need to be taken out of service 20% less often for retreading if on-condition maintenance is used instead of scheduled restoration. In this case, the effect on the availability of the vehicle would be marginal, because removing a tire and replacing it with a new one can be done very quickly. However, in cases where the corrective action requires extensive downtime, the improvement in availability could be substantial.

- by relating each failure mode to the relevant functional failure, the information worksheet provides a tool for *quick failure diagnosis*, which leads in turn to *shorter repair times.*

- the previous example suggests that greater emphasis on on-condition maintenance *reduces the frequency of major overhauls*, with a corresponding long-term increase in availability. In addition, a comprehensive list of all the failure modes which are reasonably likely together with a dispassionate assessment of the relationship between age and failure, reveals that *there is often no reason at all to perform routine overhauls at any frequency.* This leads to a reduction in previously scheduled downtime without a corresponding increase in unscheduled downtime.

  For instance, a RCM enabled a major integrated steelworks to eliminate all fixed-interval overhauls from its steel-making division. In another case, the intervals between major overhauls of a stationary gas turbine on an oil platform were increased from 25 000 to 40 000 hours without sacrificing reliability.

- in spite of the above comments, it is often necessary to plan a shutdown or an overhaul for any of the following reasons:
  - to prevent a failure which is genuinely age-related
  - to rectify a potential failure
  - to rectify a hidden functional failure
  - to carry out a modification.

  In these cases, the disciplined review of the need for preventive or corrective action that is part of the RCM process leads to shorter shutdown worklists, which leads in turn to *shorter shutdowns.* Shorter shutdowns are easier to manage and hence more likely to be completed as planned.

- short shutdown worklists also lead to *fewer infant mortality problems* when the plant is started up again after the shutdown, because it is not disrupted as much. This too leads to an overall increase in reliability.

- as explained on page 268, RCM provides an opportunity for those who participate in the process to *learn quickly and systematically how to operate and maintain new plant.* This enables them to avoid many of the errors which would otherwise be made as a result of the learning process, and to ensure that the plant is maintained correctly from the outset.

  At least four organizations with whom the author has worked in the UK and the USA achieved what each described as 'the fastest and smoothest start-up in the company's history' after applying RCM to new installations. In each case, RCM was applied in the final stages of commissioning. The companies concerned are in the automotive, steel, paper and confectionery sectors.

- the *elimination of superfluous plant* and hence of superfluous failures. As mentioned in Chapter 2, it is not unusual to find that between 5% and 20% of the components of a complex plant are utterly superfluous, but can still disrupt the plant when they fail. Eliminating such components leads to a corresponding increase in reliability.

- by using a group of people who know the equipment best to carry out a systematic analysis of failure modes, it becomes possible to *identify and eliminate chronic failures* which otherwise seem to defy detection, and to take appropriate action.

Improved Product Quality

By focusing directly on product quality issues as shown on pages 48 and 49, RCM does much to improve the yield of automated processes.

For instance, an electronics assembly operation used RCM to reduce scrap rates from 4% (4 000 parts per million) to 50 ppm.

Greater Maintenance Efficiency (Cost-effectiveness)

RCM helps to reduce, or at least to control the rate of growth of maintenance costs in the following ways:

Less routine maintenance:
Wherever RCM has been correctly applied to an existing fully-developed preventive maintenance system, it has led to a reduction of 40% to 70% in the perceived routine maintenance workload. This reduction is partly due to a reduction in the number of tasks, but mainly due to an overall increase in the intervals between tasks. It also suggests that if RCM is used to develop maintenance programs for new equipment or for equipment which is currently not subject to a formal preventive maintenance program, the routine workload would be 40% – 70% lower than if the maintenance program were developed by any other means.

Note that in this context, 'routine' or 'scheduled' maintenance means any work undertaken on a cyclic basis, be it the daily logging of the reading on a pressure gauge, a monthly vibration reading, an annual functional check of a temperature switch or a five-yearly fixed-interval overhaul. In other words, it covers scheduled on-condition tasks, scheduled restoration, scheduled discard tasks and scheduled failure-finding.

For example, RCM has led to the following reductions in routine maintenance workloads when applied to existing systems:

- a 50% reduction in the routine maintenance workload of a confectionery plant
- a 50% reduction in the routine maintenance requirements of the 11 kV transformers in an electrical distribution system
- an 85% reduction in the routine maintenance requirements of a large hydraulic system on an oil platform
- a 62% reduction in the number of low frequency tasks which needed to be done on a machining line in an automotive engine plant.

Note that the reductions mentioned above are only reductions in *perceived* routine maintenance requirements. In many PM systems, fewer than half of the schedules issued by the planning office are actually completed. This figure is often as low as 30%, and sometimes even lower. In these cases, a 70% reduction in the routine workload will only bring what is issued into line with what is actually being done, which means that there will be no reduction in actual workloads.

Ironically, the reason why so many traditionally-derived PM systems suffer from such low schedule completion rates is that much of the routine work is perceived – correctly – to be unnecessary. Nonetheless, if only a third of the prescribed work is being done in any system, that system is wholly out of control. A zero-based RCM review does much to bring situations like these back under control.

Better buying of maintenance services
Applying RCM to maintenance contracts leads to savings in two areas.

Firstly, a clear understanding of failure consequences allows buyers to specify response times more precisely – even to specify different response times for different types of failures or different types of equipment. Since rapid response is often the most costly aspect of contract maintenance, judicious fine-tuning in this area can lead to substantial savings.

Secondly, the detailed analysis of preventive tasks enables buyers to reduce both the content and the frequency of the routine portions of maintenance contracts, usually by the same amount (40% - 70%) as any other schedules which have been prepared on a traditional basis. This leads to corresponding savings in contract costs.

Less need to use expensive experts
If field technicians employed by equipment suppliers attend RCM meetings as suggested on page 269, the exchange of knowledge which takes place leads to a quantum jump in the ability of the maintainers employed by the users to solve difficult problems on their own. This leads to an equally dramatic drop in the need to call for (expensive) help thereafter.

Clearer guidelines for acquiring new maintenance technology
The criteria used to decide whether a proactive task is technically feasible and worth doing apply directly to the acquisition of condition monitoring equipment. If these criteria are applied dispassionately to such acquisitions, a number of expensive mistakes can be avoided.

Most of the items listed under 'improved operating performance'.
Most of the items listed in the previous section of this chapter also improve maintenance cost-effectiveness. How they do so is summarized below:

* *quicker failure diagnosis* means that less time is spent on each repair

* *detecting potential failures before they become functional failures* not only means that repairs can be planned properly and hence carried out more efficiently, but it also reduces the possibility of the expensive secondary damage which could be caused by the functional failure

* the *reduction or elimination of overhauls* together with *shorter worklists for the shutdowns which are necessary* can lead to very substantial savings in expenditure on parts and labor (usually contract labor)

* *the elimination of superfluous plant* also means the elimination of the need either to prevent it from failing in a way which interferes with production, or to repair it when it does fail

* *learning how the plant should be operated* together with *the identification of chronic failures* leads to a reduction in the number and severity of failures, which leads to a reduction in the amount of money which must be spent on repairing them.

  The most spectacular case of this phenomenon encountered by the author concerned a single failure mode caused by incorrect machine adjustment (operator error) in a large process plant. It was identified during an RCM review and was thought to have cost the organization using the asset just under US$1 million *in repair costs alone* over a period of eight years. It was eliminated by asking the operators to adjust the machine in a slightly different way.

Longer Useful Life of Expensive Items

By ensuring that each asset receives the bare minimum of essential maintenance – in other words, the amount of maintenance needed to ensure that what it can do stays ahead of what the users want it to do – the RCM process does much to help ensure that just about any asset can be made to last as long as its basic supporting structure remains intact and spares remain available.

As mentioned on several occasions, RCM also helps users to enjoy the maximum useful life of individual components by selecting on-condition maintenance in preference to other techniques wherever possible.

Greater motivation of individuals

RCM helps to improve the motivation of the people who are involved in the review process in a number of ways. Firstly, a clearer understanding of the functions of the asset and of what they must do to keep it working greatly enhances their competence and hence their confidence.

Secondly, a clear understanding of the issues which are beyond the control of each individual – in other words, of the limits of what they can reasonably be expected to achieve – enables them to work more comfortably within those limits. (For instance, no longer are maintenance supervisors automatically held responsible for every failure, as so often happens in practice. This enables them – and those about them – to deal with failures more calmly and rationally than might otherwise be the case.)

Thirdly, the knowledge that each group member played a part in formulating goals, in deciding what should be done to achieve them and in deciding who should do it leads a strong sense of ownership.

This combination of competence, confidence, comfort and ownership all mean that the people concerned are much more likely to want to do the right job right the first time.

Better Teamwork

In a curious way, teamwork seems to have become both a means to an end and an end in itself in many organizations. The ways in which the highly structured RCM approach to maintenance problem analysis and decision making contributes to teambuilding were summarized on page 268. Not only does this approach foster teamwork within the review groups themselves, but it also improves communication and co-operation between:
• production or operations departments and the maintenance function
• management, supervisors, technicians and operators
• equipment designers, vendors, users and maintainers.

A Maintenance Database

The RCM Information and Decision Worksheets provide a number of additional benefits, as follows:

- *adapting to changing circumstances:* the RCM database makes it possible to track the reason for every maintenance task right back to the functions and the operating context of the asset. As a result, if any aspect of the operating context changes, it is easy to identify the tasks which are affected and to revise them accordingly. (Typical examples of such changes are new environmental regulations, changes in the operating cost structure which affect the evaluation of operational consequences, or the introduction of new process technology.) Conversely it is equally easy to identify the tasks which are *not* affected by such changes, which means that time is not wasted reviewing these tasks.

  In the case of traditionally-derived maintenance systems, such changes often mean that the whole maintenance program has to be reviewed in its entirety. As often as not, this is seen as too big an undertaking, so the system as a whole gradually falls into disuse.

- *an audit trail:* Part 3 of Chapter 5 mentioned that rather than prescribing specific tasks at specific frequencies, more and more modern safety legislation is demanding that the users of physical assets must be able to produce documentary evidence that their maintenance programs are built on rational, defensible foundations. The RCM worksheets provide this evidence – the audit trail – in a coherent, logical and easily understood form.

- *more accurate drawings and manuals:* the RCM process usually means that manuals and drawings are read in a completely new light. People start asking "what does it do?" instead of "what is it?". This leads them to spot a surprising number of errors which may have gone unnoticed in as-built drawings (especially process and instrumentation drawings). This happens most often if the operators and craftsmen who work with the machines are included in the review teams.

- *reducing the effects of staff turnover:* all organizations suffer when experienced people leave or retire and take their knowledge and experience with them. By recording this information in the RCM database, the organization becomes much less vulnerable to these changes.

  For example, a major automotive manufacturer was faced with a situation where a plant was to be relocated and most of the workforce had chosen not to move with the equipment to the new site. However, by using RCM to analyze the equipment before it was moved, the company was able to transfer much of the knowledge and experience of the departing workers to the people who were recruited to operate and maintain the equipment in its new location.

- *the introduction of expert systems:* the information on the Information Worksheet in particular provides an excellent foundation for an expert system. In fact, many users regard this worksheet as a simple expert system in its own right, especially if the information is stored in a simple computerised database and sorted appropriately.

An Integrative Framework

As mentioned in Chapter 1, all of the issues discussed above are part of the mainstream of maintenance management, and many are already the target of improvement programs. A key feature of RCM is that it provides an effective step-by-step framework for tackling *all* of them at once, and for involving everyone who has anything to do with the equipment in the process.

15 A Brief History of RCM

15.1 The Experience of The Airlines

In 1974, The United States Department of Defense commissioned United Airlines to prepare a report on the processes used by the civil aviation industry to prepare maintenance programs for aircraft. The resulting report was entitled *Reliability-centered Maintenance.*

Before reviewing the application of RCM in other sectors, the following paragraphs summarize the history of RCM up to the time of publication of the report by Nowlan and Heap[1978]. The italicized paragraphs quote extracts from their report.

The Traditional Approach to Preventive Maintenance

The traditional approach to scheduled maintenance programs was based on the concept that every item on a piece of complex equipment has a 'right age' at which complete overhaul is necessary to ensure safety and operating reliability. Through the years, however, it was discovered that many types of failures could not be prevented or effectively reduced by such maintenance activities, no matter how intensively they were performed. In response to this problem, airplane designers began to develop design features that mitigated failure consequences – that is, they learned how to design airplanes that were 'failure tolerant'. Practices such as the replication of system functions, the use of multiple engines and the design of damage tolerant structures greatly weakened the relationship between safety and reliability, although this relationship has not been eliminated altogether.

Nevertheless, there was still a question concerning the relationship of preventive maintenance to reliability. By the late 1950's, the size of the commercial airline fleet had grown to the point at which there was ample data for study, and the cost of maintenance activities had become sufficiently high to warrant a searching look at the actual results of existing practices. At the same time the Federal Aviation Agency, which was responsible for regulating airline maintenance practices, was frustrated by experiences showing that it was not possible to control the failure rate of certain unreliable types of engines by any feasible changes in either the content or frequency of scheduled overhauls. As a result, in 1960 a task force was formed, consisting of representatives from both the FAA and the airlines, to investigate the capabilities of preventive maintenance.

The work of this group led to the establishment of the FAA/Industry Reliability Program, *described in the introduction to the authorizing document as follows:*

> *"The development of this program is towards the control of reliability through an analysis of the factors that affect reliability and provide a system of actions to improve low reliability levels when they exist. In the past, a great deal of emphasis has been placed on the control of overhaul periods to provide a satisfactory level of reliability. After careful study, the Committee is convinced that reliability and overhaul time control are not necessarily directed at associated topics ...."*

This approach was a direct challenge to the traditional concept that the length of time between successive overhauls of an item was an important factor in controlling its failure rate. The task force developed a propulsion-system reliability program, and each airline involved in the task force was then authorized to develop and implement reliability programs in the area of maintenance in which it was most interested. During this process, a great deal was learned about the conditions that must exist for scheduled maintenance to be effective. Two discoveries were especially surprising:

- **Scheduled overhaul has little effect on the overall reliability of a complex item unless the item has a dominant failure mode**
- **There are many items for which there is no effective form of scheduled maintenance**

The History of RCM Analysis

The next step was an attempt to organize what had been learned from the various reliability programs and develop a logical and generally applicable approach to the design of preventive maintenance programs. A rudimentary decision-diagram technique was devised in 1965, and in June 1967 a paper on its use was presented at the AIAA Commercial Aircraft Design and Operations Meeting. Subsequent refinements of this technique were embodied in a handbook on maintenance evaluation and program development, drafted by a maintenance steering group formed to oversee development of the initial program for the new Boeing 747 airplane. This document, known as MSG-1, was used by special teams of industry and FAA personnel to develop the first scheduled-maintenance program based on the principles of reliability-centered maintenance. The Boeing 747 maintenance program has been successful.

Use of the decision-diagram technique led to further improvements, which were incorporated two years later in a second document, MSG-2: Airline Manufacturer Maintenance Program Planning Document.

MSG-2 was used to develop the scheduled maintenance programs for the Lockheed 1011 and the Douglas DC10 airplanes. These programs have also been successful. MSG-2 has also been applied to tactical military aircraft; the first applications were for aircraft such as the Lockheed S-3 and P-3 and the McDonnell F4J. A similar document prepared in Europe was the basis for the initial programs for such aircraft as the Airbus Industrie A-300 and the Concorde.

The objective of the techniques outlined in MSG-1 and MSG-2 was to develop a scheduled-maintenance program that assured the maximum safety and reliability of which the equipment was capable and also provided them at the lowest cost. As an example of the economic benefits achieved with this approach, under traditional maintenance policies the initial program for the Douglas DC-8 airplane required scheduled overhaul for 339 items, in contrast to seven such items in the DC-10 program. One of the items no longer subject to overhaul limits in the later programs was the turbine propulsion engine. Elimination of scheduled overhauls for engines led to major reductions in labor and materials costs, and also reduced the spare-engine inventory required to cover shop maintenance by more than 50%. Since engines for larger airplanes then cost more than US$1 million each, this was a respectable saving.

As another example, under the MSG-1 program for the Boeing 747, United Airlines expended only 66 000 manhours on major structural inspections before reaching a basic interval of 20 000 hours for the first heavy inspections of this airplane. Under traditional maintenance policies it took an expenditure of more than 4 million manhours to arrive at the same structural inspection interval for the smaller and less complex Douglas DC-8. Cost reductions of this magnitude are of obvious importance to any organization responsible for maintaining large fleets of complex equipment. More important:

- *Such cost reductions are achieved with no decrease in reliability. On the contrary a better understanding of the failure process in complex equipment has actually improved reliability by making it possible to direct preventive tasks at specific evidence of potential failures.*

Although the MSG-1 and MSG-2 documents revolutionized the procedures followed in developing maintenance programs for transport aircraft, their application to other types of equipment was limited by their brevity and specialized focus. In addition, the formulation of certain concepts was incomplete. For example, the decision logic began with an evaluation of proposed tasks, rather than an evaluation of the failure consequences that determine whether they are needed, and if so, their actual purpose. The problem of establishing task intervals was not addressed, the role of hidden-function failures was unclear, and the treatment of structural maintenance was inadequate. There was also no guidance on the use of operating information to refine or modify the initial program after the equipment entered service or the information systems needed for effective management of the on-going program.

All these shortcomings, as well as the need to clarify many of the underlying principles led to analytic procedures of broader scope and their crystallization into the logical discipline now known as Reliability-centered Maintenance.

15.2 The Evolution of RCM2

The author and his associates began working with the application of RCM in the mining and manufacturing sectors in the early 1980s. They used a very slightly modified version of the Nowlan and Heap diagram between 1983 and 1990. During this period, the environment became more and more of an issue. In the early days, facilitators were advised to treat environmental hazards in the same way as safety hazards. However, in practice this meant that many environmental problems which did not pose an immediate and direct threat to safety were overlooked. The environment can also be a highly contentious issue which does not lend itself to subjective evaluation in the same way as safety.

As a result, in 1988 the author began working with a number of multinational organizations to develop a more precise approach to failures which threatened the environment. This culminated in the addition of question

E to the decision diagram in 1990. The use of standards and regulations as a basis for this decision removed the element of subjectivity. However, the whole issue was still accorded much the same priority as safety in recognition of the high and rising priority which society places on the environment, as discussed at length in Part 3 of Chapter 5. The addition of this question alone changed the decision diagram enough to warrant changing its name to RCM 2.

Other changes incorporated in RCM 2

When RCM 2 was launched in September 1990, a number of other changes were incorporated into the decision process which had been under development for several years. These were as follows:

- the terms 'technically feasible' and 'worth doing' were substituted for 'applicable' and 'effective'.

- the small but significant number of cases where failure-finding was either impossible or impractical led to the addition of formal selection criteria for this task. It also led to the addition of the secondary default decision process for hidden functions explained on page 186.

- question H was amended to remove a number of ambiguities.

- question S was also altered to remove possible ambiguities surrounding the meaning of the word 'safety'.

- the italicized extension was added to question O because many users tended to interpret this question too narrowly.

- questions H1/S1/O1/N1 were modified to make them easier to understand.

- the term 'scheduled restoration' was substituted for 'scheduled rework' in questions H2/S2/O2/N2 because 'rework' has a different meaning in manufacturing companies. This frequently caused confusion.

- the questions on the revised decision diagram were re-coded.

With the possible exception of the question concerning environmental consequences, none of these changes represent a significant departure from the philosophy underlying the original Nowlan and Heap decision diagram. .

The nett effect of these changes has been to make a technique which is already extraordinarily robust at the theoretical level even more robust, and to make it quicker and easier to use into the bargain.

Where RCM2 has been applied

RCM2 has been applied on more than 1 000 sites in 41 countries. The projects range from in-company awareness training for senior operations and maintenance managers to the full-scale application of RCM to all the equipment on a site. The sectors in which projects have been carried out span nearly every major field of organized human endeavor. These include mining, manufacturing, petrochemicals, utilities (electricity, gas and water), mass transport (especially railways), buildings and building services and military undertakings (armies, navies and air forces).

Space does not permit a detailed consideration of the work done in each case. However, Chapter 14 provides a general summary of the results achieved to date, together with a brief review of some of the highlights.

15.3 Other Versions of RCM and the SAE Standard

The rest of this chapter provides a summary of the evolution of RCM in general. The italicized paragraphs are extracted directly from an article entitled *"SAE's New RCM Standard"* by Netherton[2000].

In 1980, ATA [the Air Transport Association of America] produced MSG-3, Airline/Manufacturer Maintenance Program Development Document. MSG-3 was influenced by Nowlan and Heap's 1978 book, but it was intended as a continuation of the tradition begun by the earlier MSG documents. It is the document that today guides the development of initial scheduled maintenance programs for new US commercial aircraft.

However the line of thought in Nowlan and Heap's book took on a life of its own. They were commissioned to write their book by the US Department of Defense (DoD), which was looking at commercial industry for ways to make its own maintenance programs less costly. The DoD had learned that commercial aviation had found a revolutionary approach to scheduled maintenance, and hoped to benefit from their experience. Once the DoD published Nowlan and Heap's book, the US military embarked on developing RCM processes for its own use: one for the US Army, one for the US Air Force, and two for the US Navy – because the Navy's shipboard and aviation communities insisted that an RCM process that work for one would not work for the other. Support contractors and equipment vendors learned to use these processes when they sold new equipment to the US military. The processes were published in Military Standards and Military specifications (which were never updated) in the mid-1980's.

In separate but parallel work in the early 1980s, the Electric Power Research Institute (EPRI), an industry research group for the US electrical power utilities, carried out two pilot applications of RCM in the US nuclear power industry. Their interest arose from a belief that this industry was achieving adequate levels of safety and reliability, but was massively overmaintaining its equipment. As a result, their main thrust was simply to reduce maintenance costs, rather than to improve reliability, and they modified the RCM process accordingly. (So much so, in fact, that it bears little resemblance to the original RCM process described by Nowlan and Heap; it should be more correctly described as planned maintenance optimization, or PMO, rather than RCM.) This modified process was adopted on an industry-wide basis by the American nuclear power industry in 1987, and variations of this approach were afterwards adopted by various other nuclear utilities, other branches of the electricity generation and distribution industry, and parts of the oil industry.

At the same time, certain specialists in the formulation of maintenance strategies became interested in the application of RCM in industries other than aviation. Foremost among these were John Moubray and his associates. This group initially worked with RCM in mining and manufacturing industries in Southern Africa under the mentorship of Stan Nowlan, and subsequently relocated to the UK. From the UK their activities have expanded to cover the application of RCM in nearly every industrial sector spanning more than 40 countries. They have built on Nowlan's work while maintaining its original focus on equipment safety and reliability. For example, they have incorporated environmental issues into the decision-making process, clarified the ways in which equipment functions should be defined, developed more precise rules for selecting maintenance tasks and task intervals, and incorporated quantitative risk criteria directly into the setting of failure-finding task intervals. Their enhanced version of RCM is now known as RCM2.

The Need for a Standard: the 1990s
Since the early 1990's, a great many more organizations have developed variations of the RCM process. Some, such as the US Naval Air Command with its "Guidelines for the Naval Aviation Reliability Centered Maintenance Process (NAVAIR 00-25-403)" and the British Royal Navy with its RCM-oriented Naval Engineering Standard (NES45), have remained true to the process originally expounded by Nowlan and Heap. However, as the RCM bandwagon has started rolling, a whole new collection of processes has emerged that are called "RCM" by their proponents, but

that often bear little or no resemblance to the original meticulously researched, highly structured and thoroughly proven process developed by Nowlan and Heap. As a result, if an organization said that it wanted help in using or learning how to use RCM, it could not be sure what process would be offered.

Indeed, when the US Navy recently asked for equipment vendors to use RCM when building a new ship class, one US company offered a process closely related to the 1970 MSG-2 process. It defended its offering by noting that its process used a decision-logic diagram. Since RCM also uses a decision-logic diagram, the company argued, its process was an RCM process.

The US Navy had no answer to this argument, because in 1994 William Perry, the US Secretary of Defense, had established a new policy about US military standards and specifications, which said that the US military would no longer require industrial vendors to use the military's "standard" or "specific" processes. Instead it would set performance requirements, and would allow vendors to use any processes that would provide equipment that would meet these requirements.

At a stroke, this voided the US military standards and specifications that defined "RCM". The US Air Force standard was cancelled in 1995. The US Navy has been unable to invoke its standards and specifications with equipment vendors (though it continues to use them for its internal work) — and it was unable to invoke them with the US company that wished to use MSG-2.

This development happened to coincide with the sudden interest in RCM in the industrial world. During the 1990s, magazines and conferences devoted to equipment maintenance have multiplied, and magazine articles and conference papers about RCM became more and more numerous. These have shown that very different processes are being given the same name, "RCM". So both the US military and commercial industry saw a need to define what an RCM process is.

In his 1994 memorandum, Perry said, "I encourage the Under Secretary of Defense (Acquisition and Technology) to form partnerships with industry associations to develop non-government standards for replacement of military standards where practicable." Indeed, the Technical Standards Board of the SAE has had a long and close relationship with the standards community in the US military, and has been working for several years to help develop commercial standards to replace military standards and specifications, when needed and when none already existed.

So in 1996 the SAE began working on an RCM-related standard, when it invited a group of representatives from the US Navy aviation and ship RCM communities to help it develop a standard for Scheduled Maintenance Programs. These US Navy representatives had already been meeting for about a year in an effort to develop a US Navy RCM process that might be common between the aviation and ship communities, so they had already done a considerable amount of work when they began to meet under SAE sponsorship. In late 1997, having gained members from commercial industry, the group realized that it was better to focus entirely on RCM. In 1998, the group found the best approach for its standard, and in 1999 it completed its draft of the standard, and the SAE approved it and published it.

After a brief discussion about the practical difficulties associated with attempting to develop a universal standard of this nature, Netherton goes on to say:

The standard now approved by the SAE does not present a standard [RCM] process. Its title is, "Evaluation Criteria for Reliability-Centered Maintenance (RCM) Processes (SAE JA1011)." This standard presents criteria against which a process may be compared. If the process meets the criteria, it may confidently be called an "RCM process." If it does not, it should not. (This does not necessarily mean that processes that do not comply with the SAE RCM standard are not valid processes for maintenance strategy formulation. It simply means that the term "RCM" should not be applied to them.)

As mentioned in Chapter 1, the RCM process described in Chapters 2 to 10 of this book complies fully with the SAE Standard.

Appendix 1:
Asset Hierarchies and Functional Block Diagrams

Plant registers and asset hierarchies

Most organizations own, or at least use, hundreds if not thousands of physical assets. These assets range in size from small pumps to steel rolling mills, aircraft carriers or office blocks. They may be concentrated on one small site or spread over thousands of square miles. Some of these assets will be mobile, others will be fixed.

Before any organization can apply RCM – a process used to determine what must be done to ensure that *any physical asset* continues to do whatever its users want it to do – it must know what these assets are and where they are. In all but the smallest and simplest facilities, this means that a list of all the plant, equipment and buildings owned or used by the organization, and which require maintenance of any sort, must be prepared. This list is known as the *plant register.*

The register should be designed in a way which makes it possible to keep track of the assets that have been analyzed using RCM, those that have yet to be analyzed and those that are not going to be analyzed. (The plant register is also needed for other aspects of maintenance management, such as the planning and scheduling of routine and non-routine maintenance tasks, history recording and maintenance cost allocation. As a result, it should be set up and the associated numbering systems designed in such a way that it can be used for all these purposes.)

Chapter 4 explained that RCM can be applied at almost any level in a hierarchy. It also suggested that the most appropriate level is the level which leads to a reasonably manageable number of failure modes per function. 'Appropriate' levels become much easier to identify if the plant register is set up as a hierarchy which makes it possible to identify any system or any asset at any level of detail, down to and including individual components ('line replaceable items') or even spare parts.

The truck on page 85 provides one example of such a hierarchy. Figure A1.1 overleaf shows another example covering a boiler house in a food factory.

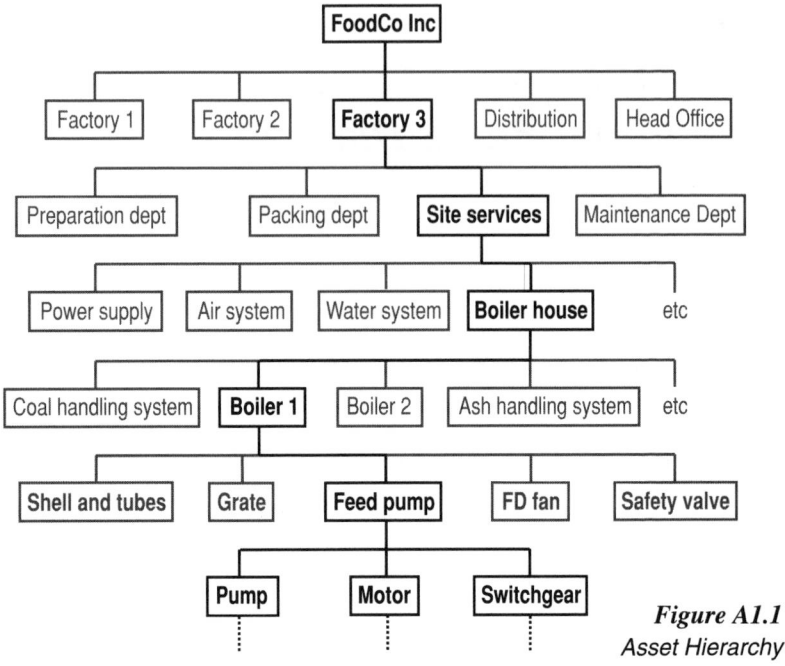

Figure A1.1
Asset Hierarchy

A list drawn up for the assets in this hierarchy, together with a hierarchical numbering system for each asset, might appear as shown in Figure A1.2.

Number	Asset
-	**FoodCo Inc**
01	**Factory 1**
02	**Factory 2**
03	**Factory 3**
0301	*Preparation Dept*
0302	*Packing Dept*
0303	*Site Services*
030301	Power Supply
030302	Compressed Air System
030303	Water System
030304	Boiler House
03030401	Coal Handling System
03030402	Boiler No 1
0303040201	Shell & Tubes
0303040202	Grate
0303040203	Feed Pump
030304020301	Pump
030304020302	Motor
030304020303	Switchgear
0303040204	FD Fan
03030403	Boiler No 2
03030404	Ash Handling System
0304	*Maintenance Dept*
04	**Distribution**
05	**Head Office**

Figure A1.2
Plant Register and
Hierarchical Numbering System

Functional hierarchies and functional block diagrams

It is possible to develop a hierarchy showing the primary functions of each of the assets in the asset hierarchy. Figure A1.3 overleaf shows how this might be done for the asset hierarchy shown in Figure A1.1.

Variations of the functional hierarchy in Figure A1.3 are used to show the relationships between functions at the same level. These are usually known as 'functional block diagrams', and they can be used to depict the relationships in a number of different ways. For instance, Smith[1993] defines a functional block diagram as 'a top level representation of the major functions that the system performs'. On the other hand, Blanchard & Fabrycky[1990], who prefer the term 'functional flow diagram', suggest that these diagrams can be prepared at many different levels. Smith tends to use the diagrams to show the movement of materials, energy and control signals through and between different elements of a system, whereas Blanchard & Fabrycky use them to depict the movement of single asset through different mission phases (such as an aircraft moving from start-up to taxi, take-off, climb, cruise, descent, landing and so on.)

A functional block diagram for the boiler house in Figure A1.1 shows that the coal flows from the coal handling plant to the two boilers, and the residue to the ash handling plant. It also shows what materials and services flow across the system boundaries. This is illustrated in Figure A1.4 on page 332, which goes on to show a more detailed functional block diagram for one of the boilers. A more complex version of these diagrams could also be used to show what control and indication signals pass across the system boundaries.

Functional hierarchies and functional block diagrams are an essential part of the equipment *design* process, because design starts with a list of desired functions and designers have to specify an entity (asset or system) which is capable of fulfilling each functional requirement.

As mentioned in Chapter 2, functional block diagrams can also be of some help when RCM is applied to facilities where the processes or the relationships between them are not intuitively obvious. These tend to be large, poorly accessible, very complex, monolithic structures such as naval vessels, combat aircraft and the less accessible parts of nuclear facilities.

However, in most other industrial applications (such as thermal power stations, food and automotive manufacturing plants, offshore oil platforms, petrochemical and pharmaceutical plants and vehicle fleets), there is usually no need to draw up functional block diagrams before embarking on an RCM project, for the following reasons:

WHAT IT IS ...

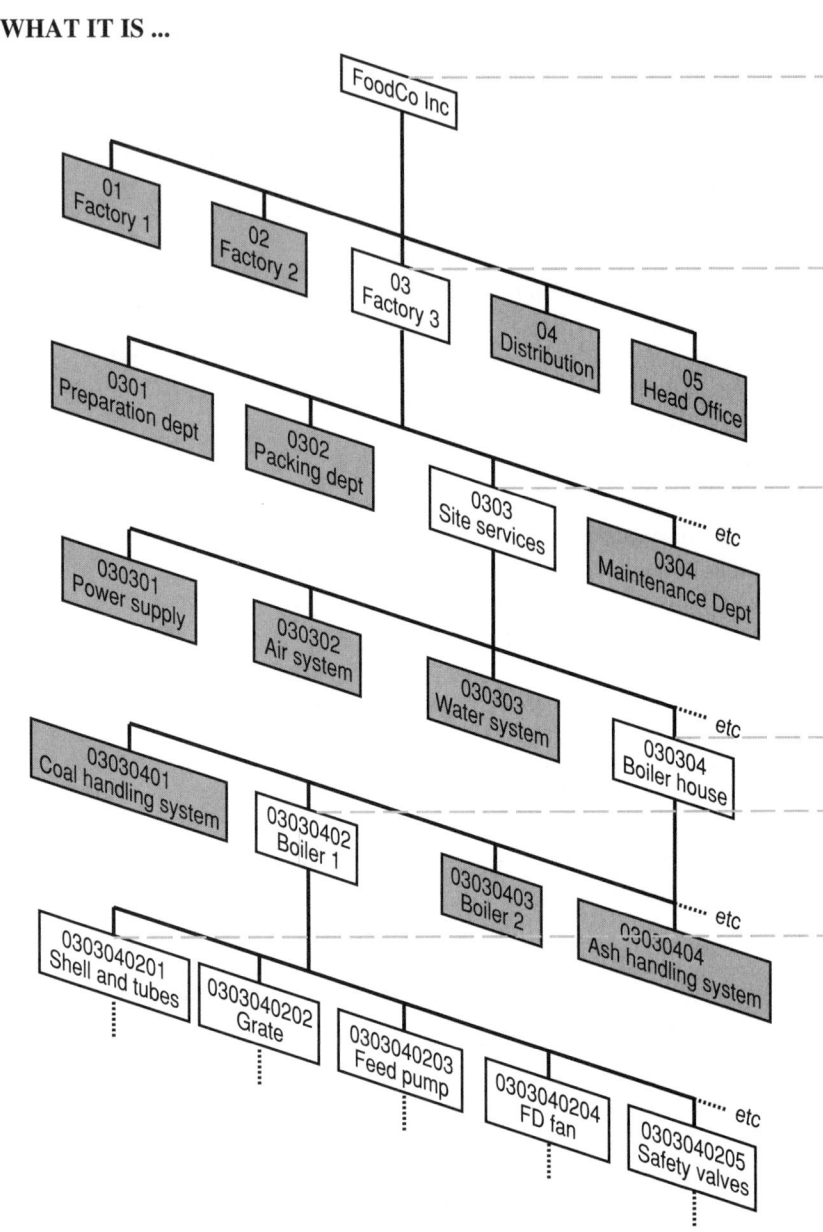

Figure A1.3
Asset Hierarchy .....

WHAT IT DOES ...

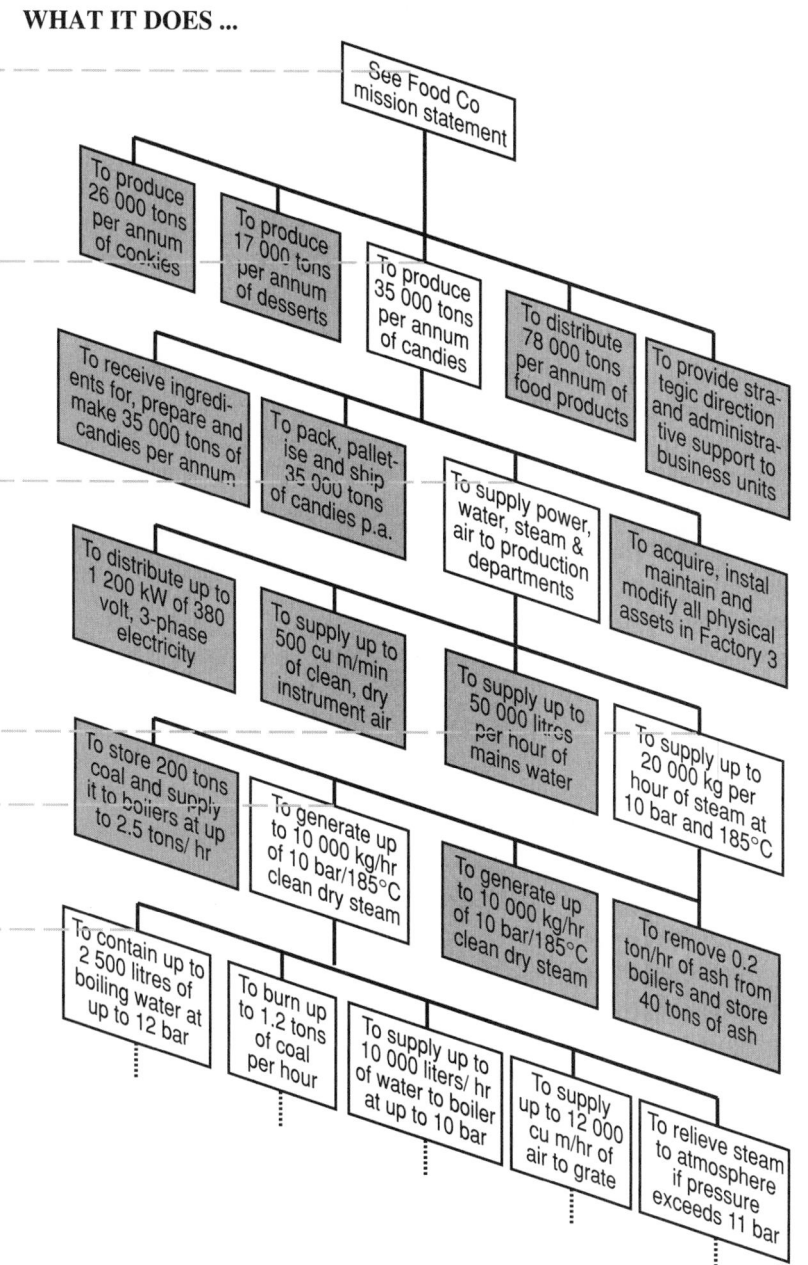

Figure A1.3 (continued)
..... *with Corresponding Functional Hierarchy*

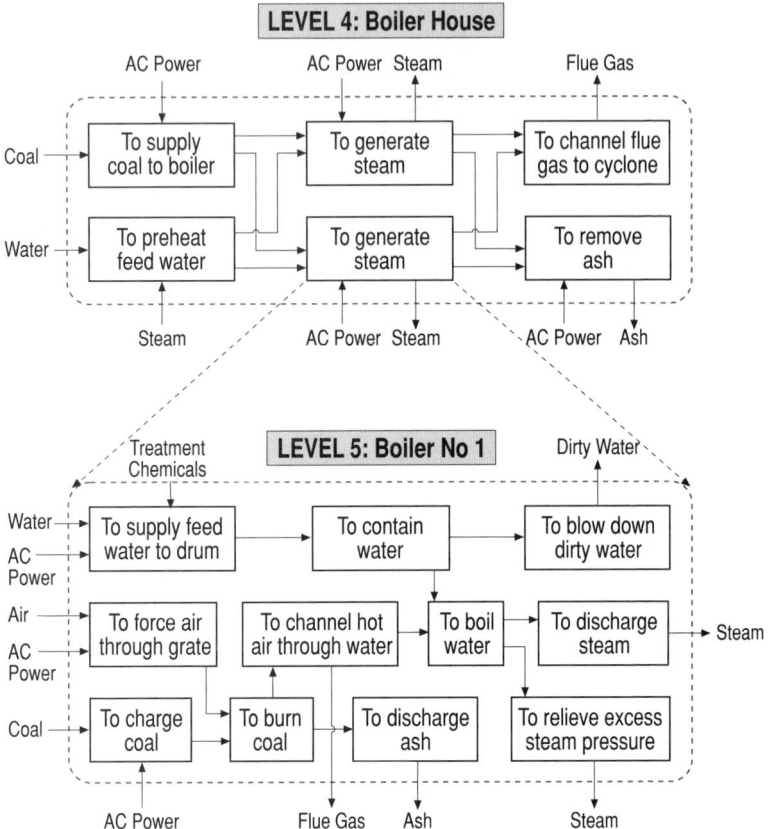

Figure A1.4: *Functional Block Diagrams*

- in most industries, the relationships between different processes is usually well enough understood by the participants in RCM review groups to make these diagrams unnecessary.

For example, the boiler house operators and maintainers would be fully aware of the fact that coal, water and air go into the boiler at one end and that steam, flue gas and ash (and occasionally dirty water) come out of the other. Most of them would probably regard the notion that these simple facts should be drawn in a diagram at best as a waste of time. As discussed at length in Chapter 2, the real challenge is not to identify the simple and usually obvious relationships between processes, but to define the desired performance relative to the initial capability for all the key elements of each system, and then to define what must be done to ensure that the system continues to deliver the desired performance.

In cases of uncertainty, equipment is usually accessible enough for it to be easy to go and see what goes where, and if not, the required information can be extracted from a set of process and instrumentation drawings. In fact, a good set of P&ID's nearly always eliminates the need for functional block diagrams entirely as a precursor to the application of RCM. In such cases, block diagrams add significantly to the time, effort and cost of the RCM process while adding nothing to its value.

• functional block diagrams only identify primary functions at each level, so they only tell part of the story. (For instance, nearly all the assets at the fourth level and below in Figure A1.1 have a secondary containment function. This cannot be shown in a functional block diagram without making it unmanageably cumbersome.)

• as explained in Part 3 of Chapter 2, the principal functions of the assets in the hierarchy *above* the level chosen for the analysis should be summarized in suitably worded operating context statements. These statements are written *only* for those assets which are relevant to the analysis in question. As a result, time is not wasted defining the functions of assets which are not germane to the asset under consideration. (If large numbers of assets are analyzed, these high-level context statements evolve into a de facto functional hierarchy for the entire organization – one that is far more detailed than a crude, single-statement-per-asset diagram.)

• assets *at or below* the level chosen for the analysis are dealt with as part of the normal RCM process. Part 7 of Chapter 4 showed that the functions of lower level assets are either listed as secondary functions in the main analysis, or dealt with as failure modes, or in the case of exceptionally complex subsystems, broken out for separate analysis.

For instance, the example of the truck shown in Figure 4.11 in Chapter 4 showed how a blockage in the fuel line could simply be treated as a failure mode of either the engine or the drive system, without needing a separate function statement for the fuel system or the fuel line.

(In the author's experience, functional block diagrams tend to be of most value to outsiders seeking to apply RCM on behalf of equipment users. Because they are outsiders, they need these diagrams – usually prepared at the expense of the owners of the assets – to improve their own understanding of the processes which they are about to analyze. The best way to avoid this expense is not to employ outsiders as analysts in the first place, but rather to train people who have a reasonable first-hand working knowledge of the plant as RCM facilitators.)

System boundaries

When applying RCM to any asset or system, it is of course important to define clearly where the 'system' to be analyzed begins and where it ends. If a comprehensive asset hierarchy has been drawn up and a decision taken to analyze a particular asset at a particular level, then the 'system' usually automatically encompasses all the assets below that system in the asset hierarchy. The only exceptions are subsystems which are judged to be so insignificant that they will not be analyzed at all, or very complex subsystems which are set aside for separate analysis.

Care is needed with control loops which consist of a sensor in one system which sends a signal to a processor in a second system, which in turn activates an actuator in a third. Chapter 4 explained that this issue can often be dealt with either by conducting the analysis at a high enough level to ensure that the 'system' encompasses the entire loop, or by analyzing control systems separately (after the controlled systems have been analyzed). However, sometimes this is not practical, in which case a decision must be made as to which system will encompass the control loop in its entirety.

Care is also needed to ensure that assets or components right on the boundaries do not 'fall between the cracks'. This applies especially to items like valves and flanges.

It is wise not to be too rigid about boundary definitions, because as understanding grows during the RCM process, perceptions about what should or should not be incorporated in the analysis frequently change. This means that boundaries may need to be extended to incorporate some subsystems, others may be dropped and yet others which are included initially may be set aside for later analysis.

(Again, the strongest exponents of rigid boundary definitions tend to be external contractors seeking to apply RCM on behalf of end-users, because system boundaries must be defined precisely in order to define the commercial scope of the contracts. The fact that the analysis is the subject of a formal contract means that boundaries have to be defined much more precisely – and much more rigidly – than is necessary from a purely technical point of view. Contracts of this type then either have to be renegotiated every time a boundary needs to be changed, or the boundary is not moved when it should be, resulting in a suboptimal analysis. The best way to avoid the time and cost associated with these commercial maneuverings is to avoid contracting out this aspect of maintenance policy formulation altogether.)

Appendix 2:
Human Error

Chapter 4 mentioned that a great many equipment failures are caused by 'human error'. It went on to mention that if a specific human error is considered to be a credible reason why a functional failure could occur, then that error should be included in the FMEA. However, human error is an enormous subject in its own right. The purpose of this appendix is to provide a brief summary of the major categories of human error, and to suggest how they might be dealt within the framework of RCM.

Principal Categories of Human Error

When considering the interaction between people and machines, Blanchard et al[1995] group the main factors under four headings:
- anthropometric factors
- human sensory factors
- physiological factors
- psychological factors.

Nearly every 'human error' can be traced to a failure or a problem which has occurred in at least one of these four areas. As a result, we review them briefly in the first part of this appendix, before looking in more detail at the fourth category.

Anthropometric factors
Anthropometric factors are those which relate to the size and/or strength of the operator or maintainer. Errors occur because a person (or part of a person, such as a hand or arm):
- simply cannot fit into the space available to do something
- cannot reach something
- is not strong enough to lift or move something
If a failure is occurring or is reasonably likely to occur for any of these reasons, it is highly unlikely that a proactive maintenance task will be found to deal with it. Note also that if a human error occurs for one of these reasons, the human error is not the root cause. The failure mode is actually poor design and the resulting human error is a failure effect.

If the consequences are such that something must be done about a failure which is occurring for anthropometric reasons, the only viable course of action is likely to be redesign. This will nearly always involve reconfiguring the asset in such a way that it becomes more accessible or easier to move. In this context, Figure A2.1 shows some dimensions which are considered by the US Navy to be adequate for reasonable human access in confined spaces.

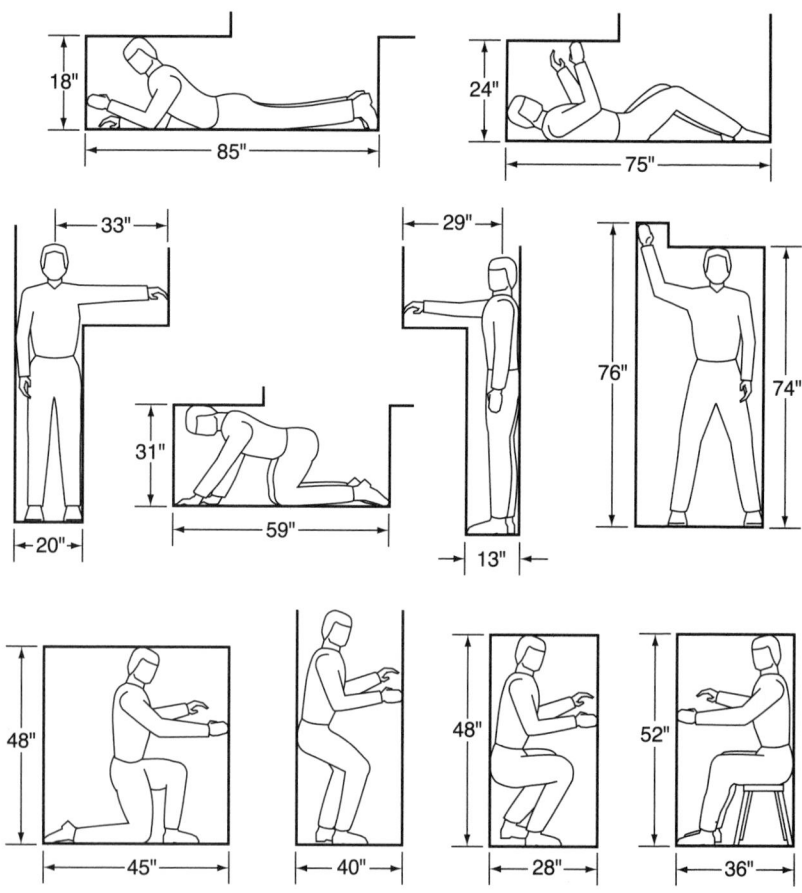

Figure A2.1:
Where people fit
(From *NAVSHIPS 94234, Maintainability Design Criteria Handbook for Designers of Shipboard Electronic Equipment.* US Navy, Washington DC)

Human sensory factors
Human sensory factors concern the ease with which people can see, hear, feel and even smell what is going on around them. In the case of operators, this tends to apply to the visibility and legibility of instruments and control consoles. For maintainers, it relates to the visibility of components in the nooks and crannies of complex systems. The volume and variability of background noise levels also affect the ability of both operators and maintainers to discern what is happening to their equipment.

Note again that if errors are occurring or thought to be likely for these reasons, the human error is not the root cause, but is the effect of some other failure. The remedies also usually entail redesigning the asset (making things easier to see, reducing noise levels).

Physiological factors
The term 'physiological factors' refers to environmental stresses which affect human performance. The stresses include high or low temperatures, loud or irritating noises, excessive humidity, high vibration, exposure to toxic chemicals or radiation, or simply working for too long – especially at a physically or mentally demanding task – without an adequate break.

Sustained exposure to these stresses leads to reduced sensory capacity, slower motor responses and reduced mental alertness. These are all manifestations of (human) fatigue, and all greatly increase the chances that the people concerned will make a slip, lapse or mistake. (These three terms are defined in the next section of this appendix.)

If errors occur or are thought to be likely for any of these reasons, the human is once again not the root cause, but the error is the effect of some other failure. Again, if the consequences warrant it, the remedy is likely to be some form of once-off change. Either the design of the physical environment can be changed in such a way that the error-inducing stresses are reduced (for instance, by reducing temperatures or by providing hearing protection), or operating procedures could be changed in a way that gives overstressed people a chance to recover (longer, more frequent or more carefully timed rest breaks).

Another environmental stress factor is a relentlessly hostile or adversarial organizational climate. While this does not necessarily have a physio-logical effect, it can lead to an increased predisposition towards psychological errors. In many cases, it boils down to excessive and inappropriate use of high task/low relationship leadership styles. Unfortunately, there is not much that RCM can do about this problem.

However, what RCM can do is alleviate – if not eliminate – the hostile relationship which so often exists between maintenance and operations people, as explained on page 268. This makes people less inclined to blame each other for errors, and more inclined to find solutions.

Psychological factors
The three sets of factors discussed so far all relate to external phenomena which cause the human to make an error. As a result, they are relatively easy to identify and to deal with (although doing so may sometimes be expensive). A far more complex and challenging category of errors are those which find their roots in the psyches of the humans themselves. As a result, these psychological factors are discussed in more detail in the next section of this appendix.

Psychological errors

Reason [1991] divides the psychological categories of human error into those which are unintended and those which are intended. An unintended error is one which occurs when someone does a task which he or she should be doing, but does it incorrectly ("does the job wrong"). An intended error occurs when someone deliberately sets out to do something, but what they do is inappropriate ("does the wrong job"). Reason divides these two categories further as follows:

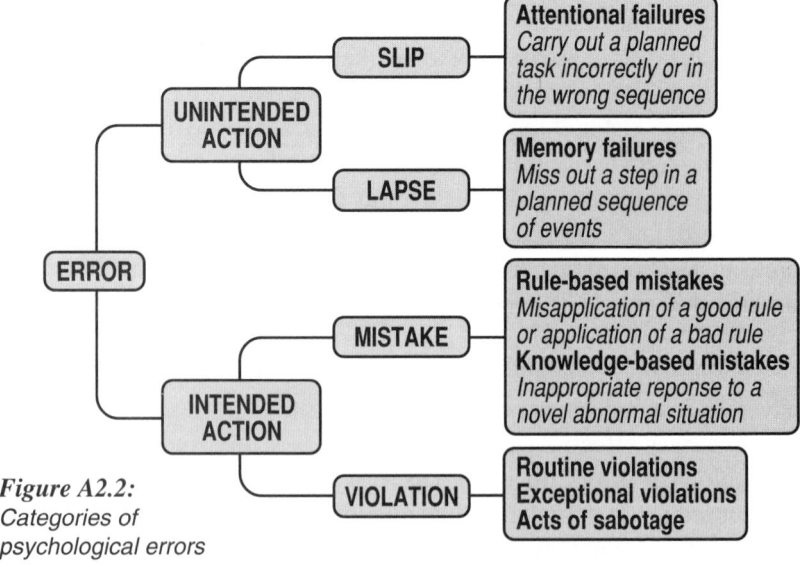

Figure A2.2:
Categories of
psychological errors

- unintended errors are subdivided into slips and lapses
- intended errors are subdivided into mistakes and violations.

These categories are illustrated in Figure A2.2 and are discussed briefly in the following paragraphs.

Slips and lapses

Slips and lapses are also known as *skill-based* errors. They occur when somebody who is fully qualified to do a job – and who may even have done it correctly many times in the past – does the job incorrectly. *Slips* occur when somebody does something incorrectly (for instance, if an electrician wires a motor incorrectly, causing it to run backwards). *Lapses* occur when someone misses out a key step in a sequence of activities (for instance, if a mechanic leaves a tool behind after working in a machine or forgets to fit a key component while reassembling it.)

These errors usually happen because the person concerned was distracted, preoccupied or simply 'absent minded'. As a result, they are usually unpredictable, although their likelihood increases if the person is working in a physically hostile environment, or if the task is exceedingly complex. However, if the environment is reasonably benign and the task is fairly simple, then this category of human errors is perhaps the only one where it is fair to describe the error as the root cause of failure.

The possibility of a great many slips and lapses can be reduced if operators and maintainers are involved directly in the RCM process (especially the FMEA). This leaves them with a much broader and deeper understanding of the effects and consequences of their actions, which in turn results in greater motivation to do the job 'right first time'. This applies especially to tasks where the consequences of failure are likely to be most serious.

Another approach to slips which occur during assembly is based on the assumption that if something can be installed incorrectly, it will be. The remedy is to go back to the drawing board and:

- redesign systems in such a way that they can only be assembled in the correct sequence
- redesign individual components in such a way that they can only be installed the right way round and in the right place.

This is the essence of the Japanese concept of *poka yoke* ('mistake proofing'). Ideally this philosophy should be applied to original designs rather than retrofitted to existing assets, because it is usually cheaper to build in good practice initially than to modify out bad practice later.

Mistakes 1: Rule-based mistakes
Rule-based mistakes occur when people believes that they are following the correct course of action when doing a task (in other words, applying a 'rule'), but in fact the course of action is inappropriate. Rule-based mistakes are further subdivided into *misapplication of a good rule* and *application of a bad rule.*

In the first case, under a given set of conditions, a person selects a course of action which seems appropriate, usually because it has been successful in dealing with similar conditions in the past – hence the term 'good rule'. However, some subtle variations on this occasion mean that the course of action, undertaken deliberately, is wrong.

For instance, a protected system might be set up in such a way that excessive pressure should cause an alarm to sound and a warning light to illuminate. However, a situation might arise where the alarm is failed, the pressure increases and the light comes on. The absence of the alarm may lead the operator to believe that the warning light on its own is only a false alarm, especially if it has a history of spurious failures. In this case, the operator may choose to take no action until the light is repaired – a course of action which has been appropriate in the past. On this occasion however, it is not the right thing to do.

The application of a bad rule means just what it says. The normally chosen or prescribed course of action is just plain wrong.

A classic example of a bad rule is a maintenance program which schedules items for fixed interval overhauls in order to deal with failure modes which conform to failure pattern E or F (see Figure 1.5 or 12.1). In the case of Pattern F especially, an action designed to improve reliability will in fact make it worse, by upsetting a stable system and inducing infant mortality.

In these cases, the 'root cause' of the failure is the rule itself or the process by which it is selected. If the rule is promulgated or selected by someone other than the person who performs the task – in other words, if the person doing the task is only following orders – then the mistake is really the effect of another failure.

The RCM process helps to *reduce the possibility of misapplying good rules* in two ways:

• the thorough analysis of failure effects, especially what could happen if a hidden function is in a failed state when it is needed, means that people are less likely to jump to inappropriate conclusions when the situation does arise (especially if they have been involved in the RCM process)

• by focusing attention on the functions and maintenance of protective devices, the RCM process greatly reduces the probability that these devices will be in a failed state in the first place.

The chances of bad habits developing are also reduced if care is taken during the FMEA to identify failure modes which give rise to spurious alarms, and to take steps subsequently to reduce them to a minimum. (In cases where the frequency and/or the possible consequences of a false alarm warrant it, the most appropriate remedy usually entails redesign.)

RCM helps to *reduce the possibility of applying bad rules* because the whole RCM process is all about defining the most appropriate 'rules' for maintaining any asset.

Of course, care must be take to ensure that the rules of RCM itself are not applied badly. This is best done by ensuring that everyone involved in the application of RCM is adequately trained in the underlying principles.

Mistakes 2: Knowledge-based mistakes
Knowledge-based mistakes occur when someone is confronted with a situation which has not occurred before and which has not been anticipated (in other words, one for which there are no 'rules'). In situations like this, the person has to make a decision about an appropriate course of action, and a mistake occurs if this decision is wrong.

In practice, the author has found that a common problem which occurs in this context is a belief on the part of senior managers and engineers that "I know, therefore my company knows". In fact, if a crisis occurs late at night when all the senior people are off-site, the requisite knowledge is useless if it is not in the mind of the person who has to take the first steps to deal with the crisis.

This suggests that first and most obvious way to avoid knowledge-based mistakes is to improve the knowledge of the people *who have to make the decisions*. In most cases, these are the operators and maintainers. Operators and maintainers are likely to make appropriate decisions more often if they clearly understand how the system works (its functions), what can go wrong (functional failures and failure modes), and the symptoms of each failure (failure effects). As mentioned several times in chapters 2, 13 and 14, this understanding is hugely enhanced if operators and maintainers are involved directly in the RCM process. The most important findings can be disseminated subsequently to people who do not participate in the analysis by incorporating the findings into training programs.

If necessary, the possibility of knowledge-based mistakes can also be reduced by designing (or redesigning) systems in ways which:

• minimize complexity, so that there is less to know

- minimize novelty, because new and alien technologies put people at the bottom of the learning curve, where mistakes are most likely to happen

- avoid tight coupling. This means designing systems in such a way that if failures do occur, consequences develop slowly enough to give people time to think and hence more opportunity to make the right decisions.

Violations

A violation occurs when someone knowingly and deliberately commits an error. Violations fall into three categories:

- *routine violations*. For instance, when people make a habit of not wearing items of protective clothing (such as hard hats) despite rules which clearly state that they should

- *exceptional violations*. For instance, if someone who usually wears a hard hat knowingly rushes outside without the hat on "because they couldn't find it and didn't have time to look for it"

- *sabotage*. This occurs when someone maliciously causes a failure.

The remedy for routine and exceptional violations usually consists of appropriate enforcement of the rules by management. However, once again, involvement in the RCM process gives people a clearer understanding of the need for safety procedures and the risks they are running if they violate them. The management of sabotage is beyond the scope of this book.

Conclusion

The most important conclusions to emerge from this appendix are that:

- not all human errors are necessarily the fault of the person who made the error. In many cases, the error is either forced by external circumstances or by inappropriate rules. So if blame is to be allocated for any error, care must be taken to identify the real source

- human error is at least as common a reason why equipment fails to do what its users want it to do as deterioration, if not more so. As a result, it should be dealt with as part of the RCM process, either as a *failure mode* when it is a root cause, or as a *failure effect* when it consists of inappropriate responses to other failures

- in the industrial context, it is only possible to come to grips with human errors if the people involved in committing the errors are involved directly in identifying them, and developing appropriate solutions.

Appendix 3:
A Continuum of Risk

Chapter 5 suggested that it might be possible to produce a schedule of tolerable risks which combines safety risks and economic risks in one continuum. It suggested that this might be made possible by combining Figures 5.2 and 5.14 in some way.

Figure 5.14, repeated as Figure A3.1, showed what an organization might decide that it can accept for *one* event that has economic consequences only.

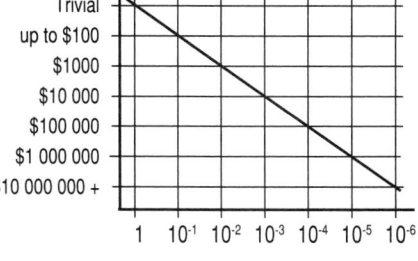

Figure A3.1:
Tolerability of economic risk

Figure 5.2 depicted what one individual might be prepared to tolerate in a specific situation from *any* event which could prove fatal in that situation, as summarized in Figure A3.1.

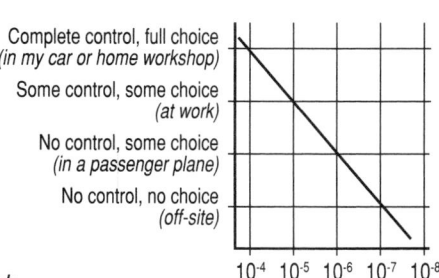

Figure A3.2: *Tolerability of fatal risk*

In fact, these two charts cannot be combined as they stand, because Figure A3.1 is based on the probability of a single event while Figure A3.2 depicts what one individual might consider to be tolerable for any event. However, with respect to the latter, part 3 of Chapter 5 went on to show that it is possible to use what one individual tolerates from *any* event in a given situation as a basis for deciding what probabilities apply to *each* event which could place him or her at risk in that situation, as follows:

The first step is to convert what one person tolerates to an overall figure for an entire site. In other words, if I tolerate a probability of 1 in 100 000 (10^{-5}) of being killed at work in any one year and I have 1 000 co-workers who all share the same view, then we all accept that on average 1 person per year on our site will be killed at work every 100 years – and that person may be me, and it may happen this year.

The next step was to translate the probability which myself and my co-workers are prepared to tolerate that any one of us might be killed by *any* event at work into a tolerable probability for *each single event* (failure mode or multiple failure) which could kill someone.

For example, continuing the logic of the previous example, the probability that any one of my 1 000 co-workers will be killed in any one year is 1 in 100 (assuming that everyone on the site faces roughly the same hazards). Furthermore, if the activities carried out on the site embody (say) 10 000 events which could kill someone, then the average probability that each event could kill one person must be reduced to 10^{-6}. This means that the probability of an event which is likely to kill ten people must be reduced to 10^{-7}, while the probability of an event that has a 1 in 10 chance of killing one person must be reduced to 10^{-5}. On a site that is divided into several areas and where each area is further divided into several sections, this process of subdividing acceptable risk could be carried out in stages, as shown in Figure A3.3.

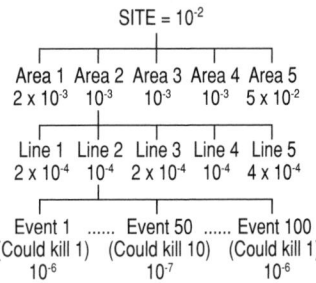

Figure A3.3:
From whole site to one event

In the example shown, an 'event' is either:

- a *single failure mode* (as defined in the FMEA) which on its own has lethal consequences. The probability allocated to this type of event defines the 'tolerable level' which is referred to when the RCM process asks the question "Does this task reduce the probability of the failure to a tolerable level?". See page 102.

- a *multiple failure* where a protected system fails and the protective device which should have rendered the system non-lethal is itself in a failed state. The probability allocated to this type of event defines the 'tolerable level' which is referred to when the RCM process asks "Does this task reduce the probability of the multiple failure to a tolerable level?" See page 122. It is also the probability used to establish M_{MF} when setting failure finding intervals. See page 179.

In complex systems, it is likely that an approach similar to a fault tree analysis would be used to allocate probabilities (see Andrew and Moss[1993]). However, in this case, we work *downwards* from a top-event probability (the probability of a fatal accident anywhere on the site) to establish objectives for each safety-oriented proactive task and to determine failure-finding task intervals, rather than *upwards* to determine a top-event probability based on an existing maintenance program.

A detailed examination of fault trees is beyond the scope of this book. The purpose of this appendix is only to suggest how it might be possible to convert risks which individual members of society might be prepared to tolerate (another manifestation of 'desired performance') into meaningful information which can be used to establish a maintenance program designed to deliver that performance.

The process described above can be used to produce a graph showing the probabilities of a single fatal event at work which would flow from the risks which one individual is prepared to accept, on the assumption that his or her judgement is accepted by everyone else on the site. This is illustrated in Figure A3.4. Note that in the next four graphs, the X-axis represents the probability of any one event occurring in any one year, (or more accurately, the annual failure rate.)

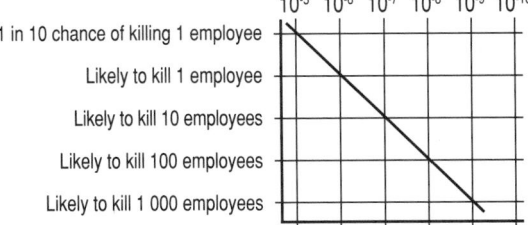

Figure A3.4:
Tolerability of one lethal event where I have some control and some choice

The same process could be applied to the situation in which the likely victims have no control but some choice about exposing themselves to the risk. The example in Figure A3.2 suggests that an airline passenger might be a typical example of someone in this situation. From the maintenance viewpoint, such people are likely to be users of mass transport systems, or people visiting large buildings (shops, offices, sports stadiums, theatres, and so on). In general, these people could be called 'customers'.

In this case, if they all tolerate the same risk as the individual in Figure A3.2 (and there are the same number of potentially life-threatening events inherent in the system), the process of apportioning risk used in Figure A3.3 could lead to the single-event probabilities shown in Figure A3.5.

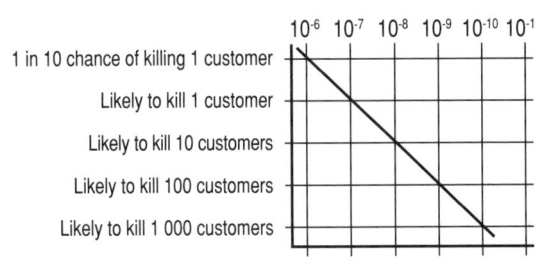

Figure A3.5:
Tolerability of one lethal event where I have no control and some choice

Similar reasoning applied to the no control/no choice scenario might yield the single event probabilities shown in Figure A3.6. (In practice, most individuals are likely to tolerate an even lower probability of being killed for this reason than is shown in Figure A3.2 – the so-called 'dread' factor. However, in most facilities, fewer events would be likely to have off-site consequences, so the probability for each event might end up about the same.)

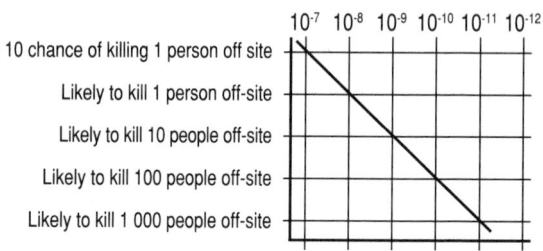

Figure A3.6:
Tolerability of one lethal event where I have no control and no choice

Once tolerable probabilities have been determined for single events as shown in Figures A3.1, A3.4, A3.5 and A3.6, it is of course possible to combine them into a single 'continuum of risk', as shown in Figure A3.7.

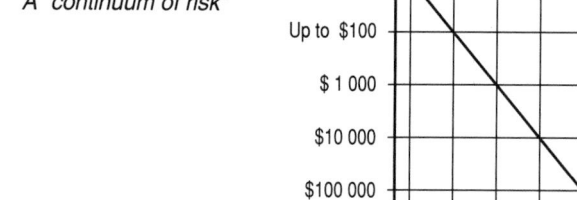

Figure A3.7:
A "continuum of risk"

Please note once again that these figures are not meant to be prescriptive and do not necessarily reflect the views of the author or any other organization or individual as to what should or should not be tolerable.

Figure A3.7 is also not intended to imply that 1 employee is worth $10 million. That figure represents a point at which two different value systems happen to coincide. The financial risks which *your* organization is willing to tolerate and the personal risks which *your* employees and customers (and society as a whole in the case of no control/no choice hazards) are prepared to tolerate may lead to a completely different set of figures in your operating context.

The key point is that the criterion upon which the whole RCM philosophy is based is what is *tolerable*, not what is practicable or what is a current industry norm (although these may coincide). Part 3 of Chapter 5 suggested that the people who are both morally and practically in the best position to decide what is tolerable are the likely victims. These are the shareholders and their management representatives in the case of financial risks, and employees, customers and the managers who have to clear up afterwards (and bear the responsibility) in the case of personal risks. As mentioned above, this appendix shows one way in which it may be possible to turn informed consensus about tolerable risk into a framework for setting targets for maintenance programs designed to deliver it.

Finally, please bear in mind that the approach outlined in this appendix is not intended to be prescriptive. If you have access to a different framework which satisfies all the parties involved, then by all means use it.

Appendix 4:
Condition Monitoring Techniques

1 Introduction

Chapter 7 explained at length that most failures give some warning of the fact that they are about to occur. This warning is called a *potential failure*, and is defined as an *identifiable physical condition which indicates that a functional failure is either about to occur or is in the process of occurring*. On the other hand, a *functional failure* is defined as *the inability of an item to meet a specified performance standard*. Techniques to detect potential failures are known as *on-condition* maintenance tasks, because items are inspected and left in service *on the condition* that they meet specified performance standards. The frequency of these inspections is determined by the *P-F interval*, which is the interval between the emergence of the potential failure and its decay into a functional failure.

Basic on-condition maintenance techniques have existed as long as mankind, in the form of the human senses (sight, sound, touch and smell). As explained in Chapter 7, the main technical advantage of using people in this capacity is that they can detect a very wide range of potential failure conditions using these four senses. However, the disadvantages are that inspection by humans is relatively imprecise, and the associated P-F intervals are usually very short.

But the sooner a potential failure can be detected, the longer the P-F interval. Longer P-F intervals mean that inspections need to be done less often and/or that there is more time to take whatever action is needed to avoid the consequences of the failure. This is why so much effort is being spent on trying to define potential failure conditions and develop techniques for detecting them which give the longest possible P-F intervals.

However, Figure A4.1 opposite shows that a long P-F interval means that the potential failure must be detected at a point which is higher up the P-F curve. But the higher we move up this curve, the smaller the deviation from the "normal" condition, especially if the final stages of deterioration are not linear. The smaller the deviation, the more sensitive must be the monitoring technique designed to detect the potential failure.

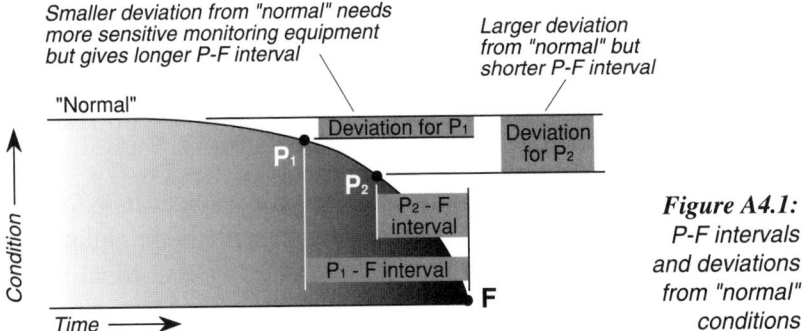

Figure A4.1:
P-F intervals
and deviations
from "normal"
conditions

2 Categories of Condition Monitoring Techniques

Most of the smaller deviations tend to be beyond the range of the human senses and can only be detected by special instruments. In other words, equipment is used to *monitor the condition* of other equipment, which is why the techniques are called *condition monitoring*. This name distinguishes them from the other types of on-condition maintenance (performance monitoring, quality variation and the human senses).

As mentioned in Chapter 7, condition monitoring techniques are really no more than highly sensitive versions of the human senses. In the same way as the human senses react to the symptoms of a potential failure (noise, smells, etc), so condition monitoring techniques are designed to detect specific symptoms (vibration, temperature, etc). For the sake of simplicity, these techniques are classified according to the symptoms (or *potential failure effects*) which they monitor, as follows:

- *dynamic effects.* Dynamic monitoring detects potential failures (especially those associated with rotating equipment) which cause abnormal amounts of energy to be emitted in the form of waves such as vibration, pulses and acoustic effects.

- *particle effects.* Particle monitoring detects potential failures which cause discrete particles of different sizes and shapes to be released into the environment in which the item or component is operating.

- *chemical effects.* Chemical monitoring detects potential failures which cause traceable quantities of chemical elements to be released into the environment.

- *physical effects.* Physical failure effects encompass changes in the physical appearance or structure of the equipment which can be detected directly, and the associated monitoring techniques detect potential failures in the form of cracks, fractures, the visible effects of wear and dimensional changes.

- *temperature effects.* Temperature monitoring techniques look for potential failures which cause a rise in the temperature of the equipment itself (as opposed to a rise in the temperature of the material being processed by the equipment).

- *electrical effects:* Electrical monitoring techniques look for changes in resistance, conductivity, dielectric strength and potential.

An enormous variety of techniques has been developed and more are appearing all the time, so it is not possible to produce an exhaustive list of all the techniques available at any time. This appendix provides a very brief summary of 96 of the techniques currently available. Some of these are well-known and well-established, while others are in their infancy or even still under development.

However, whether or not any of these techniques is technically feasible and worth doing in any context should be assessed with the same rigor as any other on-condition task. To help with this process, this appendix lists the following for each technique:
- the potential failure conditions which the technique is meant to detect (*conditions monitored*)
- the equipment it is designed for (*applications*)
- the P-F intervals typically associated with the technique (*P-F interval*)
  – for obvious reasons, this can only be a very rough 'ballpark' guide
- how it works (*operation*)
- the training and/or level of skill needed to apply the technique (*skill*)
- the advantages of the technique (*advantages*)
- the disadvantages of the technique (*disadvantages*).

Finally, before considering specific techniques, it is worth noting that a great deal of attention is being focused on condition monitoring nowadays. Because of its novelty and complexity, it is often regarded as being completely separate from other aspects of scheduled maintenance. However, we should not lose sight of the fact that condition monitoring is only one form of proactive maintenance. When it is used, it should be designed into normal schedules and schedule planning systems wherever possible, and not made subject to expensive parallel systems.

3 Dynamic Monitoring

A Preliminary Note on Vibration Analysis

Equipment which contains moving parts vibrates at a variety of frequencies. These frequencies are governed by the nature of the vibration sources, and can vary across a very wide range or *spectrum*. For instance, the vibration frequencies associated with a gearbox include the primary frequencies of rotation of the shafts (and their harmonics), the tooth contact frequencies of different gear sets, the ball passing frequencies of the bearings and so on. If any of these components starts to fail, its vibration characteristics change, and vibration analysis is all about detecting and analyzing these changes.

This is done by measuring how much the item as a whole vibrates, and then using spectrum analysis techniques to home in on the frequency of vibration of each individual component in order to see whether anything is changing.

However, the situation is complicated by the fact that it possible to measure three different characteristics of vibration. These are amplitude, velocity and acceleration. So step one is to decide which of the characteristics is going to be measured – and what measuring device is going to be used – and then step two is to decide which technique will be used to analyze the signal generated by the measuring device (or sensor). In general, amplitude (or displacement) sensors are more sensitive at lower frequencies, velocity sensors across the middle ranges and accelerometers at higher frequencies. The strength of the signal at any frequency is also influenced by how closely the sensors are mounted to the source of the signal at that frequency.

Another important characteristic of vibration is 'phase'. Phase refers to 'the position of a vibrating part at a given instant with reference to a fixed point or another vibrating part'. As a rule, phase measurements are not taken during normal routine vibration measurements, but can provide valuable information when a problem has been detected (such as imbalance, bent shafts, misalignment, mechanical looseness, reciprocating forces and eccentric pulleys and gears).

Fourier analysis also plays an important part in vibration analysis. Fourier discovered that all complex vibration curves (level against time) can be broken down into many simple sinusoidal curves (each of one frequency with one amplitude). Hence by doing a 'Fourier analysis', a complex wave can be broken down into a variety of levels (amplitudes) at a variety of frequencies. In effect, the variation level against time has been transformed into a constantly changing display of amplitude against frequency. The process by which this is done is now called a 'fast Fourier transform' (FFT).

The role of expert systems in vibration analysis is rapidly coming of age. Some systems can now find and diagnose problems as consistently as experienced vibration analysts. They are great time savers, and also enable users to compare readings with all the data from previous measurements.

The rest of this part of this chapter looks at vibration analysis in more detail.

3.1 Broad Band Vibration Analysis

Conditions monitored: Changes in vibrational characteristics caused by fatigue, wear, imbalance, misalignment, mechanical looseness, turbulence, etc.

Applications: Shafts, gearboxes, belt drives, compressors, engines, roller bearings, journal bearings, electric motors, pumps, turbines, etc.

P-F interval: Limited warning of failure

Operation: A broad band vibration system consists primarily of two parts: a transducer which is mounted on the point of measurement to convert mechanical vibrations into an electrical signal, and a measuring and indicating device called a vibration meter, which is calibrated in vibrational units. Monitors the overall reading of the vibration signature which is simply the root mean square (RMS) value of the broad band signal. It is a simple value and is suited primarily to a single sinusoidal wave rather than a complex wave. Such meters have a constant frequency response over the range 20 Hz to 1000 Hz

Skill: To use the equipment and record the vibration: a semi-skilled worker

Advantages: Can be very effective in detecting a major imbalance of rotating equipment. Can be used by inexperienced personnel. Cheap and compact. Can be portable or permanently installed. Minimal data logging. Interpretation and appraisals can be based on published condition acceptability criteria such as VDI 2056 from Germany

Disadvantages: The broad band signal provides little information about the nature of the fault. In initial spectra, spectral peaks are much lower and contribute very little to the overall broad band signature. When these spectral peaks do grow the equipment is normally in an advanced state of deterioration. Difficult to set alarm levels. Lacks sensitivity.

3.2 Octave Band Analysis

Conditions monitored and applications: As for broad band vibration

P-F interval: Days to weeks depending on application

Operation: Fixed contiguous octave and fractional octave filters divide the frequency spectrum into a series of bands of interest, which have a constant width when plotted logarithmically. The average output from each filter is measured successively, and the values are displayed by a meter or plotted on a recorder

Skill: To operate equipment and interpret results: a suitably trained technician

Advantages: Simple to use when the measurement parameters have been previously determined by an engineer: Portable: Relatively inexpensive: Good detection abilities using fractional octave filters: Recorder provides a permanent record.

Disadvantages: Limited information for diagnostic purposes: Diagnostic ability also limited by logarithmic frequency scale: Relatively long analysis time.

3.3 Constant Bandwidth Analysis

Conditions monitored: Changes in vibrational characteristics caused by fatigue, wear, imbalance, misalignment, mechanical looseness and turbulence, and to identify multiple harmonics and sidebands

Applications: Shafts, gearboxes, belt drives, compressors, engines, roller bearings, journal bearings, electric motors, pumps, turbines, and development, diagnostic and experimental work (especially on gearboxes)

P-F interval: Usually several weeks to months

Operation: An accelerometer detects the vibration and converts it into an electrical signal which is amplified, subjected to a constant bandwidth filter and then fed into an analyzer. The constant bandwidths are between 3.16 Hz and 1000 Hz and the frequency range from 2 Hz to 200 kHz. Both linear and logarithmic frequency sweeps may be selected, but linear is chosen when identifying harmonics. In order to analyze the peaks in more detail, bandwidths and frequency ranges can be changed to suit requirements

Skill: To operate the equipment: a suitably trained skilled worker. To interpret the results: an experienced technician

Advantages: Simple to use when measurement parameters have been set. Good for large frequency ranges and for detailed investigation at high frequencies. Identifies multiple harmonics and side bands which occur at constant frequency intervals. Equipment portable

Disadvantages: Relatively long analysis time. In-depth understanding of the machine harmonics and side bands required to interpret results.

3.4 Constant Percentage Bandwidth Analysis

Conditions monitored: Shock and vibration

Applications: Shafts, gearboxes, belt drives, compressors, engines, roller bearings, journal bearings, electric motors, pumps, turbines, etc.

P-F interval: Usually several weeks to months

Operation: High resolution narrow bandwidth frequency analysis is performed by sweeping through the desired frequency range (2 Hz to 20 kHz) using a constant percentage filter bandwidth (1%, 3%, 6%, 12%, 23%) which separates closely spaced frequencies or harmonics. A constant percentage filter bandwidth as narrow as 1% allows very fine resolution analyses. Continuous sweeps through the frequency range can be made in real time.

Skill: To operate the equipment a suitably trained skilled worker, to interpret the results an experienced technician

Advantages: Analysis can be done in 'real time' and is therefore faster than FFT analysis and does not suffer from certain pitfalls caused by the batch nature of FFT, such as loss of data by windowing. CPB spectra are very good for rapid fault detection. Equipment portable

Disadvantages: High skill required to interpret results.

3.5 Real Time Analysis

Conditions monitored: Acoustic and vibrational signals; measurement and analysis of shock and transient signals

Applications: Rotating machines, shafts, gearboxes etc

P-F interval: Several weeks to months

Operation: A signal is recorded on magnetic tape and played back through a real-timer analyzer. The signal is sampled and transformed into the frequency domain. A constant bandwidth spectrum is produced, measured at 400 equally spaced frequency intervals across a frequency range selectable from 0-10 Hz to 0-20 kHz. A high resolution mode can be selected, and the scan can also be adjusted to give a 'slow motion' analysis, allowing any changes in the baseband spectrum to be observed as the time window is stepped along

Skill: To operate equipment and interpret results: an experienced engineer

Advantages: Analyses all frequency bands over the entire analysis range simultaneously: Instantaneous graphical display of analyzed spectra is continuously updated: No need to wait for level recorder readout: Suited to analysis of short duration signals such as transient vibration and shock: X-Y recorders provide a permanent record

Disadvantages: Equipment not portable and very expensive: Needs high level of skill: Off-line analysis.

3.6 Time Waveform Analysis

Conditions monitored: Chipped, cracked, broken gear teeth; pump cavitation, misalignment, mechanical looseness, eccentricity, etc.

Applications: Gearboxes, pumps, roller bearings, etc

P-F interval: Usually several weeks to months

Operation: An oscilloscope is connected to a standard vibration analyzer or a real time analyzer. A vibration signal is applied to the oscilloscope vertical input. The vertical axis on the CRT is scaled in amplitude and the horizontal axis is scaled in time such as seconds or milliseconds. The oscilloscope vertical gain is adjusted until the peak or peak-to-peak value of the waveform displayed on the CRT corresponds to the amplitude reading on the vibration meter. When a machine is generating a single frequency, the time waveform is simply a sine wave with the repetition rate of the running speed of the equipment. When the equipment generates more than one frequency, a complex composite waveform is generated. Additional frequencies can be generated in the form of pulse, transient, beat, modulation, etc. which add to the complexity of the waveform. To reduce the complexity of the waveform, it is useful and even essential to use variable high pass, low pass and band pass filters

Skill: Needs considerable practice and experience to interpret complex wave forms

Advantages: Good for looking at transients, slow beats, pulses, non-linearities, sine waves, amplitude modulation, frequency modulation, instabilities, etc. Often provides more information than frequency analysis. The wave form can be used to distinguish between spectra resulting from impacts or random noise

Disadvantages: Machines that generate multiple frequencies often generate noise which makes time wave forms so complex and confusing that they are difficult to break down into component parts. To examine a wave form which might have slow beats, a long time record is required

3.7 Time Synchronous Averaging Analysis

Conditions monitored: Wear, fatigue, stress waves emitted as a result of metal-to-metal impacting, microwelding, etc

Applications: Gearbox gear teeth, roller bearings, shafts, bank of fans, rolls on a paper machine, rotating machines

P-F interval: Usually several weeks to months

Operation: Most rotating mechanical systems produce a slightly varied signal with each rotation. (Statistically this is termed 'stochastic', in comparison with identically repeated signals which are 'deterministic'.) The closer the tolerances of sliding and rolling parts, the less the variation, but nevertheless there is a variation. In many systems, this difference can be so great that it masks any changes due to a developing fault. The presence of random noise can also confuse the signal. These problems can be overcome by performing a level check at precisely the same part of the cycle rotation using a tachometer trigger pulse to initiate data capture in a data collector. A number of cycles or time records are averaged together. Signals not related to the RPM of the shaft are averaged out, leaving a very clear 'real-time' wave representing the components related to a single turning speed. The averaged wave form can be examined directly or a spectrum can be generated from it. It is devoid of random signals and will show whether one part of the cycle is changing more than another

Skill: A suitably trained and experienced skilled worker. Requires considerable practice and experience to interpret the results

Advantages: Gearboxes – specifically the individual gears – can be analyzed in detail. Very useful for analyzing equipment that has many components rotating at nearly the same speed

Disadvantages: Care must be taken with machines with roller element bearings as the bearing tones are not synchronous with the RPM and will be averaged out.

3.8 Frequency Analysis

Conditions monitored: Changes in vibration characteristics caused by fatigue, wear, imbalance, misalignment, mechanical looseness, turbulence, etc

Applications: Shafts, gearboxes, belt drives, compressors, engines, roller bearings, journal bearings, electric motors, pumps, turbines, etc

P-F interval: Several weeks to months

Operation: Data is collected from measurement points in the time domain and transformed into the frequency domain using a Fast Fourier Transform (FFT) algorithm, by either the data collector itself or a host computer. The required frequency range of the measurements is dependent on the speed of the machine. Each machine which has moving parts will produce a spectrum of frequencies. A baseline spectrum of the machine in excellent condition is compared to an actual spectrum of the same machine running at the same speed and load. Any increases over the baseline of more than one standard deviation at any forcing frequency can indicate a potential problem. One feature of frequency analysis is the "waterfall" of FFT signatures. Waterfalls are signatures taken at the same point over a time interval allowing the signatures to be trended

Skill: A suitably trained and experienced skilled worker. Requires considerable practice and experience to interpret the results

Advantages: Data collecting equipment is portable and easy to use. Expert software systems makes data interpretation easy. Using waterfall plots small changes in machine condition can be detected at an early stage

Disadvantages: Spectra resulting from impacts and random noise can look similar.

3.9 Cepstrum

Conditions monitored: Wear causing harmonics and sidebands in vibration spectra

Applications: Rolling element bearings, shafts, gears, gear meshing, belt rotation, vane and blade pass frequencies of pumps and fans

P-F interval: Several weeks to months

Operation: As a machine wears, it develops non-linearities that cause harmonics of the primary force frequencies, and sum and difference (sideband) frequencies appear in the vibration spectra. Cepstrum (pronounced "kepstrum") effectively separates the harmonics and sidebands present in the spectrum so they can be individually trended over time. In simple terms cepstrum could be defined as the FFT of the logarithmic spectrum obtained from the FFT 'a spectrum of a spectrum'. All technical words in cepstrum analyzed are reversed due to the doubling up of the transform, i.e. spectrum becomes cepstrum, frequency becomes quefrency and harmonics become rahmonics

Skill: In-depth understanding of machine behavior (harmonics and sidebands) and the expert software

Advantages: Can analyze harmonics and sidebands that usually overlap in fairly complex machines. Sidebands are easy to find in the spectra of roller element bearings. Can be performed with some expert system software.

Disadvantages: Skill and experience needed to interpret harmonics and sidebands.

3.10 Amplitude Demodulation

Conditions monitored: Bearing tones masked by noise, cracks in bearing races, eccentric or damaged gears, mechanical looseness

Applications: Steam turbines, bearings and gearboxes, low speed rotating components of paper machines, reciprocating machines, etc

P-F interval: Several weeks to months

Operation: The acceleration analog signal (time domain) is subjected to high pass filtering and then amplitude demodulation. This is where a discrete frequency often called the "carrier" in the spectrum may be modulated by another frequency called the modulator. The resulting signal is then subjected to a low frequency range spectrum analysis. The amplitude demodulation is performed in the data collector before the signal is digitized.

Skill: A suitably trained and experienced technician

Advantages: Early detection for bearing and gearbox problems (specifically bearings completely masked by noise) can easily be identified. Works well in low-speed applications such as paper machines

Disadvantages: High skill and experience needed to understand and interpret results. Difficult to implement on slow speed bearings because the stress waves are short-term transient events (less than a few milliseconds), so when the narrow pulse output from the demodulation circuit is passed through the final stage of signal conditioning (the low pass/anti-aliasing filter) a large fraction of the stress wave is filtered out, making fault detection less likely.

3.11 Peak Value (PeakVue) Analysis

Conditions monitored: Stress waves caused by metal-to-metal impacts or metal tearing, stress cracking or scuffing, spalling and abrasive wear

Applications: Anti-friction bearings and gearbox shafts and gearing systems

P-F interval: Several weeks to months depending on the application

Operation: Separates low energy faults such as those that occur in anti-friction bearings and gears, and enhances their signal causing the faults to stand above the spectral noise floor. This makes them easier to recognize. PeakVue first separates the stress waves from the vibration waveform using a high pass filter. It is then conditioned to enhance its amplitude and pulse width, making it FFT friendly. The conditioned waveform is then processed using an FFT to determine the frequency at which the stress wave occurs

Skill: Experienced vibration technician

Advantages: Reveals some faults that may have gone undetected in their earlier stages or which are buried in the noise floor (bottom) of the vibration spectrum. More consistent than demodulation. Outputs are independent of machine speeds and instrument Fmax settings. Applicable to a broad range of frequencies, from very slow speed bearings to gear meshing in excess of 1 kHz

Disadvantages: High skill and experience required to interpret results.

3.12 Spike Energy™

Conditions monitored: Dry running pumps, cavitation, flow change, bearing loose fit, bearing wear causing metal to metal contact, surface flaws of gear teeth, high pressure steam or air flow, control valves noise, poor bearing lubrication

Applications: Seal-less pumps used in the chemical and petrochemical industry, gearboxes, roller element bearings, etc

P-F interval: Several weeks to months

Operation: Some faults excite the natural frequencies of components and structures. The intense energy generated by repetitive transient mechanical impacts causes a signal to appear as periodic spikes of high frequency energy in a spectrum which can be measured by an accelerometer. A high frequency band pass filter is used to filter out low frequency vibration signals. The high frequency signals pass through a peak-to-peak detector that detects and holds the peak-to-peak amplitudes of the signal, This is called enveloping, and measurement results are expressed in 'gSE' units. Pulses with large amplitudes and high repetition rates produce high overall gSE readings. The enveloped signal can be subjected to a FFT analysis displaying a Spike Energy Spectrum. In the gSE spectrum, the fault frequency shows up as certain defect frequency and its harmonics.

Skill: A suitably trained and experienced skilled worker. Requires practice and experience to interpret the results

Advantages: Sensitive high frequency measurement parameters suited to the detection of sealless pump problems which are often difficult to detect using conventional vibration sensors such as velocity meters and accelerometers

Disadvantages: High skill and experience needed to interpret results.

3.13 Proximity Analysis

Conditions monitored: Misalignment, oil whirl, rubs, imbalance/bent shafts, resonance, reciprocating forces, eccentric pulleys and gears, etc

Applications: Shafts, motor assemblies, gearboxes, fans, couplings, etc

P-F interval: Days to weeks

Operation: In the basic mode, a signal from a transducer operates as the ordinate against a time base. With a single impulse, sinusoidal curves indicate imbalance, bent shafts, oil whirl, misalignment, adhesive bearing rubs. Two signals produce a polar diagram which provides more characteristic information than an X-Y diagram. More information can be obtained by introducing a phase indicating mark on the wave forms of the oscilloscope display. These marks are generated at the rate of one revolution by a pick up incorporated in the shaft-speed tachometer

Skill: A suitably trained and experienced technician

Advantages: Pinpoints specific problems. Can be used for balancing: Portable: Very simple to use

Disadvantages: P-F interval short: Long analysis time: Diagnostic ability limited.

3.14 Shock Pulse Monitoring

Conditions monitored: Surface deterioration and lack of lubrication causing mechanical shock waves. With data trending can identify incorrect bearing installation or replacement, using the wrong type of lubricant, poor lubricant handling or dispensing practices, or incorrect installation or maintenance of oil seals and packings, etc.

Applications: Rolling element bearings, anti-friction bearings, pneumatic impact tools, valves of internal combustion engines

P-F interval: Weeks to several months

Operation: The type and size of the bearing is entered into the analyzer. A piezoelectric accelerometer placed on a bearing housing detects the mechanical impact of shock impulses, caused by the impact of two masses (such as the rotational contact between the surfaces of the ball or roller and the raceway). The magnitude of the shock pulses depend on the surface condition and the peripheral velocity of the bearing (rpm and size). The pulses set up a dampened oscillation in the transducer at its resonant frequency. The transducer is tuned mechanically and electrically to a resonant frequency of 32 kHz. The peak amplitude of this oscillation is directly proportional to the impact velocity. As the bearing condition deteriorates from good to imminent failure, shock pulse measurements can increase up to 1000 times.

Skill: A trained and suitably experienced technician

Advantages: Relatively easy to operate. Portable. Can be used on virtually any roller element bearing. Bearing condition and lubrication status analyzed within seconds. Shock impulses are not significantly influenced by background vibration and noise. Identifies subtle changes in bearing condition or lubrication which might not be differentiated by conventional vibration analysis

Disadvantages: Needs accurate bearing size and speed information prior to taking measurement. Limited to roller element bearings.

3.15 Ultrasonic Analysis

Conditions monitored: Changes in sound patterns (sonic signatures) caused by leaks, wear, fatigue or deterioration

Applications: Leaks in pressure and vacuum systems (ie. boilers, heat exchangers, condensers, chillers, distillation columns, vacuum furnaces, specialized gas systems): bearing wear or fatigue: steam traps: valve and valve seat wear: pump cavitation: corona in switchgear: static discharge: the integrity of seals and gaskets in tanks, pipe systems and large walk-in boxes: underground pipe or tank leaks

P-F interval: Highly variable depending on the nature of the fault

Operation: Ultrasound technology is concerned with high frequency sound waves above human perception (20 Hz to 20 kHz) ranging between 20 kHz to 100 kHz. High frequency sound waves are extremely short and tend to be fairly directional, so it is easy to isolate these signals from background noises and detect their exact location. All operating equipment and most leakage problems produce a broad range of sound. As subtle changes begin to occur with deterioration, the nature of the airborne ultrasound allows these warning signals to be detected early. Ultrasonic translators convert the ultrasound sensed by the instrument into the audible range where users can hear and recognize them through headphones. The ultrasonic monitoring equipment filters out surrounding noise and other unwanted frequencies. Ultrasonic readings may be displayed visually on a VDU or a moving coil meter, as an audible signal on headphones or as traces on an electronic monitor or computer

Skill: A suitably trained skilled worker

Advantages: Quick and easy. Can be used in very noisy areas (headphones screen ambient noise). Microphones highly directional and enable the operator to detect a source of noise at long range. Equipment portable

Disadvantages: Does not indicate the size of leaks. Underground tanks can only be tested under vacuum.

3.16 Kurtosis

Conditions monitored: Shock pulses

Applications: Rolling element bearings, anti-friction bearings

P-F interval: Several weeks to months

Operation: Restricted almost exclusively to bearings where a few specific frequency ranges are examined (3-5 kHz, 5-10 kHz, 10-15 kHz). Kurtosis is a statistical analysis of the time-based (time domain) signal and looks at the fourth moment of the spectral amplitude difference from the mean level. A normal distribution has a kurtosis (K) value of 3

Skill: A suitably trained semi-skilled worker

Advantages: Applicable to any materials with a hard surface. Equipment portable. Very simple to use

Disadvantages: Limited application and significantly affected by impact noise from other sources. Considered by some users to be too sensitive.

3.17 Acoustic Emission

Conditions monitored: Plastic deformation and crack formation caused by fatigue, stress and wear

Applications: Metal materials used in structures, pressure vessels, pipelines and underground mining excavations

P-F interval: Several weeks depending on application

Operation: Audible stress waves, due to crystallographic changes, are emitted from materials subjected to loads. These stress waves are picked up by a transducer and fed via an amplifier to a pulse analyzer, then either to an X-Y recorder or to an oscilloscope. The displayed signal is then evaluated

Skill: A suitably trained and experienced technician.

Advantages: Remote detection of flaws. Covers entire structures. Measuring system set up very quickly. High sensitivity. Requires limited access to test objects. Detects active flaws. Only relative loads are required. Can sometimes be used to forecast failure loads.

Disadvantages: The structure has to be loaded. A-E activity dependent on materials. Irrelevant electrical and mechanical noise can interfere with measurements. Gives limited information on the type of flaw. Interpretation may be difficult.

4 Particle Monitoring

4.1 Ferrography

Conditions monitored: Wear, fatigue and corrosion particles

Applications: Greases: Oils used in diesel and gasoline engines, gas turbines, transmissions, gearboxes, compressors and hydraulic systems

P-F interval: Usually several months

Operation: A representative sample is diluted with a fixer solvent (tetrachloroethylene) and then passed over an inclined glass slide under the influence of a graduated magnetic field. The particles are distributed along the length of the slide according to their size. Larger particles are deposited near the entry, while fine particles are deposited near the exit of the slide. The slide, known as a ferrogram, is treated so that the particles adhere to the surface when the oil is removed. Ferrous particles are separated magnetically and are distinguished by their alignment to the magnetic fields lines, while non-magnetic and non-metallic particles are distributed in a random fashion over the entire slide. The total density of the particles and the ratio of large to small particles indicate the type and extent of wear. Analysis is done by a technique known as bichromatic microscopic examination. This uses both reflected and transmitted light sources (which may be used simultaneously). Green, red and polarized filters are also used to distinguish the size, composition, shape and texture of both metallic and non-metallic particles. An electron microscope can also be used to determine particle shapes and provide an indication of the cause of failure

Skill: To draw sample and operate ferrograph: a suitably trained semi-skilled worker. To analyze and interpret the results: an experienced technician

Advantages: More sensitive than emission spectrometry at early stages of wear: Measures particle shapes and sizes: Provides a permanent pictorial record

Disadvantages: Not an on-line technique: Time consuming, and needs some very expensive analytical support equipment: Measures generally only the ferromagnetic particles: Requires an electron microscope for an in-depth analysis.

4.2 Analytical Ferrography

Conditions monitored: Wear, fatigue and corrosion particles

Applications: Greases.: Oils used in diesel and gasoline engines, gas turbines, transmissions, gearboxes, compressors and hydraulic systems

P-F interval: Usually several months

Operation: An analytical ferrograph is used to prepare a ferrogram as described under ferrography. After the particles have been deposited on the ferrogram, a wash is used to flush away any remaining oil or water-based lubricant. Once the wash evaporates, the particles remain permanently attached to the substrate on the ferrogram. A Ferrogram Scanner scans the ferrogram in less than 20 seconds and generates standard output values that correspond to the wear mechanism. Various particles are graded by their types and shapes which reveal specific problems. For example, laminar metals (having a 'peeled look', long and thin) often indicate a problem with roller bearings. Red oxides typically are rust (likely water contamination). The software then reports the wear levels and changes in condition of the component

Skill: To draw sample and operate ferrograph: a suitably trained semi-skilled worker. To analyze and interpret the results: an experienced technician

Advantages: Available in a wide range of on-line systems. In-depth evaluation, photographic recording and data-base management. Less affected by fluid opacity and water contamination than many other techniques. Equipment expensive

Disadvantages: High level of operator experience required. Time consuming sample preparation and analysis. The need to dilute samples reduces the chance that the sample will actually be representative of actual wear.

4.3 Direct Reading Ferrograph (DRF)

Conditions monitored: Machine wear, fatigue and corrosion particles

Applications: Oils used in diesel and gasoline engines, gas turbines, transmissions, gearboxes, compressors and hydraulic systems.

P-F interval: Usually several months

Operation: A DRF quantitatively measures the concentration of ferrous particles in a fluid sample by precipitating these particles onto the bottom of a glass tube subjected to a strong magnetic field. Fibre optic bundles direct light through the glass tube at two positions corresponding to the location where large and small particles are deposited by the magnet. The light is reduced in relation to the number of particles deposited in the glass tube, and this reduction is monitored and displayed electronically. Two sets of readings are obtained for large and small particles (above and below 5 microns) which are plotted on a graph.

Skill: A suitably trained semi-skilled worker.

Advantages: Compact, portable, on-line technique, easy to operate. Less sensitive to fluid opacity and water contamination than some techniques.

Disadvantages: Measures only ferromagnetic particles: Requires further analytical ferrographic analysis when readings are high.

4.4 Mesh Obscuration Particle Counter (Pressure Differential)

Conditions monitored: Particles in lubricating and hydraulic oil systems caused by wear, fatigue, corrosion and contaminants

Applications: Enclosed lubricating and hydraulic oil systems such as engines, gearboxes, transmissions, compressors, etc.

P-F interval: Usually several weeks to months.

Operation: This instrument measures the differential pressure across three high-precision 5, 15, 25 micron screens, each with a known number of pores. As the oil passes through each screen, particles larger than the pores are trapped on the mesh surface, which reduces the open area of the screen and increases the pressure drop across the screen. Sensors measure the pressure change which is converted to reflect the number of particles larger than the screen size. This is converted in turn into ISO 4406 cleanliness codes.

Skill: To operate the portable unit: a suitably trained semi-skilled worker. To interpret the results: a suitably trained and experienced technician

Advantages: No pre-sample preparation. Equipment is portable and can be used in the field or the laboratory. An in-line version of the equipment can be used for real time continuous monitoring. Particle counts are calibrated to an ISO 4406 cleanliness standard. Most oils can be analyzed in a matter of minutes. Not affec-ted by bubbles, emulsions or dark oils that limit laser-based analyzers

Disadvantages: Provides no indication of the chemical composition of particles. Only applicable to circulating oil systems. Equipment moderately expensive.

4.5 Pore-blockage (Flow Decay) Technique

Conditions monitored: Particles in lubricating and hydraulic oil caused by wear, fatigue, corrosion and contaminants

Applications: Oils used in diesel and gasoline engines, gas turbines, transmissions, gearboxes, compressors and hydraulic systems.

P-F Interval: Usually several weeks to months

Operation: A fluid sample is pressurized between 30 and 150 psi (can go as high as 3000 psi) and allowed to flow through a selected precision calibration screen (5, 10, 15 micron) depending on oil viscosity, in a sensor assembly. Particles larger than the screen start to accumulate, restricting the flow. Smaller particles gather around the bigger particles restricting the flow even further. The result is a flow-decay time curve. The hand held computer uses a mathematical program to convert the flow-decay time curve into a particle size distribution. This is used to compute an ISO cleanliness code.

Skill: To operate the portable unit: a suitably trained skilled worker. To interpret the results: a suitably trained and experienced technician

Advantages: No pre-sample preparation. Equipment is portable and can be used in the field or the laboratory. An in-line version of the equipment can be used for continuous monitoring. Particle counts are calibrated to an ISO 4406 cleanliness standard. Most oils can be analyzed in a matter of minutes.

Disadvantages: : Provides no indication of the chemical composition of particles. Only applicable to circulating oil systems. Equipment moderately expensive.

4.6 Light Extinction Particle Counter

Conditions monitored: Particles in lubricating and hydraulic oil caused by wear, fatigue, corrosion and contaminants

Applications: Oils used in diesel and gasoline engines, gas turbines, transmissions, gearboxes, compressors and hydraulic systems.

P-F interval: Usually several weeks to months

Operation: The light extinction particle counter consists of an incandescent light source, an object cell and a photo detector. The sample fluid moves through the object cell under controlled flow and volume conditions. When opaque particles in the fluid pass through the beam it blocks an amount of light proportional to the particle sizes. The number and size of the particles in the oil sample determine how much light is blocked or reflected, and how much light passes through to the photo diode. The resultant change in the electrical signal at the photo diode is analyzed against a calibration standard to calculate the number of particles in predetermined size ranges and displays the count. From this information a direct reading of the ISO cleanliness value is determined automatically.

Skill: To operate the portable unit: a suitably trained skilled worker

Advantages: Considerably faster than visual graded filtration. Test results available within minutes. Generally the test is quite accurate and reproducible.

Disadvantages: Lacks the intensity and consistency of laser and fails to overcome reaction of the many different wavelengths of light. Accuracy dependent on fluid opacity, the number of translucent particles, air bubbles and water contamination. The count and size may also vary depending on the orientation of long, thin or unusually shaped particles in the light beam. Resolutions limited to 5 microns particle range. Provides no information on the chemical composition of the contaminants.

4.7 Light Scattering Particle Counter

Conditions monitored: Particles in lubricating and hydraulic oil caused by wear, fatigue, corrosion and contaminants

Applications: Enclosed lubricating and hydraulic oil systems such as engines, gearboxes, transmissions, compressors, etc.

P-F interval: Usually several weeks to months

Operation: The light scattering particle counter consists of three primary components; a laser light source, an object cell and a photo diode. The sample fluid moves through the object cell under controlled flow and volume conditions. When opaque particles in the fluid pass through the beam, the scattering of light is measured and translated into a particle count. From this information a direct reading of the ISO cleanliness value is determined automatically.

Skill: A suitably trained skilled worker

Advantages: Good performance in settings where conditions are controlled. High accuracy. Measures particles as small as 2 microns. Faster than the visual graded filtration – test results available within minutes. Generally the test is quite accurate and reproducible. Continuous monitoring is possible.

Disadvantages: Accuracy dependent on fluid opacity, the number of translucent particles, air bubbles and water contamination. The count and size may also vary depending on the orientation of long, thin or unusually shaped particles in the light beam. Provides no information on the chemical composition of contaminants. Dilution is often required for high particle concentrations to avoid coincidence error where several particles bunch together and appear as one large particle.

4.8 Real Time Ferromagnetic Sensor

Conditions monitored: Ferromagnetic particles caused by wear and fatigue

Applications: Oils used in diesel and gasoline engines, gas turbines, transmissions, gearboxes, compressors and hydraulic systems.

P-F interval: Weeks to months

Operation: An analog ferromagnetic sensor uses an inductive or magnetic principle to measure the quantity of ferrous particles passing the sensor. The sensor attracts the ferrous particles with an electromagnet. The particles collect around a sense coil causing a change in an oscillator frequency. The frequency is calibrated to indicate the mass of ferrous particles collected. After a measurement has been taken, the particles are released. Measurements can be trended over time.

Skill: Experienced skilled worker/technician.

Advantages: On-line technique

Disadvantages: Limited to collecting ferromagnetic particles only. Indicates mass of ferromagnetic particles only

4.9 All-Metal Debris Sensors

Conditions monitored: Ferrous and non-ferrous particles due to wear and fatigue

Applications: Designed specifically for the protection of gas turbine bearings.

P-F interval: Weeks to months

Operation: The sensor head consists of three coils wound around an insulating section of pipe. The outer stimulus coils are energized with an opposing high frequency signal. The sense coil (middle) is placed exactly at the null point between the stimulus coils. When a ferrous particle passes through the sensor, it disturbs the first field and then the second, generating a readily detectable signature in the sense coil. A non-ferrous particle generates a unique and opposite signature. The sensor will detect and measure most of the severe wear particle range. These signatures are captured and stored as time domain plots and are used real time to alert/advise operators, or to signal automatic responses from control systems

Skill: Experienced skilled worker/technician to trend results

Advantages: Detects and quantifies both ferrous and non-ferrous wear metal particles. Low probability of a false indication. On-board sensors can capture and store the time domain plots of various damage modes which can be used for identification of wear sources in near real time.

Disadvantages: Cannot determine chemical composition and size of particles.

4.10 Graded Filtration

Conditions monitored: Particles in lubricating and hydraulic oil caused by wear, fatigue, corrosion and contaminants

Applications: Oils used in diesel and gasoline engines, gas turbines, transmissions, gearboxes, compressors and hydraulic systems.

P-F interval: Usually several weeks to months

Operation: A small amount of oil (100 ml) is diluted and passed through a series of standard filter disks. Each disk is then examined under a microscope and the particles are counted manually. The results are expressed as the number of parts in a particular size range. Their statistical distribution is shown in the form of a graph. Analysis of the particles distribution profiles indicates whether wear is normal or not.

Skill: Sampling: a laboratory assistant. Examination of particle distribution profiles: an experienced laboratory technician or engineer

Advantages: Contaminants such as metal chips, pieces of seal material, or dirt can be identified visually. Relatively cheap

Disadvantages: Subjective because the operator has to determine visually the size of the particles, even though there are grid markings for reference. Setting up and examining each filter disk sample takes several hours. Specialist skills required to interpret the test results. Identification of particle elements difficult.

4.11 Magnetic Chip Detection

Conditions monitored: Wear and fatigue

Applications: Oils used in diesel and gasoline engines, gas turbines, transmissions, gearboxes, compressors and hydraulic systems

P-F interval: Days to weeks

Operation: A magnetic plug is mounted in the lubricating system so that the magnetic probe is exposed to the circulating lubricant. Fine metal particles suspended in the oil and metal flakes from fatigue break up are captured by the probe. The probe is removed regularly for microscopic examination of the captured particles An increase particle size indicates imminent failure. The debris has different characteristics (shape, color, and texture) depending on its source

Skill: To collect the sample: a suitably trained semi-skilled worker. To analyze the debris: a suitably trained and experienced technician

Advantages: Cheap. Low powered microscope only required for the analysis of the debris: Some probes can be removed without loss of lubricant

Disadvantages: Short P-F interval: High skill required to interpret the debris

4.12 Blot Testing

Conditions monitored: Wear metals, fatigue and sometimes corrosion particles, sludge, etc

Applications: Oils used in diesel and gasoline engines, gas turbines, transmissions, gearboxes, compressors and hydraulic systems

P-F Interval: A few days to a few weeks

Operation: One or two drops of oil are placed on a flat piece of blotting paper or filter paper. The oil drops spread out and dry, the large particles remain within a centre circular corona of a small radius. This removes many organometallic and detergent-dispersant additives. Further dispersion leads to oil penetration and filtration through the paper, so circular zones corresponding to the size of particles transported by the filtering oil are clearly defined. A sharply defined ring around the oil wetted area indicates the present of sludge. A period of 24 hours is needed for the oil to "blot" fully, after which the results may be analyzed photometrically. The test provides an indication when engine oil dispersants are reaching their end of their useful life. Some portable test kits have reference standards which can provide an indication of the level of sludge present

Skill: Oil blotting: a suitably trained semi-skilled worker. Analysis: a suitably trained experienced technician

Advantages: Cheap, and easy to use and set up: Provides a record: Moderately accurate indicator of oil oxidation

Disadvantages: 24 hours needed for the oil to blot: Considerable skill required to interpret the results: Only a rough indication of sludge level. Does not indicate chemical composition of particles.

4.13 Patch Test

Conditions monitored: Wear metals, fatigue and corrosion particles, sludge, etc.

Applications: Oils used in diesel and gasoline engines, gas turbines, transmissions, gearboxes, compressors and hydraulic systems

P-F interval: Days to weeks

Operation: A vacuum is used to draw a standard volume of test fluid through a 5 micron Millipore 47 mm disc filter. The degree of discoloration on the filter is compared with a standard membrane filter color rating scale and particle assessment scale to determine contamination levels. High particle levels produce a darker gray or a more highly colored spot. Free water appears either as droplets during the test procedure, or as a stain on the test filter. The patch is examined using a microscope to determine whether the system is heavily loaded with particles and to give a quick impression of the type and size of particles. An approximate (qualitative) cleanliness rating can be determined by comparing the patch to a picture chart

Skill: A suitably trained and experienced skilled worker

Advantages: Test results are dependable, repeatable and sensitive enough to detect any significant change in cleanliness. Good qualitative measure of contamination. Portable

Disadvantages: Using a microscope to count wear or contaminant particles is tedious, cannot be calibrated, and is subject to high levels of user-to-user variance.

4.14 Sediment (ASTM D-1698)

Conditions monitored: Inorganic sediment from contamination, organic sediment from oil deterioration or contamination; soluble sludge from oil deterioration

Applications: Petroleum based insulating oils in transformers, breakers, and cables

P-F interval: Several weeks

Operation: An oil sample is centrifuged to separate the sediment from the oil. The upper, sediment-free portion is decanted and used to measure soluble sludge by dilution with pentane to precipitate pentane insolubles and filtration through a filtering crucible. The sediment is dislodged and filtered through a filtering crucible. After drying and weighing to obtain total sediment the crucible is ignited at 500°C and reweighed. Loss in weight is organic and the remainder is inorganic content of the sediment.

Skill: Taking the sample an electrician. To conduct the test: a suitably trained laboratory technician

Advantages: Test quick and easy. Transformer does not have to be taken offline to monitor the insulating fluid

Disadvantages: Test suitable for low viscosity oils only, for example 5.7 to 13.0 cSt at 40°C (104°F). Test has to be conducted in a laboratory. Pentane is mildly toxic and flammable.

4.15 LIDAR (LIght Detection And Ranging)

Conditions monitored: Presence of particles in the atmosphere

Applications: Quality and dispersion of plumes of smoke from smokestacks

P-F interval: Highly variable depending on the application

Operation: Single wavelength light is directed to the area under investigation. The quantity of particulate matter is assessed by measuring backscatter. Locations are determined by triangulation based on readings taken from two points.

Skill: An experienced engineer

Advantages: A remote sensing technique which can cover large areas

Disadvantages: Very expensive: Requires a high level of skill.

5 Chemical Monitoring

A Preliminary Note on the Chemical Detection of Contaminants in Fluids

The techniques described in this section of part 5 are used to detect elements in fluids – usually lubricating oil – which indicate that a potential failure has occurred elsewhere in the system, as opposed to incipient failure of the fluid itself. The elements most commonly detected by these techniques are listed below, and they can appear as a result of wear, leaks or corrosion.

Wear metals: the following wear metals are measured in lubricating oils
- **Aluminum** from pistons, journal bearings, shims, thrust washers, accessory casings, bearing cages of planetary, pumps, gears, gear lube pumps etc
- **Antimony** from some bearing alloys and grease compounds
- **Chromium** from the wear plated components such as shafts, seals, piston rings, cylinder liners, bearing cages and some bearings
- **Copper** from journal bearings, thrust bearings, cam and rocker arm bearings, piston pin bushings, gears, valves, clutches, and turbocharger bearings. Present in brass or bronze alloys and often detected in conjunction with zinc in the former and tin in the latter
- **Iron** from cast cylinder liners, piston rings, pistons, camshafts, crankshafts, valve guides, anti-friction bearing rollers and races, gears, shafts, lube pumps and machinery structures, etc
- **Lead** from journal bearings and seals
- **Magnesium** from turbine accessory casings, shafts and valves
- **Manganese** from valves and blowers
- **Molybdenum** from wear to plated upper piston rings in some diesel engines
- **Nickel** from valves, turbine blades, turbocharger cam plates and bearings
- **Silver** from locomotive engines, solder and needle bearings
- **Tin** from bearing alloys, brass, oil seals and solder
- **Titanium** found in bearing hubs, turbine blades and compressor discs of gas turbine aircraft engines
- **Zinc** from brass components, neoprene seals.

Leaks: the following elements are associated with leaks
- **Aluminum** from atmosphere contamination
- **Boron** from coolant leaks in oil
- **Calcium** when found in fuel, generally indicates contamination by seawater.
- **Copper** from oil cooler cores - cooling water in oil
- **Magnesium** from seawater contamination
- **Phosphorus** from a coolant leak in oil
- **Potassium** from contamination by seawater in oil
- **Silicon** from contamination by silica from induction systems or cleaning fluids
- **Sodium** from anti-corrosion agents in engine cooling solutions usually as a result of a coolant leak.

Corrosion: the following elements are associated with corrosion
- **Aluminum** from engine block corrosion
- **Iron** from corrosion in storage tanks and piping
- **Manganese** sometimes found along with iron as a result of corrosion of steel

5.1 Atomic Emission (AE) Spectroscopy

Conditions monitored: Wear metals such as iron, aluminum, chromium, copper, lead, tin, nickel, and silver: oil additives containing boron, zinc, phosphorous, calcium, magnesium, or barium: extraneous contaminants such as silicon: corrosion

Applications: Oils used in diesel and gasoline engines, gas turbines, transmissions, gearboxes, compressors and hydraulic systems

P-F interval: Usually several weeks to months

Operation: AE excites the wear metal elements in the sample by raising their atomic energy states in a high voltage (15kV) temperature source. The elements are 'atomized' and emit their characteristic radiation. The resultant light energy passes through a slit to a diffraction grating which separates the individual emission lines for each element. The emission intensity at a characteristic wavelength of an element is proportional to the concentration of the element in the sample. A photomultiplier detector measures the intensity of each emission and transfers the values to a readout device (usually a computer) for additional processing and display. Standard curves are used to establish the relation between signal and element concentration values in parts per million

Skill: To draw the sample: a suitably trained semi-skilled worker. To operate the spectrometer: a suitably trained laboratory technician. To analyze the test results: an experienced chemical analyst

Advantages: Can perform sequential or simultaneous measurements (20 to 60 elements). Test takes just over a minute. Accurate to within several ppm. Low cost

Disadvantages: May fail to vaporized particles larger than five to ten microns. Cannot determine the type of wear process that may be occurring.

5.2 AE - Rotating Disk Electrode

Conditions monitored: Trace levels of wear metals, extraneous contaminants, and additive element levels in lubricants, greases and fuels

Applications: Enclosed lubricating systems in diesel and gasoline engines, gas turbines, transmissions, gearboxes, compressors, and hydraulic systems

P-F interval: Usually several weeks to months

Operation: A rotating graphite disk is immersed into a sample vessel picking up a small sample of oil, grease or fuel as it turns. The sample is introduced into a high temperature electric arc created in the gap between the disc electrode and a rod counter electrode. The sample is completely volatized, creating a plasma which emits light characteristic of the elements in the sample. The emission lines of each element are measured by an optical system, and the results are displayed on a CRT and a printer in the parts per million (ppm) range

Skill: To draw the sample and to operate the machine: a suitably trained technician. To analyze the test results: a suitably trained laboratory technician

Advantages: Simple to operate. No pre-sample preparation. Analysis takes about 30 seconds. Equipment portable. Up to 32 elements can be analyzed at the same time. No hazardous gases are produced. High precision and good repeatability

Disadvantages: Can suffer from spectral interferences. May fail to vaporize particles above 5 microns.

5.3 AE - Inductively Coupled Plasma (ICP)

Conditions monitored: Wear metals from moving parts (such as iron, aluminum, chromium, copper, lead, tin, nickel, and silver): oil additives containing boron, zinc, phosphorous, calcium, magnesium or barium: extraneous contaminants such as silicon: corrosion

Applications: Oils used in diesel and gasoline engines, gas turbines, transmissions, gearboxes, compressors and hydraulic systems

P-F interval: Usually several weeks to months

Operation: Argon gas is passed through a radio frequency induction coil and heated to a temperature of 8,000 K to 10,000 K producing a plasma. The oil sample is diluted by a low viscosity solvent such as xylene or kerosene, is nebulized and borne by the argon gas carrier into the centre of the plasma torch. The high temperature excites the metal atoms which radiate their characteristic emission lines. The lines are captured and measured by the optical system. ICP instruments are available in simultaneous or sequential measurement modes. The sequential instrument uses a movable grating and a single photo detector. Multiple (sequential) burns are necessary to acquire all elements of interest

Skill: To draw the sample: a suitably trained semi-skilled worker. To operate the spectrometer: a suitably trained technician. To analyze results: an experienced technician

Advantages: More accurate, reliable and repeatable than the rotary electrode method. A large dynamic range permits single emission lines to be used for the measurement of a range of concentration levels. Provides parts per billion (ppb) sensitivity for compounds such as metal-organics and wear metal particles less than 3 microns in size. Fast and easy to operate. No need for operator to dilute samples manually prior to analysis. Automated operation

Disadvantages: ICP spectrometer is more complex and expensive, and has a higher operating cost than the rotary disk spectrometer. It uses hazardous chemicals thus generating higher waste costs. Wear metal data generated by ICP will not correlate with data generated by other AE methods. May fail to vaporize particles above 5 microns.

5.4 Atomic Absorption (AA) Spectroscopy

Conditions monitored: Wear metals (such as iron, aluminum, chromium, lead, tin, copper, nickel and silver): oil additives containing boron, phosphorous, zinc, calcium, magnesium, or barium: contaminants such as silicon: corrosion

Applications: Oils used in diesel and gasoline engines, gas turbines, transmissions, gearboxes, compressors and hydraulic systems.

P-F interval: Usually several weeks to months

Operation: Works on the principle that every atom absorbs light of a specific wavelength. The oil sample is diluted and burned in an acetylene flame or other atomizer hot enough to dissociate the sample into its constituent atoms. The flame is irradiated by a hollow cathode lamp at the characteristic wavelength of the desired metal. The higher the concentration of the metal, the higher the absorption of the light. The degree of absorption is measured and converted into ppm values for that metal by a readout computer. Graphite furnace spectrometer uses an electrically heated hollow cylinder to contain the sample and can be used for ultra low trace wear metal levels. This can increase measurement sensitivity from 100 to 1000 times over the acetylene flame method.

Skill: To draw the sample: a suitably trained semi-skilled worker. To operate the spectrometer: a suitably trained laboratory technician. To analyze the results: an experienced chemical analyst

Advantages: Popular with smaller oil analysis facilities for determining wear metal concentrations in used oil analysis. High accuracy, precision and repeatability at low cost. AA does not suffer from spectral interference.

Disadvantages: Samples require preparation. Analysis time is longer. Requires a flammable gas. May fail to vaporize particles above 5 microns.

5.5 X-Ray Fluorescence Spectroscopy

Conditions monitored: Wear metals such as iron, aluminum, chromium, lead, tin, copper, nickel and silver: oil additives containing boron, phosphorous, zinc, cal-cium, magnesium, or barium: contaminants such as silicon: corrosion

Applications: Oils used in diesel and gasoline engines, gas turbines, transmissions, gearboxes, compressors and hydraulic systems.

P-F interval: Usually several months

Operation: An oil sample is exposed to a high energy X-Ray source which raises the energy level of the atoms in the sample. This causes the contaminants to emit characteristic secondary X-Ray energy, except that the radiation measured is the characteristic florescence of the chemical elements in the sample which is converted into their respective elemental data by a multi-channel signal analyzer.

Skill: To draw sample: a suitably trained semi-skilled worker. To operate the equipment: a suitably trained technician. To interpret the results: an experienced engineer

Advantages: Good accuracy, precision and repeatability. Current software has simplified its operation and data interpretation. Covers a wider range of chemical elements than AA or AE. Can see any particle size

Disadvantages: Requires a cryogenic cooled detector for comparable AE or AA detection limits. Longer analysis time. The analysis of lighter elements requires higher X-Ray energies and hence increased precautionary measures in the lab.

5.6 Energy Dispersive X-Ray Spectrometry

Conditions monitored: Wear metals (such as iron, aluminum, chromium, lead, tin, copper, nickel and silver): oil additives containing boron, phosphorous, zinc, calcium, magnesium, or barium: contaminants such as silicon: corrosion

Applications: Oils used in diesel and gasoline engines, gas turbines, transmissions, gearboxes, compressors and hydraulic systems.

P-F interval: Usually several months

Operation: An energy dispersive spectrometer (EDS) attachment to a scanning electron microscope (SEM) permits the detection of the X-rays produced by the impact of the electron beam on a sample, thereby allowing qualitative and quantitative analysis. The electron beam of the SEM is used to excite the atoms in the surface of a solid. These excited atoms produce characteristic X-rays which are readily detected. By utilizing the scanning feature of the SEM, a spatial distribution of the elements can be obtained.

Skill: To draw sample: a suitably trained semi-skilled worker. To do the test: a suitably trained technician. To interpret the results: an experienced engineer

Advantages: Rapid identification of particles: Very fast elemental images and line scans

Disadvantages: Not an on-line technique: Requires expensive laboratory equipment: High degree of skill to interpret the results.

5.7 Dielectric Strength (ASTM D-877 and D-1816)

Conditions monitored: The ability of insulating oil to withstand electric stress caused by conductive contaminants such as metallic cuttings, fibres, or free water

Applications: Insulating oils in transformers, breakers and cables

P-F interval: Several months

Operation: The sample container is inverted and swirled several times before filling the test cup. The test cup is filled to the top of brass electrodes and an increasing voltage applied at a rate of 3 kV/s (D-877) or 5 kV/s (D-1816) to two electrodes spaced 2,54 mm (D-877), 2 mm (D-1816) apart, until breakdown occurs. This value is recorded and trended. Five breakdowns are made with one cup filling at one minute intervals. The average of the five breakdowns is considered the dielectric breakdown voltage of the sample. High and medium voltage transformers should observe the following limit, > 25kV for in service oil, >30kV for new oil. D-877 test used for rated voltages below 230 kV, D-1816 test used for voltages rated above 230 kV.

Skill: Taking the sample: an electrician. To conduct the test: a suitably trained laboratory technician.

Advantages: Test quick and simple. Transformer does not have to be taken off-line to draw sample. A good overall indicator of transformer condition

Disadvantages: Test results dependent on sampling technique. Test sensitive to ambient temperature and humidity. Some risk involved in handling PCBs. Uses hazardous materials and equipment. Not an on-line technique.

5.8 Interfacial Tension (ASTM D-971)

Conditions monitored: Presence of hydrophilic compounds (a compound soluble in water or which attracts water to its surface)

Applications: Petroleum-based insulating oils in transformers, breakers and cables.

P-F interval: Months

Operation: Interfacial tension is determined by measuring the force needed to detach a planar ring of platinum wire from the interface between a sample of oil and distilled water. After zeroing the device (known as a tensiometer), the platinum ring is immersed in the water to a depth of 5 mm. A filtered oil sample is poured on the water to a depth of 10 mm. The oil-water interface is aged for about 30 seconds, then the container is lowered until the film ruptures. The interfacial tension is then calculated. High and medium voltages transformers should not exceed >27 dynes/cm for in-service oil and > 40 dynes/cm for new oil

Skill: Taking the oil sample an electrician. To conduct the test, a suitably trained laboratory technician

Advantages: Reliable indication of compounds soluble in water. Test takes about 1 minute. Transformer does not have to taken offline to monitor insulating oil

Disadvantages: Test dependent on sampling technique. Hazardous and flammable materials are used to conduct the test. Not an on-line technique – requires laboratory equipment.

5.9 DIAL (DIfferential Absorption LIDAR)

Conditions monitored: The chemical composition and dispersal of gases in the atmosphere

Applications: Gases emitted by smokestacks and leaks in tanks or pipelines

P-F interval: Minutes to months, depending on the application

Operation: Similar to LIDAR (see 4.15 above), except that two differential wavelengths are used. One wavelength is set to correspond to a given gas, so one wavelength is absorbed and the other reflected. The quantity of gas present is determined by measuring the amount of light reflected. The location of the gas can be determined by triangulation based on readings taken from two points.

Skill: An experienced engineer

Advantages: Can cover large areas

Disadvantages: Must be calibrated for individual gases: Very expensive and unlikely to be economic for a single site: Operating the equipment requires a high level of skill.

A Preliminary Note on the Chemical Measurement of Fluid Properties

The techniques described in this section of part 5 are used to detect incipient failure of the fluids themselves. They apply to fuels, lubricating oils and/or gases. They are used mainly to analyze the properties of the base fluid and/or the presence/ condition of additives (although some also detect contaminants). The elements most commonly detected by these techniques are listed below.

- **Antimony** from grease compounds.
- **Arsenic** from anti-corrosion or biocide agents
- **Barium** from detergent, dispersant and anti-oxidant additives for fuels and oils.
- **Boron** from anti-corrosion additive for engine coolants and as an anti-knock agent in fuels.
- **Calcium** from detergent and/or dispersant additives
- **Chromium** from an anti-oxidant in jet fuels
- **Cobalt** from natural trace levels in crude oils
- **Copper** from natural trace levels in crude oils and lubricant additives
- **Iron** from natural trace levels in crude oils
- **Lead** from anti-wear additive in some lubricants, sometimes added to fuel as anti-knock agent
- **Magnesium** from detergent and/or dispersant additives
- **Molybdenum** from natural trace levels in crude oils and as an anti-friction additive in some lubricants
- **Nickel** from natural trace levels in crude oils usually in conjunction with vanadium

- **Phosphorus** from natural trace levels in crude oils and as an anti-wear additive in some lubricants
- **Potassium** from natural trace levels in crude oils
- **Selenium** from natural trace levels in some crude oils and coal.
- **Silicon** from anti-foaming agent in some oils
- **Sodium** from natural trace levels in crude oils and seawater
- **Sulphur** from natural trace levels in crude oil and some fuels. Used as an anti-corrosion agent in gear lubricants and as anti-oxidants in lubricating oils.
- **Vanadium** from natural trace levels in some crude oils
- **Zinc** found naturally in some crude oils. Found as an anti-wear additive in automotive lubricants and as an anti-oxidant in marine lubricants.

5.10 Fourier Transform Infrared (FT-IR) Spectroscopy

Conditions monitored: Deterioration, oxidation, water content and depletion of anti-wear additives in mineral oils and synthetic lubricants

Applications: Lubricating oils from combustion engines, hydraulic systems, etc

P-F interval: Usually several weeks to months

Operation: Like atomic absorption spectroscopy, FT-IR measures absorbent light energy at specific wavelengths to determine the level of the elements in a sample. Uses a low power broadband infrared beam converted into a uniform pattern of constructive and destructive interference by a Michelson interferometer. The interference pattern is passed through a sample where it is altered by the characteristic absorbance levels of the elements of the oil and contaminants. The altered interference pattern enters a detector where it is converted into an audible frequency electronic signal, then converted into individual wavelength/amplitude data by a Fourier transform. The absorbance of the oil, additives and contaminants at their respective wavelengths is measured, generating a scalar spectrum, often called a 'fingerprint'. The sample fingerprint is compared with an unused oil sample fingerprint using intelligent software

Skill: To draw the sample: a suitably trained semi-skilled worker. To operate the spectrometer: a suitably trained laboratory technician. To analyze the test results: an experienced chemical analyst

Advantages: Does not use dangerous chemicals. Lower energy levels do not alter the molecular structure of the compounds in the sample, unlike AA. Data can be converted into ASTM equivalent parameters. Good repeatability. Total acid number (TAN) or total base number (TBN) data can be synthesized from FT-IR data

Disadvantages: Uses flammable solvent for cleaning. Different manufacturers of FT-IR equipment use different data extraction algorithms for oil condition parameters and contaminants. Only sensitive to 1 000 ppm water contamination.

5.11 Infrared Spectroscopy

Conditions monitored: The presence of gases such as hydrogen, sulphur hexa-fluoride, nitrogen, methane, carbon monoxide and ethylene; fluid degradation

Applications: As for gas chromatography

P-F interval: Highly dependent on the application

Operation: The atoms of a molecule vibrate about their equilibrium positions with different but precisely determinable frequencies. A sample, placed in a beam of infrared light, absorbs these characteristic frequencies. The absorption bands, plotted against wavelength, specify the infrared spectrum. The position of the absorption points on the wavelength scale is a qualitative characteristic and conclusions can be drawn from the intensity of the absorption bands

Skill: To operate a preset infrared spectrometer: a trained laboratory assistant. To interpret and evaluate the results: an experienced laboratory technician

Advantages: Rapid analysis: High sensitivity: Can be operated by laboratory assistant when equipment is preset: Graphs provide a permanent record

Disadvantages: Considerable experience and skill needed to analyze results: Laboratory based equipment: Wide range of applications required to justify the cost of the equipment.

5.12 Gas Chromatography

Conditions monitored: Gases emitted as a result of faults. There are over 200 gases present in electrical insulating oils of which nine are of interest. In ascending order of criticality, these are nitrogen, oxygen, carbon dioxide (CO_2), carbon mono-xide (CO), methane, ethane, ethylene, hydrogen and acetylene. Large amounts of CO and CO_2 indicate overheating in the windings; CO, CO_2 and methane indi-cate hot spots in the insulation; hydrogen, ethane and methane indicate corona discharge; methane is a sign of internal arcing

Applications: Nuclear power systems, turbine generators, sulphur hexafluoride or nitrogen sealed systems, transformer oils, breakers etc

P-F interval: Highly variable depending on the nature of the fault

Operation: A gas sample is injected through a silicone rubber septum injection port maintained at a temperature higher than the boiling point of the least volatile element in the sample. A carrier gas (usually an inert gas such as helium, argon or nitrogen) sweeps the vaporized sample out of the port and into a separation column located in a thermostatically controlled oven. Elements with a wide range of boiling points are separated by starting at a low oven temperature and raising the temperature over time to elute the high temperature elements. The separation column contains absorbent materials such as diatomaceous earth to separate the gases. Gases emerging from the column flow over a detector which can be directed into a mass spectrometer or Fourier transform infrared spectrometer to record the spectrum as eluted from the column. Different detectors are used for different separation applications

Skill: Taking the sample: an electrician. To conduct the test: a suitably trained laboratory technician. To trend and analyze the results: an electrical engineer

Advantages: High sensitivity detection (one part in 1000 million, by volume): Once the equipment has been set up, it can be operated by a laboratory assistant

Disadvantages: Adequate samples for sensitive analyses are difficult to obtain: In large systems any fault gases may be rapidly diluted: Considerable skill needed to interpret the results: Equipment is not portable: Wide range of applications required to justify purchase: Not widely used in maintenance.

5.13 Ultra-violet and Visible Absorption Spectroscopy

Conditions monitored: Changes in oil properties (alkalinity, acidity, insolubles).

Applications: Oils used in diesel and gasoline engines, gas turbines, transmissions, gearboxes, compressors and hydraulic systems

P-F interval: Several months

Operation: An oil sample is subjected to intense ultraviolet light, usually from a hydrogen or deuterium lamp, or to visible light from a tungsten lamp. Ultraviolet and visible light are energetic enough to promote the outer electrons of the sample elements to higher energy levels, causing light at specific wavelengths to be absorbed. The absorption can be monitored using a wavelength separator such as a prism or a grating monochromator. The amount of light absorbed is related to the concentration of each element. Quantitative measurements can be made by scanning the spectrum or at a single wavelength

Skill: A trained and experienced laboratory technician

Advantages: Useful for quantitative measurements

Disadvantages: The ultraviolet and visible spectra have broad features that are of limited use for sample identification. Considerable skill and experience needed to analyze the results. Equipment is laboratory-based and is expensive.

5.14 Thin-layer Activation

Conditions monitored: Wear

Applications: Turbine blades, engine cylinders, shafts, bearings, electrical contacts, rails and cooling systems

P-F interval: Months

Operation: A thin layer of atoms in the surface of the material to be monitored is made radioactive by bombarding it with a beam of charged particles. Monitoring systems are calibrated to take radioactive decay into account. Material losses of up to 1 μm can be measured up to four years after activation

Skill: To take readings: a suitably trained semi-skilled worker

Advantages: Wear can be measured during normal plant operation even with substantial intervening material

Disadvantages: Components have to be removed to be activated unless coupons can be used: Reactivation is required every four years.

5.15 Scanning Electron Microscopy (SEM)

Conditions monitored: Fractured surfaces for the presence of unusual elements

Applications: Any surface types, thin films and interfaces found in raw semiconductors, finished semiconductors, metal and steel surfaces, medical devices, ceramics, polymers, etc

P-F interval: Application dependent

Operation: A focused beam of electrons is rastered across a sample surface. This causes a secondary electron current to be emitted from the sample which varies according to the angle of incidence of the beam onto the sample. The secondary electron intensity is used to vary the brightness of a cathode ray tube which is synchronous with the raster scan, yielding a topographical image of the sample surface. Different detectors can be used to provide other information. For instance, a backscattered electron detector provides average atomic number information, while an auxiliary energy dispersive X-ray detector can identify elements such as boron and uranium.

Skill: Skilled laboratory technician

Advantages: High resolution with little sample preparation. Large depth of field allows use with rough samples. Rapid qualitative analysis of particles and small areas coupled with an energy dispersive X-ray detector

Disadvantages: More of a diagnostic technique to determine root causes of failures. Samples must be coated with a conductive film. Laboratory technique.

5.16 Scanning Auger Electron Spectroscopy

Conditions monitored: Fractured surfaces for the presence of unusual elements, elemental mapping of fine particles, corrosion and oxidation scales

Applications: Any surface types, thin films and interfaces found in raw and finished semiconductors, metal and steel surfaces, medical devices, ceramics, polymers.

P-F interval: Application dependent

Operation: A finely focused electron beam irradiates the sample and creates a core hole by ejecting a core electron from a sample atom. The resulting ion then de-excites when an electron from an upper level fills the core hole and a third electron – the Auger electron – is emitted to conserve energy. This electron has a kinetic energy characteristic of the emitting atom, which allows elements to be identified to a depth of between 2 and 20 atomic layers

Skill: Skilled laboratory technician

Advantages: SEM capabilities are usually incorporated into Auger instruments. Surface sensitive. Elemental mapping. Rapid analysis

Disadvantages: More of a diagnostic technique to determine root causes of failures. Laboratory technique.

5.17 Electro-chemical Corrosion Monitoring

Conditions monitored: Corrosion of material embedded in concrete

Applications: Structural steel pylons, gantries, etc

P-F interval: Months

Operation: Small currents are passed between the structure and a probe inserted in the ground nearby. These currents affect the potential of the structure at any point where corrosion is taking place. The changes in the potential are measured by a half-cell in contact with the ground and close to the structure. The degree of corrosion is directly related to the current required to displace the leg potential. High currents indicate the need for a physical inspection

Skill: A suitably trained technician

Advantages: Structures do not have to be excavated for inspection unless this technique reveals a real need to do so

Disadvantages: Does not measure the extent or precise location of corrosion: Ground must be moist.

5.18 Exhaust Emission Analyzers (Four-gas Analysis)

Conditions monitored: Combustion efficiency by measuring the concentrations of oxygen (O_2), carbon monoxide (CO), carbon dioxide (CO_2) and hydrocarbons (HC) in exhaust emissions. Exhaust leaks

Applications: Internal combustion engines

P-F interval: Weeks to months

Operation: A sampling probe is inserted into the exhaust pipe upstream of the catalytic convertor. Dirt and oil are removed by a prefilter and moisture by a water separator. Gas sensors pick up the gas concentrations and readings are displayed as percentages (HC in parts per million). High CO means that the engine is running rich. High O_2 indicates a lean misfire or an exhaust leak. CO_2 is at its highest at the optimum air-fuel ratio (AFR), and it drops when the AFR is too rich or too lean. High HC indicates misfires or incomplete combustion. 'Lambda' readings are also calculated on most analyzers. Lambda is the name given to the ratio of the actual AFR over the ideal ratio of 14.7. The ideal lambda reading is one, and leaner ratios are greater than one

Skill: Trained and experienced automotive mechanic

Advantages: Pinpoints emission failures. Portable

Disadvantages: Equipment needs to be taken off-line to connect to the analyzer.

5.19 Color Indicator Titration (ASTM D974)

Conditions monitored: Lubricant deterioration by determining the levels of acidity and alkalinity in an oil sample

Applications: Oils used in diesel and gasoline engines, gas turbines, transmissions, gearboxes, compressors and hydraulic systems

P-F interval: Weeks to months

Operation: The sample is dissolved into a mixture of toluene, isopropyl alcohol and water and titrated with an alcoholic base or acid solution, to the end point indicated by a color change of the added naphtholbenzene solution. The acidity or alkalinity is expressed as milligrams of potassium hydroxide needed to neutralize a gram of oil. The higher the acid or base number the greater the oil deterioration. High and medium voltages transformers should be < 0.5mgKOH/gm for new oil and < 0.1mgKOH/gm for in service oil

Skill: Laboratory technician

Advantages: Test accurate to within 15%

Disadvantages: Can only be used for petroleum based oils. Poisonous, flammable, corrosive chemicals used in the test. Cannot be used for dark oils.

5.20 Potentiometric Titration TAN/TBN (ASTM D664)

Conditions monitored: Lubricant deterioration by determining the level of acidity of an oil sample

Applications: Oils used in diesel and gasoline engines, gas turbines, transmissions, gearboxes, compressors, hydraulic systems and transformers.

P-F interval: Weeks to months

Operation: The sample is dissolved into a mixture of toluene, isopropyl alcohol and water titrated with alcoholic potassium hydroxide. The acidity is determined by measuring the change in electrical conductivity as the potassium hydroxide is added. The value is expressed as mgKOH/g. The higher the acid number, the greater the breakdown of the oil.

Skill: Laboratory technician

Advantages: Can be used for oils that are too dark to use a color change indicator. Test accurate to within 4%.

Disadvantages: Can only be used for petroleum based oils. Dangerous chemicals used in the test.

5.21 Potentiometric Titration TBN (ASTM D2896)

Conditions monitored: Lubricant deterioration by measuring alkalinity.

Applications: Oils used in diesel and gasoline engines, gas turbines, transmissions, gearboxes, compressors, hydraulic systems and transformers.

P-F Interval: Weeks to months

Operation: The sample is dissolved into a mixture of titration solvent which is titrated with perchloric acid. The potentiometric (electrical conductivity) readings are plotted against respective volumes of titrating solution. The alkalinity (base number) is calculated from the quantity of acid needed to titrate the solution expressed in milligrams of potassium hydroxide per gram equivalent (mgKOH/g). The test is a measure of an oil's ability to neutralize corrosive acids formed during operation, indicating its suitability for continued use.

Skill: Laboratory technician

Advantages: Can be used regardless of color of oil. Accurate to within 15%.

Disadvantages: Can only be used for petroleum-based oils. Dangerous chemicals used in the test.

5.22 Power Factor (ASTM D- 924)

Conditions monitored: Dielectric losses in electrical insulating oils caused by contamination and oil deterioration

Applications: Petroleum based insulating oils in transformers, breakers, and cables

P-F interval: Several weeks

Operation: A thoroughly mixed sample is poured into a clean beaker and heated to 2° below desired test temperature. The cell is removed from the test chamber and filled with the heated sample. The inner electrode is inserted into the cell together with a mercury thermometer. The electrical connections are then made to the cell. The sample is then electrically stressed by passing a voltage through the cell and the power factor is calculated. For high and medium voltage transformers the power factor limit should be less than 1% at 25°C

Skill: Taking the sample: an electrician. To conduct the test: a suitably trained laboratory technician

Advantages: Test quick and relatively simple. Transformer does not have to be taken offline to monitor the insulating fluid

Disadvantages: Uses hazardous materials and equipment. Test must be conducted in a laboratory and depends on sampling technique.

A Preliminary Note on Moisture Monitoring

Water in oil rapidly reduces machinery and component life. For instance, it can reduce roller element bearing life by as much as 100 times. It also interferes seriously with the lubricating properties of oil – for instance, a drop of water in 5 liters of oil at 85°C totally destroys zinc anti-wear additives. Water directly affects the oil itself in the following ways:
- it increases oxidation, and in so doing forms slimes and resins
- it increases conductivity, which is especially undesirable in transformer oil
- it reacts with anti-oxidants to form acids and precipitate salts
- it reacts with zinc di-alkyl di-thio phosphate (ZDDP) anti-wear additives to form hydrogen sulfide and sulfuric acid
- it promotes the growth of microbes
- it changes the viscosity of the oil
- it degrades viscosity improvers
Water also affects other aspects of the system as follows::
- it rusts and corrodes metal surfaces
- it jams valves by forming ice crystals
- it increases wear
- it gums up valves and orifices
- it shortens the life of filters
- it entrains more air, which affects bulk modulus.

5.23 Karl Fischer Titration Test (ASTM D-1744)

Conditions monitored: Water in oil

Applications: Enclosed oil systems such as engines, gearboxes, transmissions, compressors, hydraulic systems, turbines, transformers, etc.

P-F interval: Days to weeks

Operation: A measured sample is reacted with a Karl Fischer reagent which contains iodine. When iodine is present, current will pass between two platinum electrodes. Moisture entrained in the sample reacts with the iodine, perpetuating the test as long as water which has not reacted with the iodine remains. Once depleted, the electrodes are depolarized by the iodine and the test is complete. The corresponding potentiometric change is used to determine the titration end point and calculate the water concentration. The duration of the test indicates the water content. High and medium voltage transformers should not exceed 25ppm at 20°C

Skill: Laboratory technician

Advantages: Accurate for small quantities of water (parts per million). Accuracy within 10%. Test is relatively fast.

Disadvantages: Adequate samples for sensitive analyses are difficult to obtain: In large systems any fault gases may be rapidly diluted: Considerable skill needed to interpret results: Equipment not portable: Wide range of applications required to justify purchase: Not widely used in the maintenance environment.

5.24 Moisture Monitor (Vapor Induced Scintillation)

Conditions monitored: Water in oil

Applications: Oils used in diesel and gasoline engines, gas turbines, transmissions, gearboxes, compressors, hydraulic systems and transformers.

P-F interval: Several weeks

Operation: A probe with a miniature heating element is submerged into an oil sample. During the test the heating element glows at a constant temperature, causing suspended moisture in the sample to vaporize and emit a distinctive acoustic signal known as crackling. A microphone mounted near the heating element picks up this signal and electronically passes it to the data collector for analysis. The algorithm in the data collector is calibrated to convert signal threshold crossings per unit time into moisture levels in ppm or percentage. The unit is able to detect suspended moisture to as low as 25 ppm and as high as 10,000 ppm. A typical test takes 30 seconds

Skill: A trained semi-skilled worker

Advantages: No sample preparation needed. Quick and easy. Detects a wide range of concentrations. Requires only 70 milliliters of fluid to conduct the test. Contains no moving parts. Not affected by fluid's viscosity, color, density, contamination, conductivity, or flow. Portable.

Disadvantages: Equipment expensive.

5.25 Crackle Test (Human senses)

Conditions monitored: Water in oil

Applications: Oils used in diesel and gasoline engines, gas turbines, transmissions, gearboxes, compressors, hydraulic systems and transformers.

P-F interval: Days to weeks

Operation: A few drops of oil are placed on a hot plate (about 250°F). If water is present it quickly vaporizes and makes a crackling or popping sound.

Skill: A trained semi-skilled worker

Advantages: Cheap, quick and easy to use. Effective and economical

Disadvantages: Moisture under 300 - 400 ppm cannot be easily heard crackling. Test subjective, from test-to-test and user-to-user. Does not quantify the amount of water present. Requires a quiet area to hear the crackles. Danger of handling oil around a hot surface.

5.26 Crackle Test (Audio detector)

Conditions monitored: Water in oil

Applications: Oils used in diesel and gasoline engines, gas turbines, transmissions, gearboxes, compressors, hydraulic systems and transformers.

P-F interval: Weeks

Operation: A microphone is mounted adjacent to a hot plate (heating element). A few drops of oil are placed on a hot plate (about 250°F). If water is present it quickly vaporizes and makes a crackling or popping sound. The microphone picks up the sound, converts it into an electronic signal and passes it to a data collector for analysis. The algorithm in the data collector is calibrated to convert signal threshold crossings per unit time into moisture levels in ppm or percentage.

Skill: A trained and experienced technician

Advantages: Can detect moisture levels as low as 25 ppm and high as 10 000 ppm. Test takes 30 seconds. Easy to use.

Disadvantages: Danger of handling oil around a hot surface. Laboratory test.

5.27 Clear and Bright Test

Conditions monitored: Water in oil

Applications: Oils used in diesel and gasoline engines, gas turbines, transmissions, gearboxes, compressors, hydraulic systems and transformers.

P-F interval: Several days

Operation: As moisture becomes entrained in oil, the oil becomes visibly hazy – in other words, it is no longer clear and bright. However, note that some oils can dissolve significant amounts of water (depending on viscosity and the additive package) and still remain clear and bright. It is only when the oil reaches a more advanced stage of emulsification, where the oil and water combine (not mix), that it is no longer clear and bright.

Skill: Experienced technician

Advantages: No test equipment required. Cheap, quick, simple and economical.

Disadvantages: Oil color can bring error in to the test. Subjective.

6 Physical Effects Monitoring

6.1 Liquid Dye Penetrants

Conditions monitored: Surface discontinuities or cracks due to fatigue, wear, surface shrinkage, grinding, heat-treatment, corrosion fatigue, corrosion stress and hydrogen embrittlement.

Applications: Ferrous and non-ferrous materials such as welds, machined surfaces, steel structures, shafts, boilers, plastic structures, compressor receivers

P-F interval: Several days to several months, depending on the application

Operation: The liquid penetrant is applied to the test surface and sufficient time is allowed for penetration into surface discontinuities. Excess surface penetrant is removed. A developer is applied which draws the penetrant from the discontinuity to the test surface, where it is interpreted and evaluated. Liquid penetrants are categorized according to the type of dye (visible dye, fluorescent or dual sensitivity penetrants) and the processing required to remove them from the test surface (water washable, post emulsified or solvent removed).

Skill: To apply penetrant: suitably trained semi-skilled worker. Interpretation: suitably experienced technician

Advantages: Visible dye penetrant kits are very cheap (but the more expensive fluorescent kits are far more sensitive): Detects surface discontinuities on non-ferrous materials.

Disadvantages: Fluorescent penetrants require a darkened area for inspection: Highly qualified personnel required to evaluate results: Not an on-line monitoring technique: Monitors surface-breaking defects only: Cannot test materials with very porous surfaces.

6.2 Electrostatic Fluorescent Penetrant

Conditions monitored and applications: As for liquid dye penetrants

P-F interval: Slightly longer than liquid dye penetrants

Operation: As for liquid penetrant dyes, except that opposing electrostatic polarity must be induced between the workpiece and testing materials

Skill: As for liquid dye penetrants

Advantages: The polarity ensures more complete and even deposition of penetrant and developer than with ordinary penetrants, which gives greater sensitivity

Disadvantages: As for ordinary fluorescent penetrants.

6.3 Magnetic Particle Inspection

Conditions monitored: Surface and near-surface cracks and discontinuities caused by fatigue, wear, laminations, inclusions, surface shrinkage, grinding, heat treatment, hydrogen embrittlement, laps, seams, corrosion fatigue and corrosion stress.

Applications: Ferromagnetic metals such as compressor receivers, welds, machined surfaces, shafts, steel structures, boilers, etc.

P-F interval: Days to months depending on the application.

Operation: A test piece is magnetized and then sprayed with a solution containing very fine iron particles over the area to be inspected. If a crack exists, the iron particles are attracted to the magnetic flux leaking from the area caused by the discontinuity and form an indication which is then interpreted and evaluated. Fluorescent magnetic particle sprays provide greater sensitivity, but inspection should be carried out under ultraviolet light in a darkened booth.

Skill: Application: semi-skilled worker. Interpretation: an experienced technician

Advantages: Reliable and sensitive: Very widely used.

Disadvantages: Detects only surface and near-surface cracks: Time consuming: Contaminates clean surfaces: Not an on-line monitoring technique.

6.4 Strippable Magnetic Film

Conditions monitored: Surface discontinuities and cracks caused by fatigue, wear, surface shrinkage, grinding, heat treatment, hydrogen embrittlement, laminations, corrosion fatigue, corrosion stress, laps and seams.

Applications: Ferromagnetic metals such as compressor receivers, welds, machined surfaces, shafts, gears, steel structures, boilers, etc.

P-F interval: Several weeks to months.

Operation: A self curing silicon rubber solution containing fine iron oxide particles is poured into or onto the area under inspection and a magnetic field induced by a magnet. The magnetic particles in the solution migrate to cracks under the influence of the magnetic field. After curing, the rubber is removed as a plug from holes or as a coating from surfaces. Cracks appear on the cured rubber as intense black lines. Investigation of small cracks may need a microscope.

Skill: Application of solution: suitably trained semi-skilled worker. Evaluation: experienced technician.

Advantages: Can be used on areas with limited visual access: Provides a record.

Disadvantages: Detects only surface cracks: Not an on-line technique.

6.5 Ultrasonics - Pulse Echo Technique

Conditions monitored: Surface and subsurface discontinuities caused by fatigue, heat treatment, inclusions, lack of penetration and gas porosity in welds, lamination; The thickness of materials subject to wear and corrosion.

Applications: Ferrous and non-ferrous materials related to welds, steel structures, boilers, boiler tubes, plastic structures, shafts, compressor receivers, etc.

P-F interval: Several weeks to several months.

Operation: A transmitter sends an ultrasonic pulse to the test surface. A receiver amplifier feeds the return pulse to an oscilloscope. The echo is a combination of return pulses from the opposite side of the workpiece and from any intervening discontinuity. The time elapsed between the initial and return signals and the relative height indicate the location and severity of the discontinuity. A rough idea of the size and shape of the defect can be gained by triangulation.

Skill: A suitably trained and experienced technician.

Advantages: Applicable to the majority of materials.

Disadvantages: Difficult to differentiate types of defects.

6.6 Ultrasonics - Transmission Technique

Conditions monitored, applications and P-F interval: As for pulse echo technique.

Operation: A transmitter emits continuous waves from one transducer which are passed right through the test piece. Discontinuities reduce the amount of energy reaching the receiver and so their presence can be detected.

Skill and advantages: As for pulse echo technique.

Disadvantages: As for pulse echo technique: Problems of modulation associated with standing waves cause false readings to be obtained.

6.7 Ultrasonics - Resonance Technique

Conditions monitored, application and P-F interval: As for pulse echo technique. (Also used for testing the bond strength between thin surfaces).

P-F interval: As for pulse echo technique.

Operation: A transmitter is moved over the test surface and the signal observed. Resonance in the absence of discontinuities keeps the transmitted signal high. Discontinuities cause the transmitted signal to fade or disappear.

Skill, advantages, and disadvantages: As for pulse echo technique.

6.8 Ultrasonics - Frequency Modulation

Conditions monitored, applications and P-F interval: As for pulse echo.

Operation: A transducer is used to send ultrasonic waves continuously at changing radio frequencies. Echoes return at the initial frequency and interrupt the new changed frequency. By measuring the phase between frequencies the location of the defect can be determined.

Skill, advantages and disadvantages: As for pulse echo technique.

6.9 Coupon Testing

Conditions monitored: General and localized erosion and corrosion.

Applications: As for electrical resistance method, except paper mills.

P-F interval: Several months.

Operation: Coupons are usually produced from mild, low carbon steel or from a grade of material that duplicates the wall of a vessel or pipe. The coupons are carefully prepared, weighed and measured before exposure. After the coupons have been immersed in the process stream for a period of time (several weeks to several months) they are removed and checked for weight loss and pitting. From these measurements, relative metal loss from the pipe wall can be calculated and pitting can be estimated.

Skill: A suitably trained technician.

Advantages: Very satisfactory when corrosion is steady: Useful where electrical devices are prohibited: Fairly cheap: Indicates corrosion type: Very widely used.

Disadvantages: Long duration of exposure required: Response to dangerous corrosive conditions is slow: Use of coupons is labor intensive: Corrosion rate determination usually takes several weeks: Provides no allowance for unusual or temporary conditions: Coupons inadequate for pulp and paper industry.

6.10 Eddy Current Testing

Conditions monitored: Surface and subsurface discontinuities caused by wear, fatigue and stress; detection of dimensional changes through wear, strain and corrosion; determination of material hardness.

Applications: Ferrous materials used for boiler tubes, heat exchanger tubes, hydraulic tubing, hoist ropes, railway lines, overhead conductors, etc.

P-F interval: Several weeks depending on the application.

Operation: A test coil carrying alternating current at 100 kHz to 4 MHz induces eddy currents in the part being inspected. Eddy currents detour around discontinuities, becoming compressed, delayed and weakened. The electrical reaction on the test coil is amplified and recorded on a CRT or a direct reading meter.

Skill: A suitably trained and experienced technician.

Advantages: Applicable to a wide range of conducting materials: Can work without surface preparation. High defect detection sensitivity: strip chart recorder provides a permanent record.

Disadvantages: Poor response from non-ferrous materials.

6.11 X-ray Radiography

Conditions monitored: Surface and subsurface discontinuities caused by stress, fatigue, inclusions, lack of penetration in welds, gas porosity, intergranular corrosion and stress corrosion. Semiconductor discontinuities such as loose wires.

Applications: Welds, steel structures, plastic structures, metallic wear components of engines, compressors, gearboxes, pumps, shafts, etc.

P-F interval: Several months.

Operation: A radiograph is produced by passing X-rays or gamma rays through materials which are optically opaque. The absorption of the initial X-ray depends on thickness, nature of the material and intensity of the initial radiation. Film exposed to these rays becomes dark when it is developed – how dark depends on the amount of radiation reaching it. The film is darkest where the object is thinnest. A crack, inclusion or a void is observed as a dark patch.

Skill: Use of equipment: a suitably trained and skilled technician. To interpret the results: a highly skilled technician or engineer.

Advantages: Provides a permanent record: Detects defects in parts or structures not visually accessible: Most widely applied X-ray technique.

Disadvantages: Sensitivity often low for crack-like defects: Two-sided access sometimes needed.

6.12 X-ray Radiographic Fluoroscopy

Conditions monitored, applications and P-F interval: as for X-ray radiography.

Operation: The transmitted radiation produces a fluorescence of varying intensity on the coated screen instead of darker patches. The brightness of the image is proportional to the intensity of the transmitted radiation.

Skill: As for X-ray radiography.

Advantages: Quick results: Scanning capability: Detects defects in parts or structures not visually accessible: Most widely applicable technique: Low cost.

Disadvantages: No record produced: Generally inferior image quality: Less sensitive than X-ray radiography.

6.13 Rigid Borescopes

Conditions monitored: Surface cracks and their orientation, oxide films, weld defects, corrosion, wear, fatigue.

Applications: Internal visual inspection of narrow tubes, bores and chambers of engines, pumps, turbines, compressors, boilers, etc in automotive, shipbuilding, aircraft, power generation, chemical and related industries.

P-F interval: Several weeks depending on application.

Operation: Light is channelled from an external light source along a flexible fibre cable to the borescope. Very intense light (300 W) enables photographs to be taken.

Skill: A suitably trained and experienced technician.

Advantages: Inspection done with clear illumination: Parts not visible to the naked eye can be photographed and magnified.

Disadvantages: Provides surface inspection only: Resolution limited: Lens systems relatively inflexible: Operators can suffer 'optic eye' during long inspections.

6.14 Cold Light Rigid Probes

Conditions monitored, applications and P-F interval: As for rigid borescopes (also used in combustible and heat sensitive areas).

Operation: High intensity white light is channelled from a cold light supply unit (150 W) via a flexible fibre cable into a rigid borescope. The probe contains a lens relay system sheathed by glass fibres which passes the light to the working tip. No light is wasted and no heat is emitted. Forward, fore-oblique, sideways and retro-viewing versions of these probes are available. Probe diameters range from 1.7 mm to 10 mm and lengths from 8 cm to 133 cm. Parts not visible to the naked eye can be photographed and magnified or recorded by a miniature video camera.

Skill: As for rigid borescopes.

Advantages: As for rigid borescopes. No heat is generated when cold light supply used: Detailed inspection of surface finish in inaccessible areas can be obtained without dismantling: Photographs provide permanent records: Equipment portable: With the use of the video camera/endoscope technique, inspection time is reduced to a quarter of the time required for direct viewing

Disadvantages: As for rigid borescope: Probe inflexible: Not an on-line technique.

6.15 Deep-Probe Endoscope

Conditions monitored, applications and P-F interval: As for rigid borescopes. (Also used for the inspection of pipework in boilers and heat exchangers)

Operation: These are special modular endoscopes available in lengths of up to 21 m. They are made of stainless steel and screw together to provide a viewing system which can penetrate bores with severely restricted entry. Illumination is provided by a high intensity quartz halogen light

Skill: As for rigid borescopes

Advantages and disadvantages: As for rigid borescopes.

6.16 Pan-view Fiberscopes

Conditions monitored, applications and P-F interval: As for rigid borescopes

Operation: White light of high intensity from a cold light supply unit is transmitted by total internal reflection through a flexible fibre cable into a fiberscope. The fiberscope contains optical fibres bundled together to form flexible light pipes. The fiberscope has a remotely controllable prism built into its tip which can be made to view forwards or sideways as required. The instrument can be inserted using forward viewing and can be stopped to take a detailed sideways look at any passing defect by simply rotating a control knob built into the side of the eyepiece. Adaptors can be used to take photographs or mount TV viewers or cine cameras. An ultraviolet light of high intensity can also be used with fluorescent penetration to detect minute flaws in inaccessible areas

Skill: As for rigid borescopes

Advantages: As for cold light rigid probes: Flexibility makes more detailed inspections possible

Disadvantages: Not an on-line monitoring technique: Provides surface inspection only: Resolution limited: Operators can suffer from 'optic eye' during prolonged inspections: Ultra violet fiberscopes are expensive.

6.17 Electron Fractography

Conditions monitored: The growth of fatigue cracks

Applications: Metallic components subjected to cyclic stresses

P-F interval: Depends on the application

Operation: Every fracture has its own 'fingerprint', in that the history of the fracture process is imprinted on the fracture surface. By studying a replica of the fracture with an electron microscope, it is possible to establish the causes and circumstances of failure

Skill: Replica of the fracture surface: suitably trained technician. Analysis and reading: experienced engineer

Advantages: Failures can be analyzed with a high degree of certainty: No damage caused to fracture surface when replica is made

Disadvantages: Electron microscope is expensive: High degree of specialization required to read the results: Not an on-line monitoring technique: Inaccessible components must be dismantled.

6.18 Color (ASTM D-1524)

Conditions monitored: Oil color and condition

Applications: Petroleum based insulating oils in transformers, breakers and cables.

P-F interval: Weeks to months

Operation: A test tube is filled with the oil sample and placed next to the color comparator. Color is compared by revolving the color standard disk until a color match is made with the sample. The figure seen in the upper opening in the front cover gives the direct reading. For high and medium voltage transformers the color limit should not exceed 3.0 on the ASTM D-1524 color scale

Skill: An experienced electrician

Advantages: Provides rapid field screening of test samples for further testing. Transformer does not have to be taken offline to monitor the insulating oil

Disadvantages: Depends on sampling technique. Can be affected by sunlight..

6.19 Oil Appearance

Conditions monitored: Oil oxidation, water contamination, wear metal particles and particulate contamination

Applications: Lubricating oils

P-F interval: Days to weeks

Operation: Perhaps the simplest of all tests, appearance can provide distinct indications of oil condition and contamination. Most industrial oils are gold colored liquids that are bright and free of suspended solids when new. Greases, coolants and fuels also have a distinct appearance prior to use. A hazy or clouded appearance often indicates water contamination, while gradual darkening often occurs as in oil is oxidized in service. Particles as small as 40 microns can be seen by the unaided eye, providing an indication of large particulate contamination.

Skill: A trained semi-skilled worker

Advantages: Test simple, quick and cheap. No test equipment required.

Disadvantages: Subjective. Particles less than 40 microns cannot be seen by the unaided eye. Particle concentration levels and source of contamination cannot be determined. Test dependent on sampling technique.

6.20 Oil Odor

Conditions monitored: Oil oxidation

Applications: Lubricating oils

P-F interval: Days to weeks

Operation: Most oils have a bland or nondescript odor when new and develop a more pungent or 'burned' odor as they oxidize in service. An unusual odor may indicate contamination such as fuel dilution. Often, the stronger the odor, the greater the level of oxidation or contamination. This technique is also limited by the subjective nature of the observation. Some people have a more sensitive sense of smell and react differently to strong odors. In addition, vapors can collect in closed reservoirs or storage tanks, and give off a strong odor when the tank is first opened. These concentrated vapors may therefore suggest higher level of contamination or oxidation than normally exists.

Skill: Experienced semi-skilled worker

Advantages: Quick, easy and cheap. No test equipment required.

Disadvantages: Subjective.

6.18 Strain Gauges

Conditions monitored: Strain

Applications: Large civil structures such as bridges, tunnels, the load bearing elements of large buildings

P-F interval: Weeks to months

Operation: Resistance wire, foil and semiconductor strain gauges work on the principle that when an electrical conductor is stretched, its electrical resistance increases. By bonding the conductor to provide intimate mechanical contact with the surface under test, any strain on that surface will be reflected in a change of the resistance in the strain gauge. Sensitive indicating or recording equipment is needed to monitor the strains in most structures.

Skill: Operation of equipment: a suitably trained technician: Interpretation of results: a structural engineer.

Advantages: Readily attached to almost any surface.

Disadvantages: The strain gauge must be compatible with both the material under test and the environment in which it is operating.

A Preliminary Note on Viscosity Monitoring

Viscosity is an important physical property of lubricating fluids. It is essential in providing critical clearances between moving and sliding surfaces. Improper viscosity is an early indicator of general lubrication failure. Viscosity changes may also be a warning sign of many potential failure conditions. An increase in viscosity can cause sluggish valve control, pump cavitation, reduced mechanical, volumetric, and energy efficiencies and increased temperature. A decrease in viscosity can cause increased internal and external leakage, increased temperature, excessive wear due to poor lubrication and reduced control and precision.

6.22 Viscosity Monitor

Conditions monitored: Viscosity changes caused by overheating, additive failure, mixed lubricants, fuel and glycol dilution, oxidation, moisture and particulate contamination.

Applications: Oils used in diesel and gasoline engines, gas turbines, transmissions, gearboxes, compressors, hydraulic systems and transformers.

P-F interval: Several weeks to months.

Operation: A sensor is attached directly to a portable condition monitor (PCM) which controls the test sequence and displays the results. The sensor tests directly from the sample bottle and gives on-the-spot viscosity results that can be saved into the PCM for later viewing. Results can be uploaded to a desktop personal computer for trending and graphing. The fluid temperature is measured using a digital temperature probe.

Skill: A trained skilled technician.

Advantages: Fast reliable, on-site testing. Can be calibrated to ASTM viscosity standards. Measures absolute viscosity directly. Kinematic viscosity can be determined by entering specific gravity. Results can be displayed in SSU, centipoise, centistoke, or ISO viscosity grades and can be recorded at 40°C or 100°C.

Disadvantages: Equipment very expensive.

6.23 Falling Ball Comparator

Conditions monitored: Oil viscosity.

Applications: Oils used in diesel and gasoline engines, gas turbines, transmissions, gearboxes, compressors and hydraulic systems.

P-F interval: Weeks to months.

Operation: An oil sample is compared to a reference oil. Identical balls are allowed to fall freely through the sample and the reference oil. The time required to fall a specific distance provides a comparison of the viscosities. One kit provides a direct reading of the sample oil viscosity, while others require calculations.

Skill: A trained laboratory technician.

Advantages: Simple and easy to use. Accurate to within 1% in most cases.

Disadvantages: Oil sample needs to be translucent enough to see the ball as it falls, dark or oxidized oils may be unsuitable. Not a field portable technique.

6.24 Kinematic Viscosity (ASTM D445)

Conditions monitored: Oil viscosity.

Applications: Oils used in diesel and gasoline engines, gas turbines, transmissions, gearboxes, compressors and hydraulic systems.

P-F interval: Weeks to months.

Operation: This test (resistance to flow) measures the time it takes for a given volume of oil to pass through a calibrated glass capillary viscometer under a specified head (gravity) at a given temperature (usually 100°F or 38°C). The test can be used to monitor oil deterioration over time or to indicate the presence of contamination by fuel or other oils. The kinematic viscosity is the product of the time of flow and the calibration factor of the instrument. The dynamic viscosity is the product of the kinematic viscosity value and the density of the liquid.

Skill: A trained laboratory technician.

Advantages: Can be used for both transparent and opaque oils. Good repeatability. Can be used for most lubricating oils.

Disadvantages: Flammable solvents are used in the test. Not a field technique.

7 Temperature Monitoring

A Preliminary Note on Thermography

Thermography is the measurement of the radiation emitted from the surface of an object in real time, producing a visible image of the invisible infrared radiation. It is based on the principle that all objects above absolute zero (-273°F) emit infra-red radiation. Thermal imaging systems are electronic cameras that make the radiation optically visible in an image of different colors (or gray scales). These images can be recorded on conventional videotape or electronic media.

7.1 Infra-red Scanners

Conditions monitored: Electrical: current/resistance relationships from loose, oxidized or corroded connections or malfunction of the component itself. *Mechanical:* heat generated from friction caused by faulty bearings, inadequate lubrication, misalignment, misuse, and normal wear

Applications: Electrical: power distribution and high tension lines, transformers, transformer bushings, capacitor bank connections, thyristor banks, disconnects, relays and circuit breakers, meter and control connections, circuit breaker contacts, bus and fuse connections, fuse clips and stab connections, moulded case and air breakers, motor windings, thermal overloads, conductor fatigue, generator windings, generator brush riggings, generator feeders to primary, exciters, voltage regulators, motor control centres. *Mechanical:* boilers and refractories, steam piping, heat exchangers, radiators, cooling towers, diesel engines, exhaust manifolds, hydraulic systems, gas mains, bearings, bearing lubrication, conveyor belts, drive gears, drive belts, couplings, plastics, metals, gears, shafts, castings, extrusions, turbine blades, welds, buried steam lines, steam traps, brick refractories, wall and roof insulation, ducting, rotating kilns, tire defects. *Continuous processes:* glass, paper, metal, plastic and rubber manufacture.

P-F interval: A few days to several months depending on the application.

Operation: Infra-red scanners employ sets of mirrors and/or prisms rotating at high speed and a collimating lens to collect the radiation and deliver it to a few detectors. The detectors respond to the radiation by generating a current, the amount of current being proportional to the amount of radiation. This output is then processed by an on-board processor into a visible color image and presented on a view finder or monitor as a thermogram.

Skill: A suitably trained and experienced technician

Advantages: Non-contact - safe to view energized electrical system, stationary or moving processes without influencing the temperature of the object. Very sensitive, can see temperature differences as small as 0.1°F or less

Disadvantages: Equipment expensive and can be cumbersome to move around. Needs specialist to interpret the results. Camera has several moving parts.

7.2 Focal Plan Arrays (FPA's)

Conditions monitored: Electrical: current/resistance relationships form loose, oxidized or corroded connections, or malfunction of the component itself. *Mechanical:* heat generated by friction caused by faulty bearings, inadequate lubrication, misalignment, misuse and normal wear

Applications: Electrical: power distribution and high tension lines, transformers, transformer bushings, capacitor bank connections, thyristor banks, disconnects, relays and circuit breakers, meter and control connections, circuit breaker contacts, bus and fuse connections, fuse clips and stab connections, moulded case and air breakers, motor windings, thermal overloads, conductor fatigue, generator windings, generator brush riggings, generator feeders to primary, exciters, voltage regulators, motor control centres. *Mechanical:* boilers and refractories, steam piping, heat exchangers, radiators, cooling towers, diesel engines, exhaust manifolds, gas mains, hydraulic systems, bearings, bearing lubrication, conveyor belts, couplings, drive gears, drive belts, plastics, metals, gears, shafts, castings, extrusions, turbine blades, welds, buried steam lines, steam traps, brick refractories, ducting, wall and roof insulation, rotating kilns, tire defects. *Continuous processes:* glass, paper, metal, plastics and rubber manufacture

P-F interval: A few days to several months depending on the application.

Operation: A lens in the FPA focuses the radiation onto a matrix of detectors which deliver spatial and thermal resolutions that were previously unknown. Each detector is composed of many small elements. The detectors convert the radiation into electrical energy, which is amplified and processed into a visible image and presented on a view finder or monitor as a thermogram. FPA's have only one moving part – a cooler.

Skill: A suitably trained and experienced technician

Advantages: Highly versatile. Non-contact - safe to view energized electrical systems, stationary or moving processes without influencing the temperature of the object. Can see temperature differences as small as 0.1°F or less. Small and compact. Radiometric.

Disadvantages: Equipment more expensive than IR scanners. Needs specialist to interpret the results.

7.3 Fibre Loop Thermometry

Conditions monitored: Temperature variations caused by insulation deterioration, leaks, blocked cooling systems, etc.

Applications: Pipelines, engines, transformer windings, power cables.

P-F interval: Hours to months.

Operation: Light is passed down a fibre-optic cable. A certain amount of backscatter is reflected back towards the light source and diminishes the strength of the outgoing signal. There is a direct mathematical relationship between the time it takes the light to travel down the cable for a given distance, the amount of backscatter and the temperature of the cable. This relationship can be used to determine the temperature at given points along the cable.

Skill: A suitably trained and experienced technician.

Advantages: Unaffected by the presence of electromagnetic interference: Operable in hazardous environments: Can reach inaccessible locations: Combines temperature sensing and data transmission in a single component (the cable): Continues functioning even if a cable break occurs: Accurate up to 4 km.

Disadvantages: Uneconomic in small installations.

7.4 Temperature Indicating Paint

Conditions monitored: Surface temperature.

Applications: Hot spots, insulation failure.

P-F interval: Weeks to months, depending on the application.

Operation: A silicon-based paint which changes color as temperatures rise. The color starts out green, changes to blue at 204°C and turns white at 316°C. The colors do not change back again as the temperature drops.

Skill: No training required for observers.

Advantages: Simple: Permanent record of the highest temperature reached.

Disadvantages: Colors do not change back again: Only useful at two fixed temperatures: Service life of each coat only one to two years (provided it does not change color in the interim).

8 Electrical Effects Monitoring

8.1 Linear Polarization Resistance (Corrator)

Conditions monitored: Rate of corrosion in systems exposed to electrically conductive corrosive fluids.

Applications: Cooling water systems, municipal water systems, nuclear power heat exchange waters, geothermal power generating systems, desalination plants and pulp and paper mills.

P-F interval: Usually several months in most applications.

Operation: The electro-chemical polarization is based on the fact that a small voltage applied between a metal specimen and a corrosive solution produces a current. The ratio of applied voltage to current is inversely proportional to the corrosion rate, so this ratio provides a measure of the corrosion rate increase.

Skill: A suitably trained technician.

Advantages: Provides a quick and direct indication of corrosion rate and pitting tendency: Measures corrosion as it occurs: Some instruments record the corrosion condition: Automatic and portable systems available: Sensitive to corrosion rates as low as a fraction of a mil per year: Easy to interpret results.

Disadvantages: Portable equipment does not provide a permanent record: Readings must be adjusted when taken in high sensitivity corrosive media: Gives no information on total corrosion.

8.2 Electrical Resistance (Corrometer)

Conditions monitored: Integrated metal loss (i.e. total corrosion).

Applications: Petroleum refineries, process plants, gas transmission plants, underground or undersea structures, cathodic protection monitoring, abrasive slurry transport, water distribution systems, atmospheric corrosion, electrical generating plants, paper mills, etc.

P-F interval: As for linear polarization resistance.

Operation: The system is composed of a probe and an instrument to read the probe. The probe consists of a wire, strip or tube of the same metal as the plant being monitored. The electrical resistance of the probe, measured by a bridge circuit, increases as the probe cross-section decreases with corrosion. The increased resistance corresponds to total metal loss, which is easily converted to corrosion rate.

Skill: As for linear polarization method.

Advantages: When plotted against a time scale, yields both corrosion rate and total metal loss: Can be used in any environment: Portable equipment available: On-line monitoring possible: In-plant equipment provides permanent records: Interpretation normally easy.

Disadvantages: Does not indicate whether the corrosion rate at a particular time is high or low: Portable equipment provides no permanent record.

8.3 Potential Monitoring

Conditions monitored: Corrosive states (active or passive) such as stress-corrosion cracking, pitting corrosion, selective phase corrosion, impingement attack etc.

Applications: Electrolyte environments such as chemical process plants, paper mills, electrical generating plant, pollution control plants, desalination plants, etc; best suited to stainless steel, nickel-based alloys and titanium

P-F interval: Depends on the material and the rate of corrosion.

Operation: This technique takes advantage of the fact that, from the point of view of corrosion, a metal which is in a passive state (low corrosion rate) has a noble corrosion potential, while the same metal in an active state (higher corrosion rate) has a much less noble potential. The potential changes when passivity breaks down, and measurements can be made using a voltmeter of about 10 megohm input impedance and full-scale deflection of 0.5 to 2 volts.

Skill: Usually a trained technician, but sometimes needs an experienced engineer

Advantages: Monitors localized attack: Fast response to change.

Disadvantages: Small potential changes can be influenced by changes in temperature and acidity: Does not give a direct measure of corrosion rate or total corrosion: Expert assistance may be required for interpretation.

8.4 Power Factor Testing

Conditions monitored: Power loss through the insulation system caused by leakage to ground, moisture in cables.

Applications: Electrical circuits, transformer windings, high voltage transformer bushings, high and medium voltage cables.

P-F interval: Several months.

Operation: Power factor is circuit resistance divided by circuit impedance. A known voltage is applied to the winding insulation and the resulting current is measured. The cosine of the angle between the voltage and current is called the power factor. The measured current squared times the insulation resistance is called the watts loss. These values are measured and recorded when the insulation system is first installed to establish a baseline. Subsequent tests results are compared to the initial readings. As the circuit impedance changes due to aging, moisture, contamination, insulation shorts or physical damage, the power factor rises. A newly filled oil transformer should have a power factor of under 0.5% and an in-service oil filled transformer under 2%.

Skill: Conducting the tests: field technician. Analyzing the data: an engineer.

Advantages: One of the best predictive tests.

Disadvantages: Not an on-line technique.

8.5 Meggers and Other Voltage Generators

Conditions monitored: Insulation resistance.

Applications: Electrical circuits.

P-F interval: Months to years.

Operation: A known DC (250 volts to 10kV) voltage is applied to the equipment under test, resulting in a (hopefully) small current flow. If there is no current return to the test set from the equipment under test, this current must be flowing to ground. The current flowing to ground is called 'leakage current'. The insulation resistance can then be calculated using Ohms Law.

Skill: Technicians or engineers.

Advantages: A simple and very well understood technique.

Disadvantages: Test cannot be carried out on-line.

8.6 Breaker Timing Testing

Conditions monitored: Breaker contact travel, speed, wipe and bounce.

Applications: High and medium voltage circuit breakers.

P-F interval: Weeks to months.

Operation: A transducer is mechanically attached to the breaker mechanism, then electrically connected to a timing set. The breaker circuit is then operated through its entire cycle of opening and closing. The test set measures contact travel, speed, wipe and bounce. These results are compared to the last test and to the manufacturers' recommendations. Trending this information indicates whether adjustments to the breaker are necessary.

Skill: Conducting the tests: field technicians. Analyzing the data: an engineer.

Advantages: High and medium voltage breakers can benefit from this test.

Disadvantages: Not an on-line technique. Not applicable to moulded case breakers and/or low voltage breakers.

8.7 Breaker Contact Resistance Test

Conditions monitored: Breaker contact wear and deterioration.

Applications: Circuit breakers.

P-F interval: Several weeks.

Operation: A DC current, usually 10 or 100 amps is applied to the contacts. The voltage across the contacts is measured and the resistance can be calculated using

Ohm's Law. Resistances of about 200 micro-ohms are normal, although manufacturers routinely publish their own design limits. This value is trended over time to assess deterioration. Maximum limits can be obtained from manufacturers.

Skill: Conducting the tests: field technician. Analyzing the data: an engineer.

Advantages: Resistance values can be trended over time to detect potential failures before the breaker contacts deteriorate significantly.

Disadvantages: Not an on-line technique. Normal resistance meter cannot be used due to the resistance being in the order of micro-ohms. Not recognized as a true predictive technique.

8.8 Motor Circuit Analysis (MCA)

Conditions monitored: Changes in conductor path resistance caused by loose or corroded connections, loss of copper (turns) in the stator: Phase to phase inductance caused by magnetic interaction between stator and rotor: Stator inductance affected by rotor position, rotor porosity and eccentricity, stator turn, coil and phase shorting; Winding cleanliness and resistance to ground.

Applications: Electric motors (DC, AC induction, synchronous and wound rotor).

P-F interval: Several weeks to months depending on the application.

Operation: A number of tests are taken together to give a complete picture of the motor circuit condition. Algorithms and rules are used to measure the severity of any defects which may be present. The test applies low DC and AC voltage (resistance to ground test uses 500 or 1000 volts DC) at the motor control centre (MCC) power bus to measure the following: resistance to ground, circuit resistance (phase-to-phase), capacitance to ground, inductance (phase-to-phase), rotor influence, DC bar to bar and polarization index/dielectric absorption. In the conductor path of the motor circuit, the resistance of each phase is measured and compared to the other phases. Readings are usually lower for large motor circuits and higher for smaller motors. Unequal resistance in any part of the circuit unbalances the voltages in the phases, which in turn causes significant heating of the motor windings. Inductive imbalance is also measured. This indicates imbalanced magnetic fields and unequal current flows in the windings, and is most often associated with stator windings (but can be influenced by the stator iron and rotors). Increased capacitance values are normally associated with the motor. When the void between the stator and the motor casing becomes dirty and/or damp, the capacitive effect between the conductor path inside the insulation and the outer 'skin' of the insulation is increased. AC current can pass across this 'natural capacitance' and then to ground via the dirty, wet connections to the motor casing.

Skill: Conducting the tests: field technician. Analyzing the data: an engineer.

Advantages: Tests are done at low voltages and minimum current test signals which are non-destructive. Lightweight and portable equipment can be used in the field. Tests can be done at the MCC requiring no break in motor connections.

Disadvantages: Not an on-line technique. Motor circuit must be non-energized.

8.9 Electrical Surge Comparison

Conditions monitored: Turn-to-turn and phase-to-phase insulation deterioration, and reversal or open circuit in the connection of one or more coils or coil groups.

Applications: Induction or synchronous motors, DC armatures, synchronous field poles.

P-F interval: Weeks to months, dependent on motor cycle frequency and starts under loaded conditions.

Operation: A transient surge is applied at high frequency to two separate but equal parts of a winding. The resulting voltage waveforms reflected from each part are displayed on an oscilloscope. If both windings are identical, each waveform is exactly superimposed on the other, so a single trace appears on the screen. If one of the two winding segments contains a short-circuit, or a reversed or open coil, the waveforms are visibly different. If this problem is found, it is necessary to establish which segment is at fault. This can be done by comparing each segment to a third segment, and noting which combination produces the waveform deflections. Generally, shorted or missing turns cause fairly small differences in waveform amplitude. Mis-connections such as coil reversal or interphase shorts tend to cause large differences or irregularities in waveform shape. With this method it is also often possible to determine the voltage at which turn-to-turn or phase-to-phase conduction begins. If this shorting is near operating voltage, then the motor has a serious insulation fault and should be replaced as soon as possible. If shorting is not detected up to twice operating voltage plus 1000 V, the winding is considered good and the motor can be returned to service.

Skill: A trained and an experienced test operator.

Advantages: Portable. Turn-to-turn and phase-to-phase shorting often occur before deterioration of ground wall insulation giving longer P-F intervals. Most equipment can also perform high potential test (see 8.14 below).

Disadvantages: Quite complex and expensive. Cannot evaluate one coil by itself. Requires careful repetition to determine the location and severity of a fault.

8.10 Motor Current Signature Analysis

Conditions monitored: Broken rotor bar(s) or shorting rings, high resistance between bars and rings, uneven rotor-stator air gaps, rotor misposition, deteriorated or shorted rotor or stator core lamination.

Applications: AC or DC motors.

P-F interval: Several weeks to months.

Operation: This technique is based on the principle that an electric motor driving a mechanical load acts as an efficient, continuously available transducer. The motor senses mechanical load variations and converts them into electric current variations which are transmitted along the motor power cables. These current variations, though very small in relation to the average current drawn by the electric motor, can be monitored and recorded at a convenient location away from the operating equipment. Analysis of the variations provides an indication of machine condition, which may be trended over time to provide a warning of deterioration or process alteration. The test is done by placing a single split-jaw current transformer probe on one of the power leads at the motor control centre or starter cabinet. The raw waveform signal is amplified, filtered and further processed to obtain a measurement of the instantaneous load variations within the drive train and the ultimate load. In general, the current in the three phases should not differ by more than 3%. If the variation exceeds 3% for any phase, stator problems could exist. The amplitudes at line frequency can also be compared with the pole pass frequency immediately to the left of line frequency. A significant difference in amplitude between these two frequencies indicates a cracked or broken rotor bar, end ring, or slip ring, or resistance joint problems.

Skill: To clamp current transformer around one of the 3-phase power line leads an experienced electrician. To conduct the test and interpret the results: a technician with an understanding of electric motors.

Advantages: On-line measurements can be taken without breaking any electrical connections. No electrical connections are required which reduces the hazard of electrical shocks. Readings can be taken remotely and safely on large, high speed or otherwise hazardous machines.

Disadvantages: Complex due to the relatively subjective nature of interpreting the spectra (this has recently been improved from a data collection and analysis interpretation standpoint). Equipment expensive.

8.11 Power Signature Analysis

Conditions monitored: Rotors, broken bars, cracked or broken end rings, bad cage joints, bowed or bent rotors; Stators, shorted lamination, eccentricity; Single phasing, phase current and voltage balance, resistive and inductive imbalance; Torque variations, wear or deterioration of machine clearances, flow or machine output restrictions, machinery alignment; Machinery efficiencies.

Applications: AC induction motors, synchronous motors, compressors, pumps and motor operated valves.

P-F interval: Several weeks to months.

Operation: Probes are attached to motor feed lines either at the Motor Control Centre (MCC), at a breaker box or locally at the motor, to gather electric current and voltage signals while the motor is running. A signal conditioning unit conditions and filters analog signals sensed from the feed lines. Data files are compiled and analyzed using application based software tools (Fast Fourier Transform) to plot variables such as total real power, total reactive power and total power factor. Analysis of the plots enables the motor and overall system performance to be evaluated in detail. The plots can also be compared to baseline fingerprints to detect deviations.

Skill: To attach probes to live motor feed lines: an electrician. To conduct the test and interpret the results: an experienced technician.

Advantages: Tests can be done without shutting down the equipment. One of the few techniques that enables broken rotor bars to be detected under load. Allows equipment efficiencies to be determined.

Disadvantages: Skill and care required when connecting probes to live motor feed lines. Interpreting and analyzing the data takes some practice and an understanding of electric motors and the driven equipment is necessary. Limited number of industry-wide applications for comparison. Equipment expensive.

8.12 Partial Discharge

Conditions monitored: Insulation breakdown.

Applications: All types of medium voltage electrical equipment including switchgear, bus ducts, transformers, arresters, bushings, switches, motor starters, potheads, motors, generators, cable terminations, cable splices, and the cables themselves. Distribution systems and equipment > 2,000 volts AC.

P-F interval: Several weeks to months (voltage levels, the shape of the void, ambient temperature, system losses all influence how quickly the insulation fails).

Operation: A partial discharge (PD) occurs when a small void, crack, or irregularity in an insulation system causes an electric field to build up. Sensors are used to pick-up the PD. On switchgear, the sensor is connected between the grounded side of the metering CT circuit. On cables, the sensor is connected around the ground wire that connects the cable shield or placed around the insulated conductor. On motors, sensors are placed on the motor frame or around the ground connection or around the insulated motor lead. In analyzing the data, three issues are considered:

- the number of pulses per cycle and the magnitude of the pulse (field strength in pico coulombs)

- the power of the pulses (intensity)
- the rate of change over time of the power (trend analysis)

When trending data three levels of PD thresholds are set. Green means good for continued use, yellow means increased activity and red means failure is imminent.

Skill: Experienced electrical technician.

Advantages: Allows quick and more informed decisions. Can be applied to any type of electrical equipment.

Disadvantages: Current available on-line technology cannot locate the exact source of the PD while the equipment is energized. One single data point provides little or no information. Several data points are needed to trend information. No current standards available on maximum acceptable levels of PD activity, except for cables. Expert knowledge and statistical analysis required to set PD thresholds. Need special sensors off-line to determine the exact location of PD activity.

8.13 High Potential (Hi-Pot) Testing

Conditions monitored: Motor winding ground wall insulation deterioration.

Applications: AC and DC motors.

P-F interval: Several weeks.

Operation: High DC voltage is applied to the stator windings in graduated steps or ramps up to a limit, usually twice the line voltage. Test voltages are usually derived from the IEEE Standard 95. At the first sign of non-linearity in the test current or drop in insulation resistance with further voltage increase, the test voltage is recorded and the voltage removed in order to avoid complete insulation breakdown. If the insulation withstands the voltage, it is considered to be safe and the motor can be returned to service. Any trend in voltage at which non-linearity in current drop or insulation resistance occurs can be used to predict remaining life.

Skill: An experienced electrical technician.

Advantages: Tests normally correlate with surge comparison tests.

Disadvantages: Motors have to be taken out of service to conduct the test. Testing potentially destructive.

8.14 Magnetic Flux Analysis

Conditions monitored: Broken rotor bars, unbalanced phases and anomalies in stator windings such as turn-to-turn, phase-to-phase and phase-to-ground shorts.

Applications: AC induction motors.

P-F interval: Several weeks to months

Operation: A flux coil sensor is placed at the centre of the axial outboard end of the motor. (Consistent positioning of this sensor is essential for reliable and trendable data.) The signal received from the sensor is transformed into the frequency domain using an FFT analyzer. A trend of certain magnetic flux frequencies will indicate electrical asymmetries associated with the rotor and stator windings. Most of the peaks in a flux coil spectrum occur at frequencies which have some relationship to running speed. Broken rotor bars increase the sideband activity around running speed. Unbalanced supply voltage (which causes motor heating and eventually leads to premature deterioration of the stator windings) shows no change except around the peak occurring at line frequency +1 x RPM. One of the first faults a winding will encounter is turn-to-turn shorts, which then migrate into phase-to-phase or phase-to-ground shorts. A winding fault can be indicated around the 3 x running speed sideband of line frequency. A variation of this technique is used to detect turn-to-turn shorts by looking at the family of 'slot pass' frequencies from measurements taken with a flux coil. Flux measurements are taken as mention above, and the resulting signature is analyzed at the 'slot pass' frequencies. The principle slot pass frequency occurs at the product of the number of rotor bars and running speed. The technique involves comparing spectra over time to determine when changes occur.

Skill: To record the spectrum: an electrician/technician with an understanding of motors. To interpret the results: an engineer.

Advantages: One of the few techniques that can detect faults associated with electrical insulation of electric motors while the motor is on-line.

Disadvantages: High degree of skill and knowledge of electric motors required to interpret results.

8.15 Battery Impedance Test

Conditions monitored: Cell deterioration.

Applications: Emergency power and DC control power batteries.

P-F interval: Several weeks.

Operation: As a battery ages and begins to lose capacity, its internal impedance rises. A battery impedance set injects an AC signal between the terminals of the battery. The resulting voltage is measured and the impedance calculated. Two comparisons can then be made: first, the impedance is compared with the last reading for that battery; and second, the reading is compared with other batteries in the same bank. Each battery should be within 10% of the others and 5% of its last reading. A reading outside these values indicates a cell problem or capacity loss. There are no set guidelines and limits for this test. Each type, style and configuration of battery has its own impedance, so it is important to take a baseline reading early in the battery's life.

Skill: Field technician.

Advantages: The test can be performed without removing the battery from service, as the AC signal is low level and 'rides' on top of the DC of the battery.

Disadvantages: Test could take a long time on large battery banks.

9 A Note on Leaks

With the exception of ultrasonic leak detection, a topic which has not been covered in much detail in this Appendix is leaks, especially in underground storage tanks. This is because a publication which provides a comprehensive description of 36 different leak detection methods is already available. It is called "Underground Leak Detection Methods - A State of the Art Review", and is in the form of a report prepared in 1986 by Shahzad Niaki and John Broscious of the IT Corporation in Pittsburgh and commissioned by the Hazardous Waste Engineering Research Laboratory, Edison, New Jersey. Copies of the report are available from the National Technical Information Service, a division of the United States Department of Commerce based in Springfield, Virginia, USA.

Glossary

Applicable: See "technically feasible"

Conditional probability of failure: The probability that a failure will occur in a specific period provided that the item concerned has survived to the beginning of that period.

Desired performance: The level of performance acceptable to the owner or user of a physical asset or system

Effective: See "worth doing"

Environmental consequences: A failure mode or multiple failure has environmental consequences if it could breach any corporate, municipal, regional, national or international environmental standard or regulation which applies to the physical asset or system under consideration

Evident failure: A failure mode that will on its own become evident to the operating crew under normal circumstances

Evident function: A function whose failure will on its own become evident to the operating crew under normal circumstances

Failure consequences: The way (or ways) in which a failure mode or a multiple failure matters.

Failure effect: What happens when a failure mode occurs

Failure-finding task: A scheduled task that seeks to determine whether a specific hidden failure has occurred (check whether an item has failed)

Failure management policy: A generic term that encompasses on-condition tasks, scheduled restoration, scheduled discard, failure-finding, run-to-failure and redesign

Failure mode: A single event that causes a functional failure

Function: What the owner or user of a physical asset or system wants it to do

Functional failure: A state in which a physical asset or system is unable to perform a specific function to a level of performance that is acceptable to its owner or user

Hidden failure: A failure mode that will not become evident to the operating crew under normal circumstances if it occurs on its own

Hidden function: A function whose failure will not become evident to the operating crew under normal circumstances if it occurs on its own

Initial capability: The level of performance of which a physical asset or system is capable at the moment it enters service

Multiple failure: An event that occurs if a protected function fails while its protective device or protective system is in a failed state

Non-operational consequences: A failure mode has non-operational consequences if it is not hidden and does not have safety, environmental or operational consequences, but only requires repair

On-condition task: A scheduled task used to determine whether a potential failure has occurred (Check whether an item is failing or about to fail)

Operating context: The circumstances in which a physical asset or system is expected to operate

Operational consequence: A failure mode or multiple failure has operational consequences if it could adversely affect the operational capability of a physical asset or system (output, product quality, customer service, military capability, or operating costs in addition to the cost of repair)

P-F Interval: The interval between the point at which a potential failure becomes detectable and the point at which it degrades into a functional failure (also known as 'failure development period' or 'lead time to failure')

Potential failure: An identifiable condition that indicates that a functional failure is either about to occur or is in the process of occurring

Primary function(s): The function(s) that constitute the main reason(s) why a physical asset or system is acquired by its owner or user

Protective device or **protective system:** A device or system that is intended to avoid, eliminate or minimise the consequences of failure of some other system

Redesign: Any action taken to change the physical configuration of an asset or system (modification), to change the operating context of the asset or system, to change the method used by an operator or maintainer to perform a task, or to change the capability of an operator or maintainer (training)

Routine: Same as 'scheduled'

Run-to-failure: A failure management policy that permits a specific failure mode to occur without any attempt to anticipate or prevent it

Safety consequence: A failure mode or multiple failure has safety consequences if it could injure or kill a human being

Scheduled: Performed at fixed, predetermined intervals.

Scheduled discard: A task that entails discarding an item or component at or before a specified age limit regardless of its condition at the time

Scheduled restoration: A task that restores the initial capability of an item or component at or before a specified age limit, regardless of its condition at the time

Secondary function: Functions which a physical asset or system has to fulfil apart from its primary functions, such as those needed to fulfil regulatory requirements and those which concern issues such as protection, control, containment, comfort, appearance, structural integrity and energy efficiency.

Technically feasible: A task is technically feasible if it is physically possible for the task to reduce, or enable action to be taken to reduce, the consequences of the associated failure mode to an extent that might be acceptable to the owner or user of the asset

Worth doing: A scheduled task is worth doing if it reduces (avoids, eliminates or minimizes) the consequences of the associated failure mode to an extent that justifies the direct and indirect costs of doing the task.

Bibliography

American Society of Testing Materials. *Annual Book of ASTM Standards*. Philadelphia Pennsylvania. ASTM. 1995

Andrews JD & Moss TR. *Reliability and Risk Assessment*. Harlow, Essex. Longman. 1993

Blanchard BS, Verma D & Peterson EL. *Maintainability*. New York. Wiley. 1995

Blanchard BS & Fabrycky WJ. *Systems Engineering and Analysis*. Englewood Cliffs, New Jersey. Prentice Hall. 1990

Blanchard BS. *Logistics Engineering and Management*. Englewood Cliffs, New Jersey. Prentice Hall. 1986

Berry JE. "Detection of Multiple Cracked Rotor Bars on Induction Motors using both Vibration and Motor Current Analysis" *P/PM Technology*. **9 (3)** 1996

Bowers SV. "Integrated Strategy for Predictive Maintenance of AC Induction Motors". *Predictive Maintenance Technology National Conference*. Indianapolis, Indiana. 4 - 6 December 1995

Cox SJ & Tait NRS. *Reliability, Safety and Risk Management*. Oxford. Butterworth Heinemann. 1991

Dalley R. "Wear Particle Analysis". *Predictive Maintenance Technology National Conference*. Indianapolis, Indiana. 4 - 6 December 1995

Davis D J: "An Analysis of Some Failure Data". *Journal of the American Statistical Association*, **47 (258)** 1952

DelZingaro M & Matthews C. "Using Power Signature Analysis to Detect the Behavior of Electric Motors and Motor-driven Machines". *P/PM Technology*. **8 (5)** 1995

Gaertner J P. *Demonstration of Reliability-Centered Maintenance*. Palo Alto, California: Electric Power Research Institute. 1989

Gleick J. *Chaos – Making a New Science*. New York. Penguin. 1987

James R. "Basic Oil Analysis". *Predictive Maintenance Technology National Conference*. Indianapolis, Indiana. 4 - 6 December 1995

Jones RB. *Risk-based Management*. Houston, Texas. Gulf. 1995

Kane CF: "Predictive Maintenance Technologies Can Help Prioritize Maintenance Dollars". *Northwest Indiana Business Roundtable & Trade Show*. Merrillville, Indiana. 3 October 1996

Maintenance Steering Group - 3 Task Force . *Maintenance Program Development Document MSG-3*. Washington DC: Air Transport Association (ATA) of America. 1993

Mercier J-P. *Nuclear Power Plant Maintenance.* Maisons-Alfort, France. Editions Kirk. 1987

Mohr G. "Technology Overview - Ultrasonic Detection". *P/PM Technology.* **8** (5) 1995

Moubray J M . "The Responsible Custodianship of Physical Assets". *The Tenth Annual Canadian Maintenance Management Conference,* Toronto, Canada. 2 - 5 November 1998

Moubray JM. "Maintenance Management – A New Paradigm". *Third Annual Conference of the Society of Maintenance & Reliability Professionals.* Chicago Illinois. 2 - 4 October 1995

Moubray J M . "Maintenance and Product Quality". *International Conference on Total Quality,* Hong Kong: 16 - 17 November 1989

Moubray J M. "Maintenance and Safety - a Proactive Approach". *Annual Conference of the Accident Prevention and Advisory Unit of the UK National Health and Safety Executive,* Liverpool, UK; 19 May 1989

Moubray J M. "Developments in Reliability-centered Maintenance". *The Factory Efficiency & Maintenance Show and Conference,* NEC, Birmingham, UK; 27 - 30 September 1988

Moubray J M. "Maintenance Management - The Third Generation". *The 9th European Maintenance Congress,* Helsinki, Finland; 24 - 27 May 1988

Moubray J M. "Reliability-centered Maintenance". *A Conference on Condition Monitoring,* Gol, Norway; 2 - 4 November 1987

Nelson W. *Applied Life Data Analysis.* New York: Wiley. 1982

Netherton D: "SAE's New Standard for RCM". *Maintenance* 15 (1) 3 - 7, 2000

Niaki S and Broscious J A. *Underground Tank Leak Detection Methods – A State-of-the-Art Review.* Springfield, Virginia: National Technical Information Service, US Department of Commerce. 1985

Nicholas J R Jr (1995): "AC & DC Motor Circuit Testing and Predictive Analysis". *Predictive Maintenance Technology National Conference.* Indianapolis, Indiana. 4 - 6 December 1995

Nowlan F S & Heap H. *Reliability-centered Maintenance.* Springfield, Virginia: National Technical Information Service, US Department of Commerce. 1978

Oakland JS. *Total Quality Management.* Oxford. Butterworth Heinemann. 1989

Perrow C. *Normal Accidents.* New York. Harper-Collins. 1984

Reason JT. *Human Error.* Cambridge, England. Cambridge University Press. 1990

Resnikoff H L. *Mathematical Aspects of Reliability-centered Maintenance.* Los Altos, California: Dolby Access Press. 1978

Robinson Dr JC & Piety Dr KR. "Peak Value (PeakVue) Analysis: Advantages over Demodulation for Gearing Systems and Slow-speed Bearings". *Northwest Indiana Business Roundtable & Trade Show.* Merrillville, Indiana. 3 October 1996

Rose A. "How to Set Up an Electrical Predictive Maintenance Program". *Predictive Maintenance Technology National Conference.* Indianapolis, Indiana. 7 - 9 November 1994

Royal Navy RCM Implementation Team. *Naval Engineering Standard 45 (NES45): "Requirements for the Application of Reliability Centred Maintenance Techniques to HM Ships, Royal Fleet Auxiliaries and other Naval Auxiliary Vessels" (Restricted Commercial):* Foxhill, Bath. 1999

Sandtorv H & Rausand M. "RCM - Closing the Loop between Design Reliability and Operational Reliability". *Maintenance,* 6(1), 13 - 21. 1991

Shiffrin CA. "Aviation Safety Takes Center Stage Worldwide". *Aviation Week & Space Technology.* **145 (19).** 1996

Smith AM. *Reliability-centered Maintenance.* New York. McGraw-Hill. 1993

Smith DJ. *Reliability, Maintainability and Risk.* Oxford. Butterworth Heinemann. 1993

Snow DA (editor). *Plant Engineer's Reference Book.* Oxford. Butterworth Heinemann. 1991

Society of Automotive Engineers, Inc (SAE): *Surface Vehicle/Aerospace Standard JA1011: "Evaluation Criteria for Reliability-centered Maintenance Processes":* Warrentown, Pennsylvania. 1999

Tissue BM. *SCIMEDIA - Analytical Chemistry and Instrumentation.* Website <http://www.scimedia.com>. 1996

Toms LA. *Machinery Oil Analysis.* Pensacola, Florida. Published by author. 1995

US Naval Air Command. *NAVAIR 00-25-403: "Guidelines for the Naval Aviation Reliability Centered Maintenance Process".* US Department of Defense Publications: Philadelphia Pennsylvania: 1996

US Navy (Engineering Specifications & Standards Department). *Mil-Std-2173: Reliability-Centered Maintenance Requirements for Naval Aircraft, Weapon Systems and Support Equipment.* Lakehurst, New Jersey. US Department of Defense. 1986

van der Horn G & Woyshner W. "Electric Motor Predictive Maintenance - A Comprehensive Approach". *Predictive Maintenance Technology National Conference.* Indianapolis, Indiana. 4 - 6 December 1995

Weaver C. "Time Waveform Analysis" *P/PM Technology.* **8 (5)** 1995

White G. "Vibration Data Collectors and Analyzers". *Predictive Maintenance Technology National Conference.* Indianapolis, Indiana. 4 - 6 December 1995

White G. "Designing and Installing a Monitoring System". *Predictive Maintenance Technology National Conference.* Indianapolis, Indiana. 4 - 6 December 1995

Xu Ming & Le Bleu J. "Condition Monitoring of Sealless Pumps using Spike Energy™". *P/PM Technology.* **8 (6)** 1995

Index